The Anticipation of Converging Industries

Clive-Steven Curran

The Anticipation of Converging Industries

A Concept Applied to Nutraceuticals and Functional Foods

Springer

Clive-Steven Curran
Fresenius Medical Care AG & Co. KGaA
Bad Homburg
Germany

ISBN 978-1-4471-5169-2 ISBN 978-1-4471-5170-8 (eBook)
DOI 10.1007/978-1-4471-5170-8
Springer London Heidelberg New York Dordrecht

Library of Congress Control Number: 2013936537

Printed on acid-free paper

Springer is part of Springer Science+Business Media (www.springer.com)

Foreword

Industries are often perceived to be rather 'monolithic blocks' characterized by common value propositions, customer bases, and product offerings. Within these industries those companies will be most successful, which can work the system most effectively. In most areas this is particularly dependent on (1) the capability to manage manufacturing (or service offering) capacities to produce a reliably good product (or service) at the lowest possible costs. Furthermore, (2) marketing and sales forces need to be able to convince potential buyers of their need to purchase the company's offering at the highest possible price. Many companies are lost, however, when they suddenly need to deal with radically changing technological foci, market dynamics, or regulatory frameworks. These factors are fundamental aspects of blurring boundaries between hitherto separate industries.

This phenomenon is called industry convergence or industry fusion. What used to be true or smart in a company's traditional environment may prove detrimental in a new industry. One primary example of converging industries has been the area of information technologies, consumer electronics, and telecommunication (ICT). For a long time, e.g., the telecommunication industry was in most countries dominated by a state-owned monopolist that would only very reluctantly let other companies take over market shares or react slowly to changing customer needs. On the other hand, consumer electronics companies had already experienced the opportunities and threats of a more competitive environment. While a mere two decades ago people would have to own half a dozen products to take pictures, place phone calls, listen to music, or work on electronic documents, today's children are growing up in a world where nothing is more natural to them than that all of those functionalities are incorporated in one device weighing around 100 g. In the same time span a lot of corporate mergers, divestitures, and strategic maneuvers, especially in ICT, but also in industries such as the food or the pharmaceutical industry have tried 'to seize the day' of convergence. Companies such as Apple have spent billions of dollars expecting a success in marketing convergence products. Their past financials speak a clear language in respect to the overall success and importance of these investments—between 2008 and 2012 alone Apple's total net sales more than quadrupled from $37 billion to $157 billion. More importantly, the contribution of more traditional computer products has long been outweighed by smartphones and similar devices in Apple's revenue statistics. On the other hand, many companies have lost billions of dollars when trying to turn convergence into an opportunity for the firm.

As the correct timing for major maneuvers is difficult to achieve, corporate foresight activities have gained importance over the more recent past. Clive-Steven Curran adds to this important field his approach to anticipate converging industries. Whilst a large part of the literature focuses on the more qualitative variants of generating intelligence about the future, his research utilizes publicly available data to identify early patterns of blurring boundaries between scientific, technological, market, or even industrial areas. He compares keyword-based analyses of scientific publications and patents with classification-based analyses of patents in the area of ICT and Nutraceuticals and Functional Foods. The latter is an equally active new industry segment involving the pharmaceutical and food industries. Building on a thorough literature review and deriving comprehensive and concise definitions of hitherto fuzzy concepts in this rather new research field, Curran analyzes approximately 1.8 million patents and a few thousand scientific publications. Identifying a set of triggers and drivers for industry convergence, he develops a very convincing method for the anticipation of converging industries.

This work's impressive theoretical foundation in combination with its clear practical applicability has proven appealing to academics and practitioners alike. Parts of it have been presented at international conferences including the International Society for Professional Innovation Management (ISPIM) Conferences in Vienna (Austria), Tours (France), Bilbao (Spain), and Hamburg (Germany), the International Conference on Management of Technology (IAMOT) in Cairo (Egypt), the R&D Management Conference in Ottawa (Canada), the Portland International Center for Management of Engineering and Technology (PICMET) Conference in Portland (USA), and the Academy of Management (AOM) Annual Meeting in Chicago (USA). At these conferences he has been nominated for Best Student Paper Award at the XX ISPIM Conference (Tours 2008) and awarded the Brad W. Hosler Outstanding Student Paper Award, PICMET'09 (Portland 2009). In addition, the dissertation on which this book is based has received the ISPIM-Wiley Innovation Management Dissertation Award (2011). Furthermore, Clive-Steven Curran's inspiring work has been published in leading international journals, such as the International Journal of Innovation Management and Technological Forecasting and Social Change. I would like to close by expressing my conviction that this book will receive its well-deserved broad dissemination and by wishing you an inspiring time reading it.

Prof. Dr. Jens Leker

Preface

When I started working at the University of Münster my idea of what my research topic would be was quite different from what is presented here. After looking into different topics in the broader realm of innovation management two main aspects have triggered my excitement about the anticipation of converging industries. First, the research devoted to different forms of forecasting/foresight and distilling information from the vast amount of publicly available data, most notably scientific publications and patent documents. With a natural sciences background this quantitative data appealed to me more than the more qualitative, psychological aspects of innovation management. Second, with a specific interest in the pharmaceutical industry and its related industries I came across some strategic maneuvers that did not seem to make sense within the traditional, narrower boundaries of their industry. A feasible explanation I found was the few works on 'convergence' and immanent principles that largely concentrated on the ICT (information technologies, consumer electronics, and telecommunication) industry. In consequence, this yielded my hypothesis that anticipating such convergence on the basis of publicly available data would be possible.

The present book was written as the result of my doctoral research project at the Institute of Business Administration at the Department of chemistry and pharmacy, University of Münster. However important an initial spark may have been, there would be no book if not for the overwhelming support I have experienced from so many people. Consequently, it is an honor and my great pleasure to extend my heartfelt thanks to them.

First, I would like to express my gratitude to my thesis supervisor Prof. Dr. Jens Leker, who has provided me with one of the most important ingredients in my endeavor—the freedom to operate—whilst at the same time being supportive of my stream of research and open to new approaches and methods. Second, I would like to thank Prof. Dr. Hans-Ulrich Humpf for serving as the co-advisor.

Third, I am indebted to Dr. Walter Buckel and his colleagues at FIZ Karlsruhe/ STN International for their support with and access to STN AnaVist and the respective databases. Fourth, I owe special thanks to Prof. Dr. Stefanie Bröring for infecting me with the 'convergence virus' and for several collaborative papers. Fifth, such a project is always considerably influenced by feedback from editors, reviewers, and conference participants, even if the individual contribution is impossible to specify. Thank you for your open-minded and yet critical discussion of several aspects of my research.

As much as a Doctorate on the one hand needs to be a 'lonesome, entrepreneurial' task, as much is it on the other hand, dependent on one's environment. In my case I consider myself extremely lucky to have been able to conduct my research at the Institute of Business Administration. A diverse team of colleagues with various educational backgrounds in natural sciences as well as economics has always led to inspiring discussions and a lot of fun—be it the weekly football session or the occasional campus coffee. Thank you all! Specifically, I am grateful to Anne Große Daldrup, who is the heart and soul of the institute. Furthermore, thanks to Gudrun Maria John and Nina Preschitschek for conducting some of the patent searches and for co-authoring papers with me. Nina, to you also, special thanks for some very fruitful last-minute discussions.

Dr. Verena Thiemann (nee Holtorf) and Dr. Benjamin Niedergassel deserve my thank you for constituting a great team on 'the third level of the institute' as well as on our several conference trips. Additionally, to Ben I want to direct very special thanks. He was not only a great colleague in many projects, publications, and teaching assignments, but also a perfect co-inhabitant of our common office and a very thorough and dedicated proof reader of my manuscript. His input, thoughts, and our endless hours of discussion had a major impact on this research project. Moreover, he was a great friend challenging me when I was too tempted to take short-cuts as well as when I was overly skeptic about the meaning of my results.

Finally, I am greatly indebted to my family and friends who have always been supportive and provided me with many a much needed distraction from my academic task. Most importantly, this book would never have been possible if not for the incredible love, patience, encouragement, and unlimited support of my wife Anna. To you and to our two wonderful children, Liam James and Nell Barbara, I dedicate this book.

Contents

Chapter 1
Introduction

1.1 Research Problem and Relevance of the Study

"The value of the products of our imagination depends of course on the number, accuracy, and clearness of our impressions; on our judgment and taste in selecting or rejecting the involuntary combinations, and to a certain extent on our power of voluntarily combining them" [1, p. 46].

Many scholars active in the field of *futures analysis* [cf. 2, p. 3] would likely dispute a connection of their forecasts to Darwin's 'products of imagination'. Forecasts are supposed to be mainly quantitative pictures of a likely future, based on clear principles and strict methodologies—giving little room to opinions or even imagination.[1] An extant body of literature covers various applications of forecasting, from meteorology-based weather forecasts to forecasts of nations' economic growth. Within business management, forecasts are often employed as a part of strategic or corporate foresight, which include heuristic and interactive methods as well as questionings.[2] The lively debate on principles and methods of futures analysis features various topics, such as a focus on experts versus on publicly available data, or the use of statistical testing within predictions of the future.[3] While foresight's combination of methods already points more in the direction of Darwin's quotation, anticipation goes even further in being based less on an exact mathematic formula with calculated probabilities and more on judgment and a combination of different observations.[4]

Futures analysis techniques are often used to generate information on emerging technologies and the predicted impact they are going to have [cf., e.g., 12, 13–15]. Emerging technologies may not always have decisive effects, but especially when a combination of new technologies arises it can substantially alter products,

[1] For a delineation of forecasting from foresight, futures analysis and anticipation employed in the present study, see Sect. 3.1.2.

[2] [cf. 3, p. 283f] for a classification of methodologies used within foresight approaches.

[3] See, e.g., [4–6], for differing opinions on the use of experts or other sources for predictions of the future. For a discussion on the value of statistical testing in forecasting see, e.g., [7–11].

[4] Two other important facets of anticipation, as introduced in Sect. 3.1.2 are the monitoring and the immanently active stance to it.

C.-S. Curran, *The Anticipation of Converging Industries*,
DOI: 10.1007/978-1-4471-5170-8_1,

manufacturing processes or even whole industries. For centuries, industries have undergone evolution, decisive or radical changes or even dematerialized completely. Within the last few decades, they have been facing a new and powerful phenomenon—the convergence of industries. Although this blurring of boundaries does not have to be based on emerging technologies[5] (or technologies at all), it often will. Especially in science and technology-driven industries like the computer, telecommunication, or the pharmaceutical industries, a convergence of technologies will often be a decisive part of industry convergence [cf. 16, pp. 2ff.].[6]

Convergence has in the past mainly been discussed in connection with the information technology, consumer electronics, and telecommunication industries.[7] This blurring of boundaries between hitherto disjunct industries had consequences on different levels. While consumers may not realize the evolutionary track of their new Smartphone, this product is a result of technology convergence as well as market convergence. Driven by new technological possibilities and at the same time by consumers' demands for multi-purpose devices, convergence products may also affect whole industries. A recent example of the forceful dynamics resulting from convergence was Hewlett-Packard's[8] announced acquisition of Palm at a value of $1.2 billion. It was explicitly targeted at enabling HP "to participate more aggressively in the fast-growing, highly profitable smartphone and connected mobile device markets" [cf. 17].[9]

Recently, scholars have also put more attention on convergence at the outposts of the chemical and pharmaceutical industries as well. The chemical industry is, for example, facing increasing competition from the agricultural industry in the area of (bio)polymers. They used to be synthesized from petroleum by chemical companies, but within the last years agricultural companies are increasingly producing them from renewable raw materials [cf. 18, pp. 1ff.]. Similarly, the pharmaceutical industry is affected by fading boundaries between pharmaceutical drugs and food products in the area of Nutraceuticals and Functional Foods (NFF).[10] Beyond M&A (mergers & acquisitions) activities these developments are also triggering new regulations and changes in consumer preferences.[11]

---

[5] Furthermore, the potentially massive effects of emerging technologies as well as convergence in general bear a strong resemblance. Hence, a large part of the literature on emerging technologies can be related to cases of convergence as well. See also Sect. 3.1.1.

[6] See also Sect. 2.1 for different loci of convergence and Sect. 2.4 for a description of the sequential process of convergence on which this study is based.

[7] Overall, the body of literature on convergence is comparably scarce. For a literature review on different aspects of convergence see Chap. 2.

[8] Hewlett-Packard Company (HP).

[9] Hewlett-Packard Development Company is a subsidiary of HP. See also Sect. 2.3.1 for a more thorough discussion of convergence, its drivers and effects in the ICT industries.

[10] See Sect. 4.3 for a detailed description of convergence around NFF.

[11] Allowedly, convergence is not only driving changes in consumer preferences, but is also driven by them [cf. 16, pp. 75ff.]. See Sect. 2.2 for a delineation of different triggers and drivers in respect to convergence and Sect. 4.3 for a detailed description of convergence around NFF.

Consequently, industry convergence poses a massive challenge to different stakeholders, mainly academic researchers, company managers, and policy makers. Especially for companies it often proves difficult to adjust their traditional "industry recipes"[12] [19, p. 60] to a new environment where they are often facing new technologies, competitors, customers, and even regulations. In the light of such substantial alterations and developments, a way to **anticipate convergence** at an early point in time is of natural interest to managers as well as (academic) scientists and policy makers involved in areas of potential future convergence. It would enable firms to form strategic alliances or acquire new technologies already at an early stage and to prepare for the challenges and pitfalls of the new segment in advance. Academics could engage in research collaborations or amend their research foci and policy makers could start the often lengthy procedures to introduce new or alter existing regulations as well as direct funding incentives accordingly.[13] Surprisingly, till date, methods to anticipate convergence are lacking, although a clear demand for them has been voiced, especially in the light of many failed half-hearted convergence predictions [cf. 20, p. 157].

1.2 Research Questions and Objective

Thus, the study at hand will try to fill this research gap by introducing a comprehensive approach to anticipate converging industries on the basis of publicly available data (i.e., mainly scientific publications and patent documents). A further shortcoming of the literature on convergence and forecasting is the lack of clearly delineated definitions of the different steps and aspects of convergence. Therefore, this study will not only fill the gap regarding a concept for the anticipation of converging industries but also enrich the body of literature with a more detailed set of definitions and analyses of industry convergence. Furthermore, it will focus on developments around the new inter-industry segment of NFF and scrutinize the state of convergence in respect to it. As the larger part of the convergence literature has concentrated on the ICT industries, the analyses of scientific and patent publications around NFF will be compared to an analysis in ICT.

Hence, in the course of this study the following **research questions** will be derived from a discussion of the resource-based view (RBV) as a **theoretical** foundation for the analysis of converging industries:

1. Q.1:*Can convergence be anticipated on the basis of weak signals in publicly available data?*

[12] Spender relates to the *industry recipe* as "the rationality which managers use to resolve the firm's strategic uncertainties." See also [16, p. 2].

[13] See also Sect. 4.2 for a discussion of the implications of convergence (anticipation) for academics, companies, and policy makers.

2. Q.2: *Can the assessment of different levels of convergence be based on analyzing samples derived from different data sources?*
3. Q.3: *Is a keyword-based anticipation of convergence possible and can this also be achieved employing broader classes of articles or patents?*

In addition, the following research questions will be derived from the **empirical** examples of NFF and related fields[14]:

1. Q.4: *Are the food and the pharmaceutical industries converging in the area of NFF?*
2. Q.5: *Are the personal care and pharmaceutical industries converging in the area of Cosmeceuticals, perhaps even into one broad NFF/Cosmeceuticals area?*
3. Q.6: *Does the analysis of activities in different countries lead to different results?*
4. Q.7: *Are patterns of convergence generalizable across industry boundaries?*

These research questions will be addressed in the course of the study to fulfill the **research objective** of identifying underlying patterns in converging industries focusing on two differentiating factors. First, the sequential process of convergence can be divided into four different steps: science, technology, market, and industry convergence. Second, patterns will be unraveled in respect of environmental differences regarding industries and countries. Thus, it should yield **implications for theory and practice** alike. The research streams on RBV, convergence, and forecasting/foresight could be profiting from theoretical implications. At the same time, the present study could contribute to the tool set of academics, managers, and policy makers by developing a comparably simple, transparent, and robust early warning approach.

1.3 Outline of the Study

As depicted in Fig. 1.1, the study is segmented into seven chapters. After a brief discussion of the research problem, questions, and objectives chapter, the remaining chapters are briefly described in the following.

Constituting the first part of the theoretical foundation, Chap. 2 focuses on the phenomenon to be anticipated—the construct of industry convergence. Therefore, it is introduced through a discussion of its fundamental aspects and then defined and delineated in the context of the present study. Subsequently, a classification groups triggers and drivers mentioned in the scattered pieces of the literature. Followed by a discussion of the most important examples of convergence, the chapter concludes with a brief analysis of the convergence process.

---

[14] See Sects. 4.2 and 4.3 for the reasoning in respect to the derived research questions.

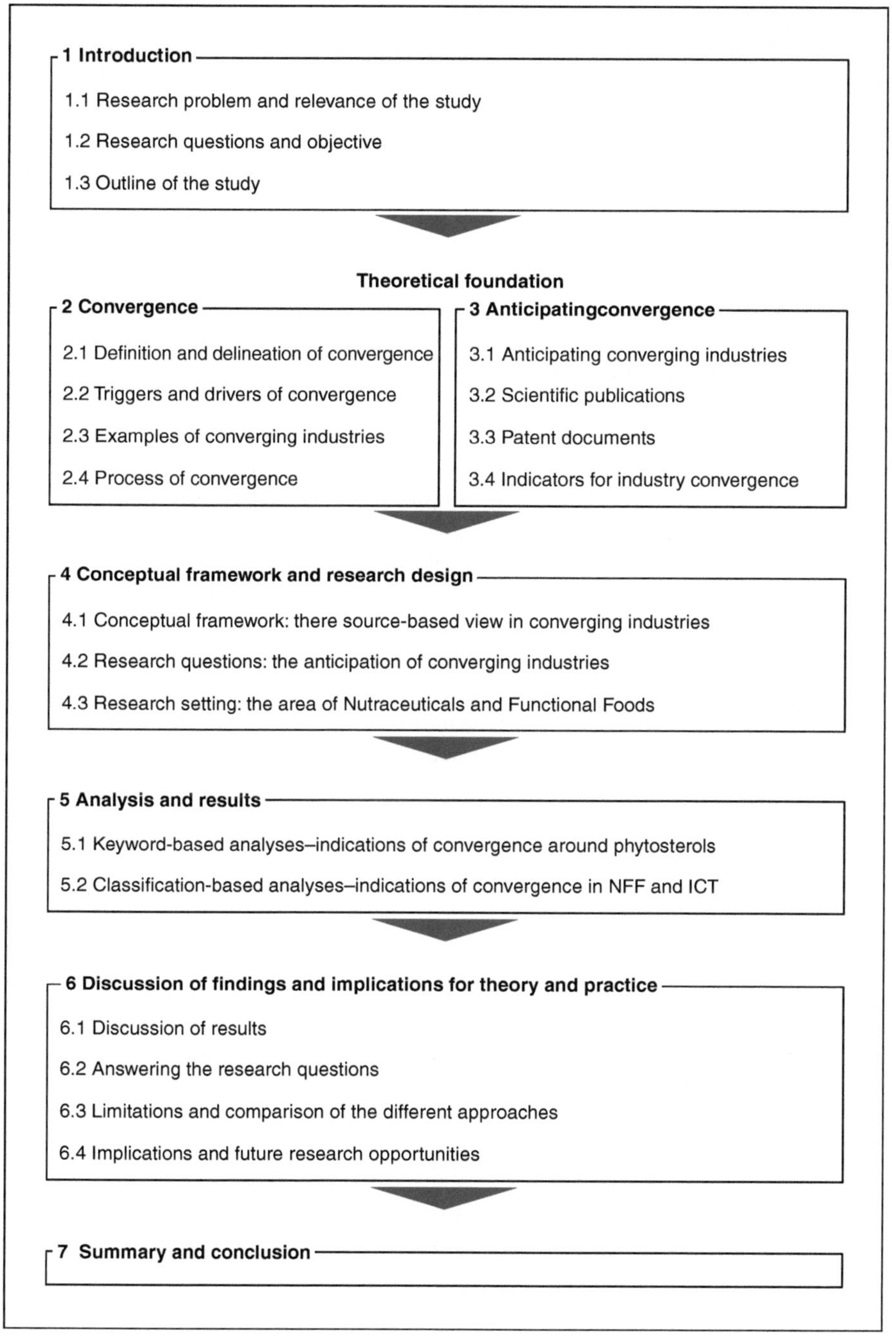

Fig. 1.1 Structure of the study

Concentrating on the concept for anticipating convergence, Chap. 3 represents the second part of the theoretical foundation. Based on the sequential model described in the preceding chapter, it outlines the respective concept and highlights potential fields of analysis. Although it is not intended to provide complete overviews, it will briefly discuss forecasting/foresight, scientific publications and bibliometrics as well as patents and their analyses. In the course of these discussions particular emphasis will be put on the applicability within this study's approach. The chapter closes with an introduction of indicators used.

Chapter 4 characterizes the conceptual framework for this study on the basis of the RBV as a theoretical concept. Central research questions are derived from this framework prior to the analysis of Nutraceuticals and Functional Foods as the setting for the empirical analyses.

In Chap. 5 the results from these analyses are presented from four different samples based on two different search strategies. The first two samples concentrate on phytosterols as a class of substances employed in NFF and are generated based on keyword searches. The latter two samples are derived classification-based and cover the greater area of NFF as well as the area of ICT.

Subsequently, Chap. 6 discusses the findings of this study critically. This discussion starts with a detailed discussion of individual important results, before using these results to answer the seven research questions. Based on a description of the overall limitations of the different approaches, they are compared and their suitability for this study's concept discussed. This is followed by the presentation of theoretical implications for the research streams on convergence, forecasting/foresight, and the RBV. Finally, practical implications and potential future research opportunities are pointed out.

Chapter 7 summarizes and concludes the main findings of this study.

References

1. Darwin, C.: The Descent of Man, and Selection in Relation to Sex, vol. 1. John Murray, London (1871)
2. Rohrbeck, R., Arnold, H.M., Heuer, J.: Strategic Foresight in multinational enterprises—a case study on the Deutsche Telekom Laboratories. Paper presented at the 1st ISPIM-Asia International Conference, New Delhi, India, 9.-12.01.2007
3. Mietzner, D., Reger, G.: Practices of strategic foresight in biotech companies. Int. J. Innovation Manage. **13**(2), 273–294 (2009)
4. van der Meulen, B.: The impact of foresight on environmental science and technology policy in the Netherlands. Futures **31**(1), 7–23 (1999)
5. Tichy, G.: The over-optimism among experts in assessment and foresight. Technol. Forecast. Soc. Chang. **71**(4), 341–363 (2004)
6. McMaster, M.: Foresight: exploring the structure of the future. Long Range Plan. **29**(2), 149–155 (1996)
7. Armstrong, J.S.: Significance tests harm progress in forecasting. Int. J. Forecast. **23**(2), 321–327 (2007)
8. Hauer, E.: The harm done by tests of significance. Accid. Anal. Prev. **36**(3), 495–500 (2004)

9. Ord, K.: Comments on significance tests harm progress in forecasting. Int. J. Forecast. **23**(2), 331–332 (2007)
10. Stekler, H.O.: Significance tests harm progress in forecasting: comment. Int. J. Forecast. **23**(2), 329–330 (2007)
11. Goodwin, P.: Should we be using significance tests in forecasting research? Int. J. Forecast. **23**(2), 333–334 (2007)
12. Porter, A.L., Ashton, W.B., Clar, G., Coates, J.F., Cuhls, K., Cunningham, S.W., Ducatel, K., van der Duin, P., Georgehiou, L., Gordon, T., Linstone, H.A., Marchau, V., Massari, G., Miles, I., Mogee, M., Salo, A., Scapolo, F., Smits, R., Thissen, W.: Technology futures analysis: toward integration of the field and new methods. Technol. Forecast. Soc. Chang. **71**(3), 287–303 (2004)
13. Daim, T.U., Rueda, G., Martin, H., Gerdsri, P.: Forecasting emerging technologies: use of bibliometrics and patent analysis. Technol. Forecast. Soc. Chang. **73**(8), 981–1012 (2006)
14. Day, G.S., Schoemaker, P.J.H.: Avoiding the pitfalls of emerging technologies. In: Day, G.S., Schoemaker, P.J.H., Gunther, R.E. (eds.) Wharton on managing emerging technologies, pp. 24–52. Wiley, New York (2000)
15. van Lente, H., van Til, J.: A combined roadmapping-cluster approach for emerging technologies. Int. J. Foresight Innov. Policy **3**(2), 121–138 (2007)
16. Bröring, S.: The front end of innovation in converging industries: the case of nutraceuticals and functional foods. DUV, Wiesbaden (2005)
17. Hewlett-Packard development company: HP to acquire palm for $1.2 Billion. http://www.hp.com/hpinfo/newsroom/press/2010/100428xa.html (2010). Accessed 29.04.2010
18. Preschitschek, N., Curran, C.-S., Leker, J.: Are plastics turning organic? Antecedents of convergence in the chemical industry. Paper presented at the The XXI ISPIM Conference—The dynamics of innovation, Bilbao, Spain, 06.-09.06.2010
19. Spender, J.-C.: Industry recipes: an enquiry into the nature and sources of managerial judgement. Blackwell, Oxford (1989)
20. Schnaars, S., Thomas, G., Irmak, C.: Predicting the emergence of innovations from technological convergence: lessons from the twentieth century. J. Macromarketing **28**(2), 157–168 (2008)

Chapter 2
Convergence

After this short introduction of the motivation for a concept of anticipating converging industries this chapter will focus on the phenomenon to be anticipated. Only few studies have thus far participated in laying a solid theoretical foundation to the industry convergence construct. To partly heal this major shortcoming the present study will try to contribute to this foundation in so far as it is relevant for anticipating convergence. First the term's historic applications will be discussed followed by its fundamental aspects a review of definitions by preceding researchers then its definition delineation in the present study. Second an array of different triggers drivers is derived from literature before compiling them into a comprehensive classification scheme. Third in a case study approach different examples of industry convergence are characterized on the basis of the derived classification. Finally this chapter is concluded by a brief analysis of the convergence process itself its individual steps. In summary this chapter will address the questions:

- What is convergence?
- Why does convergence happen?
- Where does convergence happen?
- How does convergence happen?

In doing so, this chapter will introduce delineations and definitions for the central aspects of this study, while at the same time providing a literature review of the thus far rather scarce and scattered literature on convergence.

C.-S. Curran, *The Anticipation of Converging Industries*,
DOI: 10.1007/978-1-4471-5170-8_2, © Springer-Verlag London 2013

2.1 Definition and Delineation of Convergence

2.1.1 Concept of Convergence

Although ‘convergence’ has been attested “broad implications for economic welfare” [1, p. 1079], it has at the same time been coined “an often used but rarely defined”[1] [3, p. 5] buzzword [cf. 4, p. 2; or 5, p. 13] by authors aiming at developing a comprehensive definition for it. This section will focus on uses of and definitions for this term and delineate it from other terms used in the same context. The use of ‘convergence’ in the English language dates back to the seventeenth century. It is of Latin origin “convergere” (inclining together) and its dictionary definition encompasses basic definitions like “(of lines) tend to meet at a point” or “(of a number of things) gradually change so as to become similar or develop something in common” [6, p. 378]. Furthermore, individual disciplines use it in more specialized meanings. Biologists speak of convergence when talking about unrelated organisms that “develop superficially similar characteristics under similar environmental conditions” and meteorologists think of meeting airflows or ocean currents [6, p. 378]. It is used as well in mathematics, natural and social sciences as in computing and economic sciences. More generally it describes the concept of at least two discernible items moving towards union or uniformity or the merging of distinct technologies, devices or industries into a unified whole. A common point in all of these definitions is the emphasis on convergence being a process rather than a steady state. The converging items are ‘inclining’, ‘meeting’, ‘developing’, ‘moving’ or ‘merging’. This longitudinal dimension will also be highlighted in Chap. 3, where the concept for an anticipation of convergence is developed.

Historically it had been applied in mathematics and natural sciences first, before describing a convergence of US and Soviet systems in the political sciences in the middle of the twentieth century [cf. 7, p. 58]. Stieglitz argues the term’s application in so different environments lead to structurally different definitions [cf. 3, pp. 21ff.]. In mathematics ‘convergent’ is defined as “(of a series) approaching a definite limit as more of its terms are added” [6, p. 378]. Thus, it is easily measurable or quantifiable. Consequently, Stieglitz speaks of a **quantitative-exact convergence** definition, which is also applicable to macroeconomics. There it is used in connection with growth rates in productivity or *per capita* income in different societies that are supposed to tend to a common number, based on the influences of exogenously determined parameters. These definitions all share the quantitative perspective, where convergence can be objectively identified and even measured, as it is solely based on numbers and facts. Contrastingly, biology and the social sciences employ a **qualitative-heuristic convergence** term. In biology, exogenous factors influence initially genetically different organisms to develop

[1] Amongst those who criticize the lack of a clear definition are also [2, p. 2224], who mention a wide range of definitions from “easily understood to scientific approaches”.

similar physiological characteristics in a similar environment. As these characteristics are present in completely distinct organisms and may be structurally different, they cannot be measured in a mathematical sense. The same holds true for the social sciences, where convergence is being discussed as a result of similar problems in similar environments leading to similar solutions in different societies or groups [cf. 3, p. 22]. Accordingly, Sorokin argues that the US and the Soviet systems would be beginning to exhibit institutional similarities due to the necessities and the influence of a growing degree of industrialization [cf. 3, p. 22]. Stieglitz sees a central difference between the quantitative and the qualitative convergence definitions in respect to the degree of forecastability of developments. Whereas the quantitative-exact view implies a high degree of determinism, this would be a lot less the case in respect to the qualitative-heuristic view. There, serendipitous reactions to environmental impacts make a precise forecast impossible [cf. 3, p. 23].[2]

The first use of convergence in a managerial context is attributed to Rosenberg, who in his article on technological changes in the machine tool industry in the second half of the nineteenth century coins the expression of "technological convergence" in contrast to "sequences of parallel and unrelated activities" [8, p. 423]. He employs the phrase to describe processes that are commonly used by different (unrelated) industry sectors and different stages of tool production. While pre-industrial economies had seen a necessity for rather specialized skills to assemble any final product, these techniques had become more and more generic during industrialization. Thus, completely different machinery and metal-using sectors—from firearms to bicycles—had begun to share a common technological basis. 8 years later, Oettinger for the first time invented a hybrid term to describe an area of convergence between computers and telecommunications—"communications" [cf. 9, p. 74], claiming that both areas had become so inseparable that it was time to fuse their names as well. Farber and Baran, in their 1977 Science article, add to the evolving talk about convergence in computers and telecommunication and predict "a large number of new industries" to emerge [cf. 10, p. 1166]. This academic viewpoint was at the same time complemented by a growing industrial awareness for convergence. The Nippon Electric Company (now NEC) estimated a convergence between computers and communications to occur by 1990 [cf. 11, p. 26]. In 1978 MIT (Massachusetts Institute of Technology) researcher Negroponte became the pioneer in depicting convergence. He used three circles labeled with the three industries and technologies of computers, broadcasting and printing that he described to be overlapping nearly completely by 2000 [cf. 7, pp. 58f.; 11, p. 25; 12, p. 25]. Shortly afterwards, Apple's then CEO (Chief Executive Officer) Sculley

[2] Consequently, this study largely follows the quantitative-exact approach, with limitations regarding the degree of determinism. Convergence within the present study is not viewed as an unavoidable result of outside impacts. Even when taking an industry insider perspective, managers and other stakeholders can exert substantial influence on the course of events. See Sect. 2.2 for an overview of triggers and drivers as well as Sects. 2.3 and 4.3 for a discussion of the role of individual actors within possible cases of convergence.

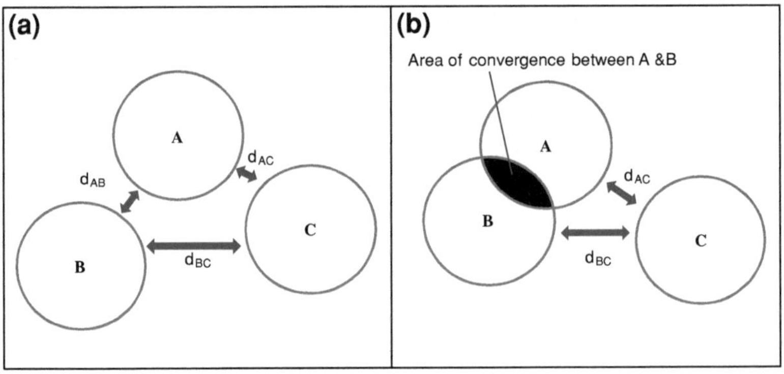

Fig. 2.1 The process of convergence, depicted in two states (time t_0 and t_1). **a** *Status quo* at time t_0. **b** *Status quo* at time t_1 (*Source* Adapted from [13, p. 28])

used a similar figure to embed his idea of Apple's future position. He used boxes to illustrate seven individual industries from media/publishing to computers, office equipment and distribution representing the situation in 1980. For the year 2000 he expected overlaps between these seven areas, with new industry segments like "Interactive News" where the old ones had overlapped. He labeled the whole figure "convergence" [cf. 7, p. 59]. Figure 2.1 depicts a similar concept. Areas A, B and C (Fig. 2.1a) represent different and initially disjoint fields like science disciplines, technologies, markets or industries. Their circular geometry is only an arbitrary illustration; others have used other forms such as ellipses. Each area contains all the items belonging to this particular field. Depending on the level of analysis, such an area might be seen from an individuals' perspective and thus be constituted by all the scholars conducting research in the respective field. Or it might be regarded as a field in the sense of all the scientific publications in it. Consequently, the distance between each of the areas cannot be calculated on a general basis.[3] But basically Fig. 2.1b shows a difference in the distances between the three areas. Whilst A&B have begun to overlap, the distances between A&C (d_{AC}) and B&C (d_{BC}) are more or less unchanged.

Besides any needs for further specification, Negroponte's illustration is an easily comprehensible way to graphically explain what happens in converging fields. This is of even greater interest, as the breadth of what is to be understood as convergence varies greatly by author [cf. 3, p. 5]. Some limit their definitions to just one industrial field [cf. 14, p. 9]. Nyström, e.g., in her study on convergence in the Finnish telecommunication industry, initially defines it as the "merging of the three sectors of telecommunications, media and information technology" [12, p. 8].[4] Only later in her text she acknowledges the fact that this would be too

[3] Chapter 3 will provide a more detailed discussion of possible measures to calculate the size of each of the areas and the distances between them.

[4] Nyström in her definition refers to a green paper by the European Commission [15, p. 1].

narrow an explanation and that convergence is affecting other industries as well [cf. 12, p. 25]. Accordingly, Stieglitz criticizes that the market-oriented strategy stream views technological convergence solely in connection with developments in the ICT (information technology, consumer electronics and telecommunication[5]) industries [cf. 3, p. 15].[6] This concentration on ICT is being driven by three factors. First, in many studies industry convergence—or even convergence in general—is equaled with technological convergence. Possibly the most prominent technological convergence is the digitization of information in any form. And the momentum of the digitization has been a central enabler of changes in the ICT industries. Second, it has seen huge mergers and acquisitions which in turn caught the attention of scholars and the general public. Third, the changes that are happening around devices, services and technology in ICT affect everyone. People realize that their phones have changed as much as the way in which they communicate with other people, buy their goods or conduct their daily business. Thus it appears to be quite natural to focus on developments around these industries. It is reflected by the number of studies on convergence in ICT. In contrast to this ICT-focused approach, other scholars are discussing convergence with respect to different industries and on a more general level.[7] Section 2.3 will delve deeper into ICT and other examples of converging industries.

But if convergence does not refer to one single case, a concrete definition becomes even more necessary. In Rosenberg's early seminal work a foundation for the commonplace (mis-)understanding of industry convergence equaling technological convergence is laid. While in many cases like the ICT industries, technological convergence and industry convergence will be strongly related, they are still not interchangeable [cf. 12]. The OECD (Organisation for Economic Co-operation and Development) in a 1992 report on developments in telecommunications and broadcasting acknowledged that "the term 'convergence' has become something of a buzzword in the information industry" and that processes on a further three levels would need to be taken into account: networks, services and corporate organization [cf. 5, p. 13]. But it entails even more facets, including science convergence, demand convergence and convergence in regulations. Consequently, definitions mirror these different levels of convergence. And even 'levels' might be a misleading word. While some of the different forms of

[5] A plethora of different definitions exists for the abbreviation 'ICT'. Within this study, it is used to encompass the three broad sectors of computers and information technology (I), consumer electronics and media appliances (C) and telecommunication devices and technology (T). They have in common the generation, processing and storing of (by now) digital information. On a more content-related note, also media and broadcasting should be taken into account when discussing developments in ICT (see, e.g., [16]). Other abbreviations describing similar developments include TIME (telecommunications, information technology, media and entertainment) and TMC (telecommunications, media and consumer electronics).

[6] Others, like [14, 17, p. 9], share this criticism.

[7] These include but are not limited to [4, 7, 14, 16, 18–26].

convergence are clearly inclusive of some others, a few are somehow detached from most forms. The present study is conducted on the basis of an understanding of a sequential structure of science, technology, market and industry convergence.[8]

To develop a comprehensive definition as a foundation for the analyses in this study, it is important to delineate prior works regarding these facets. One part is representing different *loci* of convergence[9]:

- **Science** Different scientific disciplines or areas
- **Technology**[10] Central technologies of different application areas
- **Market** Markets for different sets of demands
- **Corporate organization** Companies being unattached to each other and active in different markets
- **Industry** Set of companies with (partly) different technology bases, different application fields and different target groups

All of them are possible answers to the question *On what level is convergence happening?* and are partly related or dependent on each other. The nature of their hierarchical interdependencies is the starting point for the sequential process of convergence, which is being introduced in Sect. 2.4. An additional part of facets is answering the question *What is the result of convergence?* concentrating on the type of convergence result:

- **Product** Combination of hitherto disjoint products functionalities
- **Service** Different services combined into one
- **Demand** Customers' demands traditionally being met by different companies/industries
- **Regulation** Regulation for different services or products in their respective applications

Traditionally, convergence has mainly been discussed in connection with products or services, but some papers also address the demand type of

[8] See Sect. 2.4 for a detailed discussion of this sequential structure.

[9] A short description for the pre-convergence status of each of the loci is given in parentheses.

[10] As a convergence of networks (e.g., networks for cellular phones, cable television and broadband) represents only a form of technological and/or market convergence, it is not included in the selection of *loci*. Müller [27, pp. 40ff.], distinguishes further between technology convergence and functional convergence as sub-types of market convergence. In his view technological convergence describes a merger of different technologies hitherto used. Following his line of argument, functional convergence leads to a combination of different functions (and mostly products) into one new device. As this study is aiming at developing a concept for the anticipation of convergence, only a clear distinction between different *loci* of convergence will enable a stepwise employment of the concept (see Sect. 2.4 for a more detailed argumentation for this sequential structure).

convergence. The European Commission speaks vaguely of "interacting with society" [15, p. ii]. Pennings and Puranam label it "demand side convergence" and discern between two archetypes. The first refers to converging customer demands in different parts of the world. This, in their view, leads to a "uniform set of needs" and global markets [cf. 25, p. 6]. The second describes demand convergence as it is understood within the present study, as a growing tendency of customers towards 'one-stop shopping'—the wish to buy products or services satisfying multiple needs. This **demand-side convergence** is complemented by what Pennings and Puranam call **supply-side convergence** [cf. 25, pp. 6ff.]. Following a similar perspective, Bröring concludes that only the co-occurrence of technology-driven input side convergence and demand-driven output side convergence leads to the emergence of new inter-industry segments between or instead of at least two distinct industries [cf. 20, p. 80]. Bröring also stresses the importance of convergence in regulation [cf. 20, pp. 79f]. She views it as a possible trigger as well as a possible outcome of convergence processes. On the one hand, changes in regulation can lead to changes in industry structure.[11] On the other hand, new areas of convergence may make changes or even an initial introduction of regulation inevitable.[12] Products are probably the most naturally accepted outcome of convergence. Smartphones, e.g., have reached a status where they incorporate not only phone and text message functionalities, but also commonly cameras, radios, web browsers, e-mail and even GPS (Global Positioning System). Thus, consumers can conduct an array of different tasks with their smartphone, instead of carrying about a dozen different devices. Such a convergence outcome is present in peoples' everyday life as well as very physically graspable. So, a large share of the limited amount of literature on convergence contains product convergence examples cf., e.g., [1, 3, 4, 7, 12, 14, 20, 24–26, 28–33]. But additionally to products as outcomes, new services may also be a result of convergence. This tendency to underestimate the importance of the service sector is somewhat similar to discussions within the broader innovation management literature [cf. 34, p. 1]. For instance, Tidd and Hull criticize that we know "comparatively little" about organization of new 'product' development in service sectors [cf. 35, p. 3]. And this is the case, although an increasing amount of the real economy is based on services rather than products [cf. 36, pp. 1f.], especially in the so called 'developed nations'. Statistics for the USA, the UK or Australia list a

[11] An example for such a convergence-enabling change in regulation is the ICT area. Privatization of formerly state-owned companies led to a reshuffling of companies and their portfolios in most countries. See Sect. 2.3 for more details on this.

[12] Biotechnology is a good example for such a regulation-demanding convergence. New technological possibilities may lead to a myriad of legal and moral questions and challenges. This is especially the case when dealing with gene technology and issues such as embryonic stem cells or the cloning of animals or humans. Differing from country to country, policy makers have set up more or less tight guard railings. See Sect. 2.3 for more details on this.

share of nearly 80, 75 or 70 %, respectively, of the gross domestic product (GDP)[13] provided by the service sector [cf. 37].[14] Hence, new ideas, insights or combinations may lead to either new products or services—or even both. Staying with examples from ICT, phone calls may be placed through more traditional means like landline phones or cell phones, but may also be conducted via VoIP (voice over internet protocol) services. The service of letting two or more people talk to each other is increasingly being detached from certain, dedicated devices. Consequently, many of the studies and reports on convergence have acknowledged the importance of service convergence cf., e.g., [1, 5, 7, 12, 14, 15, 20, 24, 32, 33].

2.1.2 Literature Review on Definitions of Convergence

As briefly mentioned in Sect. 2.1.1, most of the papers and books on convergence use their own definition—if the term is defined at all. A rough selection of the most important ones will be provided in the following before developing a comprehensive one on their basis. Included will be also the term ‘fusion’ or ‘industry fusion’ that has been used by some scholars to describe similar ideas. This overall definition will serve as the foundation for the development of an anticipation concept in Chap. 3.

Amongst those that define convergence, there are those that focus solely on ICT. Accordingly, Fransmann describes convergence as “paradigmatic changes [...] in the information, communication and entertainment areas” [33, p. 35]. Similarly, Borés et al. restrict their delineation to ICT, but are apparently less convinced that it actually happens. They view convergence as “a process by which the telecommunications, broadcasting, information technologies and entertainment sectors [...] may be converging towards a unified market” [38, p. 1]. Interestingly, the breadth or scope of definitions does not show any clear trend over time. Neither a generalization nor a specialization tendency can be extrapolated from the different contributions over time. For example Yoffie employs a broader definition earlier than Fransmann or Borés et al. He emphasizes the importance of a shared functionality as well as a set of base technologies: “the unification of functions—the coming together of previously distinct products that employ digital technologies” [39, p. 2]. This output-oriented take on functionality is shared by Gill, who

[13] GDP = gross domestic product; value of all final goods and services produced within one nation (or group of nations) in one given year.

[14] Interestingly, there are some noteworthy exceptions to this general rule. For example, Norway’s service sector share amounts only to about 54 % and Ireland’s to 49 %. Whereas most countries with high per capita GDP exhibit a share of the agricultural sector somewhere between 1 and 2 %, this number is 5 % for Ireland. For the whole world, the distribution is 4 % agricultural sector, 32 % industrial sector and 64 % services sector. Many poor nations are a lot more dependent on the agricultural sector, with Liberia marking the highest number of nearly 77 % (5 % industry and 18 % services).

views convergence as "the addition of disparate new functionalities to existing base products" [cf. 40]. According to Stieglitz, a distinction should be made between convergence as a state and convergence as a process. [cf. 3, p. 23]. After discussing a selection of market convergence types, he concludes with his definition of market convergence as a process of qualitative change leading to the combination of two or more formerly disjoint markets [cf. 3, p. 32]. Most studies inherently deal with convergence as a process, but only few acknowledge the decisive difference. This perspective also has important implications for measuring convergence. If it is a process, it does not make much sense to try to assess a degree of convergence at one point in time. Only when following a longitudinal approach, differences in the degree of overlap between the involved areas can be calculated. Section 3.4 will discuss possible measures for estimating degrees of convergence. It will be followed by a more comprehensive concept for measuring and anticipating convergence, based on longitudinal data collection.

Bally sums up earlier works to a quite comprehensive definition of technological convergence: "the growing together of technologies, which fundamentally alters the boundaries of previously distinct industry or market sectors and merges them into a new competitive environment" [14, p. 13]. In contrast to this, Lind defines it within his study only with respect to market convergence as being "a confluence and merging of hitherto separated markets, removing entry barriers across the market and industry boundaries" [16, p. 1]. He emphasizes the difference between his definition and the widespread belief that industry convergence would lead to separate industries merging into one big converging industry [cf. 4, p. 2; 16, p. 1]. In a later paper he adds an alternative, very broad definition: "a market/industry definition generated by technological change" [4, p. 1]. Complimentary to technology and market convergence, Bröring focuses on industry convergence leading "to an overlap of two formerly separated industries starting to produce similar products" [20, p. 76]. In the following she also incorporates fusing technologies that form [4] new products or product categories. Albeit adding to this industry perspective, Choi and Valikangas broaden it to a blurring of "boundaries between industries by converging value propositions, technologies and markets" [41, p. 426]. Although the OECD report on telecommunications and broadcasting brands convergence a buzzword "which is high on concept, but low on content" [5, p. 13], it defines it simply as the "blurring of technical and regulatory boundaries between sectors of the economy" [5, p. 13]. Thus it follows the line of industry convergence, while at the same time stressing more unusual influencing factors such as regulation.

Another group of scholars detaches its definitions from too industry- and technology-focused viewpoints. Pennings and Puranam emphasize that "convergence between previously disjointed markets can be viewed as the erosion of boundaries that define and isolate industry-specific knowledge" [25, p. 3]. But they do not specifically define convergence. Hacklin follows this approach and defines (besides other forms of convergence): "Knowledge convergence denotes the emergence of serendipitous co-evolutionary spillover between previously unassociated and distinct knowledge bases, giving rise to the erosion of established boundaries that isolate industry-specific knowledge" [11, p. 57]. Interestingly, knowledge is already a broader term than technology and leads to the question of convergence on levels uncoupled with commercial activities. While most definitions focus on markets, industries and technologies in connection with their application in new products or services, some scholars also see the importance of extending this construct to changes in knowledge bases or scientific disciplines. Accordingly, Guilhon sees convergence as "the process by which the application of the same scientific concept allows the putting together of two or many fields of activity" [42, p. 33]. Here the independence of technological aspects plays a crucial role. The notion of science convergence will be discussed further in Sect. 2.4.

Finally, while stating that convergence "eludes precise definition", a Green Paper of the European Commission describes it as "the ability of different network platforms to carry essentially similar kinds of services, or the coming together of consumer devices such as the telephone, television and personal computer" [15, p. 1]. This is an example for the kind of definitions provided in most texts on convergence. Most of them are either rather fuzzy (like the OECD definition) or they explicitly focus on a rather limited scope. Most likely Nyström's final definition is the one closest to the scope of this study. After a detailed overview of different definition approaches she concludes: "Convergence is a change process initiated by technological, socio-economical and organizational forces, removing or changing traditional industry borders, framing and enabling new resource constellations and eventually leading to industry convergence in the form of sub-industries, new business ecosystems and new markets" [12, p. 52]. Against this background, this study considers convergence on all levels (e.g. science, technology, market or industry), in respect to all results (product/service, demand, regulation), all application fields,[15]and inclusive of what has been termed 'fusion'.

2.1.3 Delineating Convergence from Fusion

Bierly and Chakrabarti refer to a state of 'industry fusion' as "industries that are without clear boundaries and are continually combining with other industries" [19, p. 35]. They argue that while many industries have not yet entered the "industry fusion stage", they will in the near future. Other industries, like the ICT, the pharmaceutical or the financial services sectors are already in the middle of this process. They further develop their concept of managing a company affected by industry fusion and thus add to the aforementioned definition. Such companies will have to employ new knowledge from different areas, will have to face new and very strong competitors from different sectors and hence will engage in mergers, alliances and licensing deals. Especially the challenge to deal with "so many knowledge areas" [19, p. 41] and to successfully integrate new knowledge in the company is nearly unimaginable—if not done in connection with close collaboration of any form with other companies [cf. 18, pp. 10f.; 19, pp. 35ff.].

[15] For an overview of fields prominently affected by convergence see Sect. 2.3.

While Bierly and Chakrabarti do not delineate fusion from convergence, Adner and Levinthal do. To them convergence is a process, "in which the common domain is an application domain in which one of two antecedent technologies is already applied" [43, p. 64]. Fusion is something that two technologies undergo, with "the resulting technology" being "applied to a new domain" [43, p. 65]. In contrast to their opinion, 'fusion' has etymologically a slightly different meaning. Just like 'convergence' it is of Latin origin ("fundere"—pour, melt), but fusion is defined as "the process or result of joining two or more things together to form a single entity" [6, p. 704]. Whereas the earlier is mainly being used for objects that move towards each other and form one single entity somewhere else, the latter is covering complete melts of objects in the very same spot. Adding to a basic perception of fusion, Kim distinguishes between two main types of technology fusion. First, fusing across technologies (either basic and applied or applied and applied), relating mainly to the employment of a different technology base within a given technology. Second, fusing across industries (either among neighboring or non-neighboring sectors), which describes technology fusion in the sense of blurring industry boundaries. Whereas the first definition would not be coined technology fusion by most, it is easily understandable against the background of Kim's study. Discussing different possibilities to reflect fading boundaries in university education, technology fusion according to his definition still requires universities to set up blended teaching schemes with courses from a wider array of disciplines [44, pp. 222f.].

2.1.4 *Defining Convergence and Fusion*

A definition within this study should entail convergence as well as fusion, as both are grounded on a similar idea—the blurring or fading of boundaries. When identifying early warning signals, every possible point of information should be taken into account. Thus, all *loci* of convergence will be included, whereas drivers and triggers are only an additional feature of the definition. They will be added at the end of Sect. 2.2. Furthermore, as will be discussed in Sect. 2.4, convergence does not necessarily imply the development of a new industry. In some cases industry areas or, e.g., scientific disciplines may create a new area of collaboration, without changes in the anteceding areas and even nothing more than loosely coupled to them. In addition to Fig. 2.1, these areas will in most cases be heterogeneous forms with some of the tentacles beginning to move towards each other (Fig. 2.2). This may mean a movement from both[16] areas, but as well only one of the fields may be evolving their expertise further in such a direction.

[16] When speaking of 'two areas' or 'both', this is only the smallest possible number. Convergence may as well happen between three or more areas, of course. But for increased legibility 'two' will always represent 'at least two'.

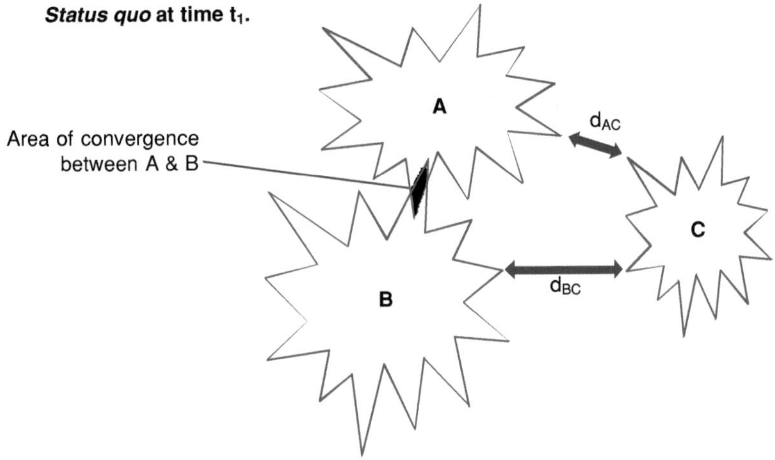

Fig. 2.2 Convergence of non-circular areas *A&B*, still completely distinct from area *C* (*Source* Own figure)

To differentiate convergence from fusion this study delineates the terms as follows:

Convergence describes a process, where objects move or stretch further from their prior and discrete spots, to a new and common place.

Fusion describes a process, where objects begin to merge with each other in the very same place of at least one of the objects.

Hence, the difference lies in the movements of the objects. For example, industries would be labeled 'converging', if they (or parts thereof) began to merge with each other in a new field. This field would have to not have been part of any of the hitherto distinct industries. In contrast to this, industries would be labeled 'fusing', if the new replacing segment or sub-segment would substitute (parts) of the prior segments.

Accordingly, Nutraceuticals and Functional Foods can be interpreted as an example of convergence, as they are replacing neither the demand for nutrition nor the demand for full-scale medication.[17] On the other hand, smartphones are a fusion[18] product of cameras, cellular phones and portable computers (etc.). All of the attached demands had been fulfilled prior to the fusion process. Afterwards a new sub-segment of the established industries was created in the spot of the hitherto disjoint devices. Figure 2.3 depicts the difference with an illustrative example. Figure 2.3a (at time t_0) consists of the two disjoint areas A&B separated

[17] See Sect. 4.3 for a more thorough explanation of NFF as an example for convergence.

[18] This delineation is not reflected in other researchers' works. As discussed, most appear to not discern between both terms at all. For studies explicitly dealing with fusion see, e.g., [18, 19, 44–46].

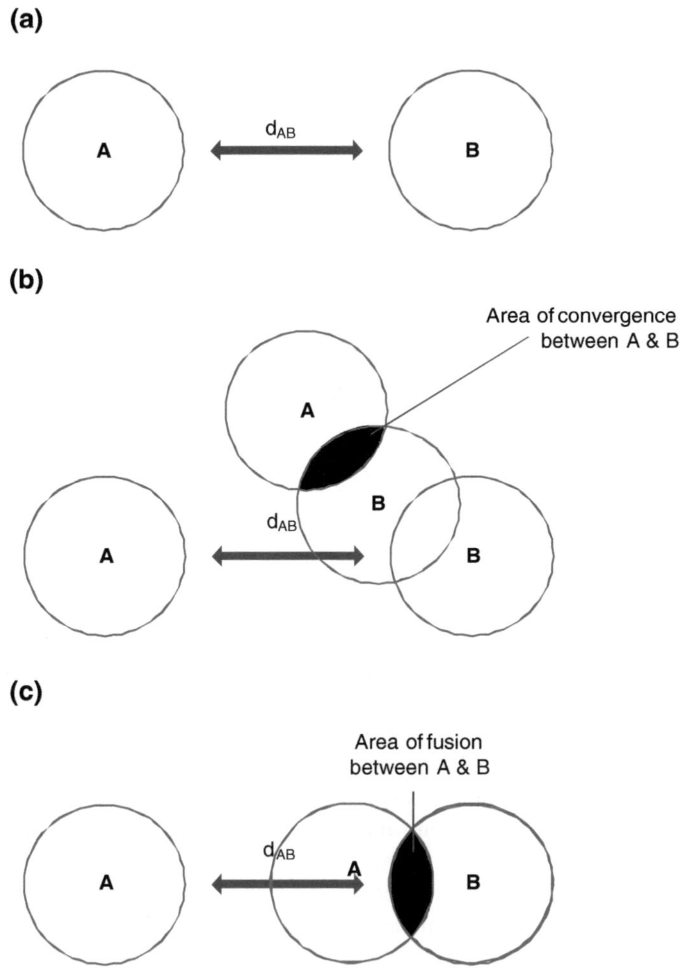

Fig. 2.3 The process of convergence (**a/b**) graphically delineated from the process of fusion (**a/c**). **a** *Status quo* at time t_0. **b** *Status quo* at time t_1-convergence. **c** *Status quo* at time t_1-fusion. *Source* Own figure

by the distance d_{AB}. Upon reaching a convergence status at time t_1, both areas (or parts thereof) have moved to a common new spot, demonstrated by their old positions, depicted in the background (Fig. 2.3b). Contrasting this situation, Fig. 2.3c shows a fusion scenario. B has not moved at all, but is rather merging with A in its own spot.

Consequently, building on the notion of Nyström and the aforementioned convergence/fusion[19] delineation, this study defines convergence in a managerial context as:

Convergence [fusion] *is a blurring of boundaries between at least two hitherto disjoint areas of science, technology, markets or industries. Through this convergence [fusion], a new (sub-)segment is being created in a new spot [the same spot] as a merger of (parts) of the old segments. It is marked by an increase of interchangeability and connectedness between the respective areas, as can be seen in collaboration, licensing, patenting or publishing behavior.*

2.2 Triggers and Drivers of Convergence

2.2.1 Differentiating Between Triggers and Drivers

After arriving at a definition for convergence and fusion within this study and before looking at examples of convergence and its progress, it is imperative to question what causes it.[20] Causes of convergence are commonly referred to as drivers or triggers. First of all, it should be distinguished between triggers and drivers. Incidents and developments may be seen as clear **triggers** of convergence, but they may not necessarily **drive** its further development. Whereas literature tends not to discern between the two,[21] this chapter will try to delineate the one from the other. The Oxford Dictionary of English lists as one definition for 'driver' "a factor which brings about or sustains a particular phenomenon" [6, p. 531], even though its original Germanic meaning is "urge (a person or animal) to go forward" [6, p. 531]. In contrast to this, 'trigger' is listed to mean "an event that is the cause of a particular action, process, or situation" [6, p. 1884]. Hence, within this study a distinction will be made between:

---

[19] As the present study aims at developing a monitoring concept for a blurring of boundaries, both cases are of high interest—convergence as well as fusion. As the concept will not discern between convergence and fusion, fusion will always implicitly be addressed when convergence is mentioned—if not stated otherwise.

[20] Barriers for convergence processes could be discussed together with the triggers and drivers. But this will not be explicitly done within this study. Most barriers mentioned in literature thus far are somewhat inverted drivers, if addressed at all see, e.g., [12], including organizations' inability to adjust to convergence effects or regulation stopping firms from entering new markets. Thus, potential barriers are to a large extent inherently addressed when talking about triggers and drivers. Furthermore, this study does not view convergence as a kind of natural force happening if not stopped. On the contrary, it interprets convergence as something that has to be actively driven in order to finally prevail.

[21] Pennings and Puranam, for example, even combine both terms to: "We attempt to identify a range of drivers that trigger the onset of convergence." [25, p. 3].

*A **trigger** denominates an event or situation that **sets off** a certain sequence of reactions.*
*A **driver** denominates a series of events or a situation that **sustains** a certain sequence of reactions.*

For instance, when looking at the launch of a rocket, the **trigger** would be the ignition spark, whereas the launch would be **driven** by the constant burning of the propellant. Following this delineation, triggers of convergence include inventions, instant regulation changes or demand shifts, the unavailability of necessary raw materials and other sudden events. Drivers of convergence can be seen in slower and more constant developments. These can be based on technological changes, modifications of business *foci* and *loci*, policy alterations or changes in customer structures. It can be argued that the convergence processes in ICT are triggered (amongst others) by inventions on the technological side, rather bold moves of single companies—like Apple's decision to enter the cellular phone market—and policy shifts such as the privatization of hitherto state-owned telecommunication monopolists. But what drives the ICT convergence is much more a constant digitization, miniaturization and the tendency of customers to buy multi-purpose devices. Section 2.3 will shed more light on triggers and drivers for convergence in individual cases.

2.2.2 Literature Review on Triggers and Drivers

Concentrating on a classification of triggers and drivers, most scholars have mentioned a comparable set of triggers/drivers.[22] But nearly all of the contributions have thus far only very briefly addressed the question of antecedents of convergence [cf. 11, p. 38], if at all. A clustering of prior research into streams of literature proofs especially difficult in this context as most authors use their own terminology and classification schemes. Notwithstanding this challenge, the remainder of this chapter will try to reflect the most important contributions to literature's discussion of influences on convergence. Starting with rather simple and mono-causal views, it will become more complex before channeling the different works into one more comprehensive perspective. On a general level, the listed examples include causes based on:

- Technology
- Business models
- Market structure
- Regulation and legal issues
- Societal developments

[22] As most researchers do not discern between triggers and drivers, this chapter will start with listing examples from the literature without a further classification. At the end an overview will be provided, delineating the mentioned triggers from the potential drivers of convergence.

A first part of the literature provides rather narrow, nearly mono-causal drivers/triggers. While most articles feature possible gains from the convergence phenomenon, Bainbridge's work on scientific convergence stands out for its particularly positive tone towards fading boundaries and even "unifying science". This may as well be rooted in a neglect to discuss possible negative effects as in the focus on (far from commercialization) fundamental research. Consequently, he speaks of mathematics as a universal language and new tools 'helping' researchers to turn to adjacent fields for possible solutions to their challenges [cf. 47]. Similarly positive are Yasunaga et al., working for the Japanese Ministry of Economy and examining ways to promote and further technology convergence. Whilst not really providing further insight into potential triggers and drivers, they are discussing possibilities to sustain convergence through the use of grants and road-mapping [cf. 48].

As the first to use convergence in a managerial context, Rosenberg focused solely on technology as a cause for convergence. In his view, the increasingly ubiquitous use of machinery and metal working technologies led to a close relationship of industries hitherto completely distinct [cf. 8, p. 423]. Von Tunzelmann attributes trends of convergence in telecommunications and computing to the similarities between hardware and function. Computers began to be as much connected through networks as the telecommunications sector switched from analogue technology to computerized processing [cf. 28, p. 125].[23] Taking a little broader perspective, Kim lists three reasons for fusion on industry level: constantly growing complexity of customer demands, surmounting demand limits and blending manufacturing with advanced services (such as renting or consulting). Interestingly, all of these are rooted in the demand sphere [cf. 44, p. 223].

A second part of the literature introduces also a second set of causes. For instance, Bierly and Chakrabarti list the increasing rate of technological diffusion (caused by easier and wider access to information) as well as the globalization of markets (leading to shorter product life cycles and a need for more effective innovation activities) as the main reasons for convergence [cf. 18, p. 7; 19, p. 35]. Concentrating on developments affecting the computer games industry, Ip identifies drivers of convergence to increasingly stem from market convergence and the complexity of content [cf. 49, p. 220]. Interestingly, most researchers who see changing demands as a main driver for convergence do not differentiate between B2B (business to business) and B2C (business to consumers) customers.[24] A deviant approach is voiced by Myring. In his opinion—due to differing economies of scale and technological expertise between both groups—technological developments are often driven by businesses first and only later by consumers [cf. 32, p. 7].

---

[23] Also, Duysters and Hagedoorn, argue in a similar fashion [cf. 30, p. 356].

[24] Business-to-business (B2B) delineates business transactions between businesses (e.g., wholesaler/retailer or manufacturer/wholesaler) from transactions between businesses and consumers (B2C) or governments (B2G).

Gambardella and Torrisi see technological convergence being prompted by an increasing importance of "generic technologies". In spite of their applicability in different markets or products, the authors identify the problem of many companies to be able to get used to new "value networks". They further emphasize the role of the individual company in highlighting companies' influences. According to their line of argument, more complex products are not always the result of inevitable changes in technological knowledge bases, but sometimes rather a way to generate more income through selling multi-technology devices [cf. 50]. Finally, Bally classifies the influencing factors as either being internal or external factors. In this classification, technologies, innovation and management constitute internal factors, whereas technological change, regulation and customers' demands represent the main external factors. As compelling as this dichotomy is, the delineation of 'technologies' from 'innovations' and from 'technological change' remains somewhat unclear [cf. 14, pp. 9f.].

Yoffie as a representative of the third group of authors focuses his explanation of the convergence phenomenon on three main drivers. Next to the widely accepted drivers (de-)regulation and technological change, he puts large emphasis on what he calls "managerial creativity". Against the background of a strong impetus of technological change in computers and communications, large multinational corporations like Sony or International Business Machines (IBM) had still failed in "forcing convergence" due to missing creative products or services. In contrast to this, smaller start-ups were more successful in off-setting convergence through innovative products [cf. 2, 39]. Although Kaluza et al. follow Yoffie's introduction of three main drivers, they discern between drivers and triggers. According to them, the effects of dramatically reduced telecommunication costs were "triggered by the breakup of AT&T (American Telephone and Telegraph Company)" [2, p. 2225], leading to further convergence in ICT. In spite of largely adhering to Yoffie's classification and acknowledging the particular importance and acceptance of "digital technology" as a driver, they criticize the lacking consent among scholars on the central drivers behind convergence [2, p. 2245]. Contradictory to this assessment is Rikkiev and Mäkinen's take, who see external environmental forces as well as internal company actions as consistently listed by several authors [31, p. 375].

Pennings and Puranam differentiate between "sudden surges of convergence", such as those caused by deregulation and delayed convergence developments. In their opinion, socio-economic developments and changing demographics are particularly slow triggers/drivers. Technological drivers' pace is depending on the structural parameters and the characteristics of the technological innovation [cf. 25, p. 3]. Again, Katz identifies three classes of drivers: technological developments, "evolution of business thinking" and regulatory changes. For the blurring of boundaries in ICT, technological change (especially digitization) is one of the central drivers. But in his view, convergence in ICT has moved from being mainly supply-side driven to a more demand-side phenomenon. Convergence started on the basis of a mere possibility to change technology before being a lot more dependent on users' interests. Katz discusses regulation not so much as a driver but rather as a barrier to convergence. In ICT, e.g., regulation distinguished for a long time between telecommunication and enhanced services, forcing their separation. Furthermore, mono- or at least oligopolizing the telephone market barred other possible service providers from entering the market [cf. 1, pp. 1082ff.].

In conclusion, developing further this notion of three classes, Nyström discerns between organizational, technological and socio-economic drivers. According to her line of argument, firms have the possibility to cause convergence by seeking new knowledge and new solutions and combining them into something innovative across technological boundaries. Additionally, firms will in most cases be more likely to engage in developments outside their traditional fields of expertise, if their recent past has not been as successful as that of competitors. Technological drivers are said to consist of technological change, digitization and the growing importance of the internet. Finally, Nyström lists socio-economic drivers, including liberalization, regulation and globalization. Interestingly, she does not discuss regulation as only a barrier to convergence (with de-regulation or liberalization as drivers), but also as a driver. Regulations may also be aimed at fostering convergence developments (such as promoting internet services or multipurpose devices) [cf. 12, pp. 29ff.; 51, p. 1]. Similar to this more complex discussion, Choi and Valikangas regard convergence as one of ten strategic innovation themes, caused not only by deregulation and new customer demands, but also by a "ubiquity of information". While this is part of other researchers' discussion about globalization, it adds another perspective to it [cf. 41, pp. 425f.]. It is a combination of globalization (borders become less important and economic systems are becoming more equal), digitization (information flows are increasingly dependent on the flow of bits and bytes) and also changes in consumers' perception of information. New developments in the information business are concentrating on providing basically all the information at anytime and anywhere.[25]

The last stream of the literature develops and discusses multi-causal concepts, classifying triggers/drivers along different parameters. Accordingly, Hacklin draws in several works the conclusion that drivers are to be found inside and outside of the respective industries. As one of the scholars dwelling most intensively on triggers and drivers, he argues that "convergence occurs when innovations emerge at the intersection of established and clearly defined industry boundaries, thereby sparking off an evolutionary development with much broader impact" [52, p. 723] (pointing at a trigger). A combination of technological trajectories leads to outcomes with improved performance when compared to the sum of the performance of each part. He also sees a cause for technology as well as industry convergence in the "discontinuity in a globalized industry" [53, p. 317], resulting from a global market, new innovation opportunities due to more and more intersections between technologies, opportunities to establish innovation collaborations, amendments in

[25] See also Sect. 2.3.1 for a more detailed description of changes around information technology, media and communication.

regulation and customers' need for more complex solutions and problem solving [cf. 11, pp. 36ff.]. Referring to Theilen, he lists possible triggers as "new technologies, societal change, globalization, liberalization, deregulation, legal change and market-implied changes of existing value chains[26]" [53, p. 318], but also better opportunities for product-bundling. He delineates technological drivers (the evolution of underlying technologies) from economic drivers (deregulation) in the field of ICT [cf. 53, p. 317]. In his nomenclature—and similar to Bally's classification—endogenous drivers on the firm level are to be separated from exogenous drivers on industry level.

Malhorta and Gupta discern between demand-side and supply-side convergence. Whilst the first is set off by customers identifying substitutive or complementary linkages between products from different industries, the second is initiated by companies delivering goods to two industries. Both the drivers may occur alone or in a combination of both [cf. 24, p. 1]. Although there is arguably a dichotomy between supply- and demand-side drivers, especially their elucidation of what is to be understood as supply-side drivers is rather weak. Taking a more holistic approach, Weaver in his work on the security industry acknowledges technological drivers as being amongst the most important. At the same time, he argues that these are complemented by other causes, such as regulation, standards, business models, demand shifts and the structure of industry channels, which all "tend to be overlooked" a little [cf. 26, pp. 2ff.]. He follows the dichotomy of demand- and supply-side convergence drivers and elaborates further on them. At the same time, he is caught in a 'chicken and egg' problem, mixing antecedents and phenomena of convergence [cf. 55, 56]. In spite of these inconsistencies, he adds a few more aspects to the current discussion. As part of technological drivers, there are also the results of 'Moore's Law',[27] leading to more and more functionalities on an ever smaller chip and the setting and adoption of universal standards. Compellingly, Weaver sees internet protocols as examples for a *de facto* communication standard. Furthermore, he adds firms' internal factors, including entrepreneurial activities and strategic actions [cf. 26, p. 8; 55, p. 5].

Finally, Bröring further refines Malhorta and Gupta's classification and divides drivers into two different groups. First, **"technology-driven input-side convergence"** [20, p. 81] evolves from new technologies, which are applied across different industry sectors. Accordingly, one of the industries may develop a new

[26] See also [54].

[27] Moore's Law describes a historical trend in computing and is named after Intel co-founder Gordon E. Moore. In his seminal 1965 paper "Cramming more components onto integrated circuits" and later works he predicted that the number of transistors on one single computer chip would double about every 2 years. This forecast has proven to be correct for more than 40 years and is expected to be still valid for the next years. In 1965 an average chip would hold approximately 50 transistors. This number has now reached 2 billion with Intel's newest Itanium processor range "Tukwila". Closely related to the number of transistors are also related parameters like the size of chip components, the number of calculations per time, memory capacities and the number of pixels on an average digital camera chip [cf. 57, 58].

technology and then this is employed in other industries as well. In this case a new technology **substitutes** an incumbent technology in a different industry. It is referred to as **complementary**, if both industries are developing individual technologies that are combined only later, to form new solutions or products. Second, "**market-driven output-side convergence**" [20, p. 82] is said to occur once customers start to "treat products of different industries in the same way, i.e. products which originally do not stand in any competition to each other start to become substitutes" [20, p. 82]. Just as regarding technology-driven input-side, the output-side convergence can be classified further into substitutive or complimentary, depending on the reaction of customers [cf. 21, p. 489; 22, pp. 165f.]. Third, another dimension is added with respect to regulations. Changes in regulations can be triggers or outcomes of convergence. Contrastingly, existing regulation may actually block or slow down convergence. Interestingly, Bröring argues that complete industry convergence will only happen when both input- and output-side occur [cf. 60, pp. 275f.].[28] In one paper she introduces different *loci* of convergence as triggers and drivers in their own right. Consequently, she voices her opinion that technology convergence is the "ultimate trigger for industry convergence, whereas market convergence is reinforcing it" [60, p. 276].[29] This concept of input-side and output-side bears analogy to the discussion about technology push or market pull as drivers for innovation by Mowery and Rosenberg or Howells [cf. 61, pp. 102ff.; 62, pp. 1209ff.]. While technological developments enable firms to supply the market with new or enhanced products, changes in customer structures and behavior may lead to demand-driven output-side convergence. Reasons for a decisive change on the demand-side include a growing interest in satisfying multiple needs with one transaction as well as socio-demographic developments [cf. 25, pp. 3ff.].[30]

2.2.3 *Classification of Triggers and Drivers*

To summarize the prior literature review, Table 2.1 represents a classification of triggers and drivers and their respective sources in literature. Most of the above-mentioned literature focused on individual cases and specific industry settings, largely parts of the greater ICT sector. Bearing in mind that it is not always easy or at all possible to sharply distinguish between triggers and drivers, this is supposed to be a more generalized approach.

[28] This is in line with Hauschildt and Salomo's argument that it takes technology push and market demand for successful innovations [cf. 59].

[29] See Sect. 2.4 for an introduction into different *loci* of convergence and their interconnectedness.

[30] These encompass developments such as growing customer bases in thus far developing countries, relatively free and global trade or a "homogenization of customer segments due to changing demographics".

Concentrating on the meta-level, four different clustering schemes appear to be reasonable. As discussed at the beginning of Sect. 2.2, a first distinction should be made between triggers and drivers. While it should be possible in most cases to judge on the pace and stage of an impact (sudden and early = trigger; slower, longer-lasting and later in the process of convergence = driver), this dichotomy alone seems to rough of a classification. Nevertheless, each cause should be assessed in respect to its positioning. The same holds true for the next two schemes. As compelling as the distinction between endogenous (or internal) and

Table 2.1 Classification of triggers and drivers in converging industries

	Triggers	Drivers
Technological	• Combination of technological trajectories yields increased performance [53] • Similarity between hardware and function [28]	• Ubiquitous use of technology [8] • Digitization [1, 2, 12, 41, 51] • Technological developments [1, 14, 20, 26] • Generic technologies [50] • Increased rate of technological diffusion [18, 19] • Ubiquity of information [41] • Technological change [14, 38, 39, 55, 56] • Complexity of content [49] • Growing importance of the internet [12] • More intersections between technologies [11]
Managerial	• Managerial creativity [39] • Management's actions [14, 26, 55] • Entrepreneurial activities [26, 55] • Rather unsuccessful past of company [12, 30]	• Evolution of business thinking [1] • More complex products as means to generate more income [50] • Innovation [14] • Blending manufacturing with advanced services [44] • Business models [26] • Establishing innovation collaborations [11]
Industrial	• Breakup of companies [2] • Standards (introduced by industry) [26] • Innovations emerging at intersection of different industries [52]	• Companies delivering goods to two industries [24] • Market-implied changes of value chains [53] • Internet protocols as quasi standards [26, 55] • Structure of industry channels [26] • Market convergence [49]
Political	• De-monopolizing markets [1] • Deregulation [25] • Legal changes [11, 54] • Standards (imposed by governments) [26]	• (De-)regulation/liberalization [1, 11, 12, 14, 20, 26, 38, 39, 41] • New regulations aimed at fostering convergence [41] • Grants promoting convergence [48]

(continued)

Table 2.1 (continued)

	Triggers	Drivers
Societal		• Customers' demands [1, 14, 26, 41] • Globalization of markets [18, 19, 41, 53] • Changing demographics [25] • Consumers' perception of information [41] • Customers identifying linkages between products from different industries [20, 24] • Socio-economic developments [25] • Growing complexity of customer demands [1, 11] • Surmounting demand limits [44]

Source Own table. For sources of individual triggers and drivers the aforementioned studies are listed. They are grouped with a common terminology, if a comparable stance can be assumed. Where no reference is provided, it is an additionally identified trigger/driver. Individual bullet points are sorted by year of oldest reference. The delineation between drivers and triggers is not always provided by original authors. Where no difference was made, it was conducted by this study

exogenous (or external) drivers[31] is, this only makes sense if it is defined in greater detail. Is the internal perspective that of an industry, an individual firm, or even a specific technology? For instance, following Bally's classification, technologies, innovation and management would constitute internal factors (similar to Hacklin's endogenous, firm level drivers). Whereas management actions may convincingly be seen as completely firm-dependent, technology and innovation should also include an outside perspective. Following increasing trends towards "Open Innovation" cf., e.g., [63], this would largely neglect the importance and also the direct influence of innovation partners beyond the factory gate. While these differentiations add an important level of analysis to any discussion on drivers of convergence, their general applicability remains questionable.

Malhorta and Gupta as well as Bröring discuss a differentiation between supply-side (or input-side) and demand-side (or output-side) drivers. Again, it is necessary to specify the entity receiving input or producing output. As convergence is a phenomenon not to be expected on the level of individual companies, it seems to be more appropriate to apply this classification on industry level. First, reducing drivers to be either input- or output-related would leave the industry and company influences unacknowledged. Second, especially when looking at more complex value chains, it could prove to be difficult to generalize across all participants in that industry. The value chain of chemical products, e.g., starts with oil-drilling and cracking before oil is being transformed into base chemicals, specialty chemicals and then eventually final products, such as fast moving consumer goods

[31] For increased legibility, 'driver' will be used for the remainder of this study to describe both, triggers and drivers, if not stated otherwise.

(FMCG). Depending on the parameters, a 'chemical industry' perspective would mostly contain the producers of base and specialty chemicals [cf. 63, pp. 123ff.]. Within the chemical industry the use of renewable resources instead of petrochemicals is increasingly being discussed.[32] A specialty chemicals producer could switch to renewable feedstocks through either purchasing base chemicals made from renewable or through purchasing them from outside the chemical industry (e.g., from companies in the agricultural sector). In the latter case this would have to be coined an input-side driver, whereas in the earlier case it would come from within the industry. Accordingly, an analysis of output-side drivers would pose similar problems—in spite of the same underlying trend and thus driver. Thus, these dimensions add an important angle to analyses of convergence drivers but appear to fall short of a generalizable and comprehensive classification of causes.

Building on the five areas introduced at the beginning of Sect. 2.2.2, this study proposes to ask where possible drivers can evolve. Building on the fact that all developments within economic systems are somehow set off by human individuals (or groups of individuals) this translates into the question: "Who is the source of this particular convergence driver?" Adhering to the five areas these drivers are:

- Technology: scientists/researchers (**technological drivers**)
- Business models: managers (**managerial drivers**)
- Market structure: companies (**industrial drivers**)
- Regulation and legal issues: governments (**political drivers**)
- Societal developments: societies/consumers (**societal drivers**).

2.2.3.1 Technological Drivers

Technological developments are grounded on new technological possibilities (such as miniaturization) or evolutions (e.g., digitization). These developments are not a somehow unavoidable result of technology but rather the outcome of research and development activities. They are conducted mainly in academia, semi-academic research labs or governmental agencies and industry. Furthermore, especially in areas like software development, unaffiliated individuals with good expertise may be another important source.

2.2.3.2 Managerial Drivers

Several authors emphasize the role of individual business action. Yoffie calls it "managerial creativity" [39, p. 9], Katz "evolution of business thinking" [cf. 1, pp. 1082ff.], Weaver "business models" [cf. 26, pp. 2ff.] and Bally simply "management" [cf. 14, pp. 9f.] (to name only a few). Strategic actions driving

[32] Cf., e.g., Nordhoff et al., who critically discuss the influence of raw material prices (renewable as well as petrochemical) on their selection in the chemical industry [64, pp. 1505ff.].

convergence are sometimes massive corporate mergers like that of Time Warner and America Online (AOL) or IBM and Rolm,[33] but at the same time they can be much more silent developments. The decision of Apple's managers (and probably particularly Steve Jobs) to introduce Apple's design and functionality stance to the mobile phone industry, has certainly fortified a blurring of boundaries around smartphones. As most managers' decisions are rather short- to medium-term oriented cf., e.g., [65, pp. 1469ff.], managerial drivers will in most instances really be drivers rather than triggers. Only if a company is already facing a possible convergence will most managers be prepared to make bold moves to engage in it.

2.2.3.3 Industrial Drivers

The differentiation between managerial and industrial drivers may be a little difficult, as developments on industry level will often be initiated by individual companies' decisions. But at a certain point in time (where industry structure and positioning is a given fact, for the time being), the parameters of the respective industries will determine their scope of decisions and actions. For example, the interest of food companies to sell Functional Foods is to a considerable extent based on their higher margins compared to conventional food products. Rather slim margins in food production are a characteristic that only a whole industry sector could try to change.[34] This leads to the conclusion that certain enablers of convergence are rooted on an industrial level and thus also pose challenges to the concerned companies.

2.2.3.4 Political Drivers

Political drivers encompass regulation, liberalization (or de-regulation), standardization (if not conducted on industry level), legislature, government funding and the way governmental authorities deal with debated issues.[35] Without the decision in most countries to open the telecommunication markets to non state-owned companies, the initiation of a convergence between the communication sector and the computer sector would not have been possible. On a similar note, regulation around NFF plays a substantial role in the question whether they can successfully compete with conventional foods and pharmaceutical industry's drugs.[36]

---

[33] See Sect. 2.3.1 for a more thorough discussion of these activities within ICT.

[34] See Sect. 4.3 for an overview of the situation in the NFF industry.

[35] Especially when it comes to disputed new technologies, governmental practice can be even more important than legislature. This is even more so the case, where non-governmental organizations (NGOs) or the society at large may be opposing new developments and where decisions are of a particularly long-term orientation.

[36] This will also be discussed in greater detail in Sect. 4.3 on the NFF industry.

2.2.3.5 Societal Drivers

Societal drivers are probably the broadest selection of different impacts. They include changing demographics, trends of global thinking and flexibility and changes in public's values and political views. At least in the so-called developed nations people's life-span is constantly increasing. On the one hand, people in these ageing societies may have another 20 years of average life expectancy (and thus of buying products and services) after retiring from their jobs. On the other hand, reaching 80 or 90 years leads to a sharp increase in illnesses caused by the shear age of the body.[37] Accordingly, the demand for products and services targeted at older age consumers increases, including medical devices, pharmaceuticals and functional foods [cf. 66, pp. 125ff.].[38] This trend is strengthened further by a more dominant desire to be healthy, young and beautiful.[39] A possible antagonistic trend in the food industry, rooted in people's values and beliefs, may be seen in the purchase of organic food products. And at the intersection between both trends is a market for "natural functional foods" [cf. 66, pp. 125ff.]. Societal drivers also include developments caused by society as a group of consumers, leading to demand shifts. For example, consumers' tendency towards one-stop shopping leads to unified systems in cellular phones. A rise in the share of people living alone and spending a high percentage of their time in front of computers leads to a different attitude towards what, where and how to buy, providing a strong commercial basis for internet stores or auctioning platforms like eBay.

After discussing the classification of triggers and drivers on a more theoretical level, Table 2.1 gives an overview of the developed matrix and possible examples. As most authors do not discern between triggers and drivers, the classification has to be based on this study's understanding of their respective line of argument. But although they have been freely moved between both classes, the majority of causes can be regarded as being rather drivers than triggers. This is easily understandable on the basis of the extent of changes initiated by convergence. In most cases they will be much too massive to be caused by one "once in time" trigger and will need other drivers to sustain the impetus of change. In any way, convergence has thus far never been caused by just one trigger or driver, but rather a broader selection. Strikingly, in contrast to other scientific fields, there is no increasing specialization visible in the discussion of causes for convergence. When looking at the use of different drivers over time no clear pattern is identifiable. Rather specified terms like 'more complex products as means to generate more income' (Gambardella

[37] While the overall process of ageing and its relation to an increased mortality is very complex, major factors include cells being lost without replacement and mutations in our chromosomes. This organismal senescence is characterized by a declining ability to mitigate stress results and unhealthy conditions, to a considerable part caused by environmental factors.

[38] Demand shifts caused by an ageing society will also be briefly discussed in the context of developments at the intersection of nutrition and pharmaceuticals in Sect. 4.3.

[39] While the content of this desire and its implications could well be discussed in a study of its own, the mere existence is as undisputed as the massive effects on buying behavior and leisure.

and Torrisi) have been introduced in 1998, whereas the way more generic 'technological change' seems to have been introduced by Bally in 2005. To shed some more light on real cases of convergence and the way in which they happen, Sect. 2.3 will discuss a selection of examples from different industries and initiated by different triggers and drivers.

2.3 Examples of Converging Industries

The preceding paragraphs have provided the terminology and definitions for the most important individual aspects of convergence. In a case study approach, this chapter will substantiate the largely theoretical discussion in the preceding paragraphs with an analysis of the blurring of boundaries in the prototype of converging industries—the ICT sector. This is supposed to shed some more light on the practical relevance of industry convergence. Additionally, the influence of the different triggers and drivers and the degree of convergence realization are briefly discussed.[40] Exemplary company examples are included in this discussion, where applicable. Followed by a few more exotic manifestations,[41] this chapter is specifically aiming at elaborating different parameters of convergence.

2.3.1 *The ICT industry*

As discussed earlier, most scholars have thus far concentrated on media-, information technology- or electronics-related industries for studying effects of industry convergence. Most of the remaining works employ developments in the ICT industry at least as one of their examples. But as also other sectors have experienced their fair share of convergence-related activities, why should so many researchers focus on the intersection of computers, communication and consumer electronics? Partly, this may be due to the strong technological focus, the search for signs of industry reactions and the necessity of comprehensible and illustrative examples, marking a considerable portion of the literature.[42] But also the extent of influences on peoples' every-day life is probably unrivalled by other occurrences of convergence. First, a decisive customer surplus is often a result of new combinations of products or services. Second, the omnipresence of (digitalized) information in the form of text, pictures, sounds, or a combination of these is

---

[40] Especially the degree of convergence in NFF will be discussed in-depth in Chap. 6 after the presentation of results of the measuring of convergence.

[41] Being the second major example of convergence and the basis of the present study, developments in NFF will be discussed in greater detail as the research setting in Sect. 4.3.

[42] See Sect. 2.1.1 for a discussion of reasons of literature's (over-)representation of ICT examples. See also Chap. 6 for an assessment of the state of convergence in the ICT industry.

altering the global economy. Third, for some developments even new economic measures need to be developed. Products like Google's search engine (or any of its competitors, for that matter) or the Adobe Reader experience nearly unlimited market dissemination, partly due to apparently being free of charge. Thus, the only way in which their importance for metrics such as GDP can be measured is due to their indirect effects in enabling Google to sell advertising or Adobe to sell other connected software tools [cf. 67, p. 165].

When looking at convergence possibilities at the outskirts of the ICT cluster, there is a myriad of boundaries to be faded. The industry sectors consist of a broad array of different segments. While the information technology sector is about computers and their production, software and service providers play a significant role as well. In consumer electronics, e.g., developments regarding (digital) cameras, music storage and reproduction, television sets and radios are involved. Communication deals with landline phones, cellular phones and access to the internet as well as other forms of modern communication. Finally, there are other sectors affected, like the media industry, publishing and printing and to some extent completely different sectors: biotechnology, nanotechnology or chemistry.

Strikingly, convergence involving the computer and the communication sector has been discussed at least since the early 1970s. As early as 1971 Oettinger spoke of "computers and communications [having] long since become inseparable" [9, p. 74].[43] Retrospectively, this was a time at which a merger between both sectors was far from being finalized (if it had really begun at all). Indeed, while some classify it as a clear trend observed by many and as a well suited example [cf. 53, p. 318], others question the connection between an undisputed technological convergence in ICT and blurring market boundaries [cf. 50, p. 445]. Notwithstanding this discussion, many areas have in the meantime merged into common markets. In addition to the emergence of smartphones, voice and data streams are increasingly managed by the same providers and delivered through one single line to the end customer.

In 1965, at the time of Intel-cofounder Moore's groundbreaking paper on the expected evolution of integrated circuits, most processes were still based on analog technologies. In his article he envisions a future with "such wonders as home computers [...] and personal portable communications equipment" [57, p. 1] and discusses different approaches to miniaturize and reduce the costs of electronics equipment. In the remainder of the paper he derives time-dependent cost curves for the number of components per integrated circuit and predicts the number of components (at minimum costs) to be exponentially rising for at least a decade.[44] Two decisive technological drivers of convergence in ICT are listed here. First, digitization enables a free transfer of information (no matter whether pictures, sounds or text) through various transport mechanisms. Second, miniaturization

[43] See also Sect. 2.1.1.

[44] With a few alterations this prediction has been named "Moore's Law" after him. See also Sect. 2.2.2.

secures the applicability of integrated circuits in a wide array of increasingly small equipment. Digitization most likely plays the central role [cf. 2, p. 2224]. It could even be classified a prerequisite for convergence in ICT, instead of just one of many triggers and drivers. Without the switching from analog to digital means of information transportation, the different branches would have had no common 'language' to use. To that extent, grounding products and services on bits and bytes may be regarded a 'generic technology' See, e.g., [68, p. 228], as it can be employed similarly across different industry boundaries. Indeed, digital technology probably is a generic technology. And much more, as the notion 'language' would suggest. Consequently, it is not only a technology to send information from A to B in certain settings. As a result, these products or services have now been replaced into a competitive environment, addressing similar demands and markets.

Furthermore, digitization actually changed underlying paradigms of, for instance, telecommunication and computers. When the internet began to receive increased attention during the 1990s, people used their telephone lines and analog modems to connect to it. Nowadays, more and more consumers and companies use their digital broadband access to administer their telephoning activities. Also, VoIP handsets have been introduced by mobile device vendors and the underlying technology is being switched to packet-switched data transmission [cf. 14, pp. 14ff.; 52, p. 728; 53, p. 321]. This technology in combination with efforts to grant free access to "Wireless Local Area Networks" (WLAN) or even "Wireless Metropolitan Area Networks" (WMAN), could lead to substantial losses for telecommunication providers concentrating on established airtime business models. Next to free and wireless internet access in many hotels, restaurants and shops, several cities and even countries around the world have free WLAN access in place. For instance, all of Estonia's 45,000 km^2 are covered by WLAN, partly financed through advertisement, partly through a kind of levy on purchases [cf. 69]. Also, cities like Swindon (UK) [cf. 70] or Milpitas (USA) [cf. 71] provide free WLAN access to their population. On the other hand, some very ambitious projects have been called off due to a lack of investors or sponsors, such as in San Francisco (USA) [cf. 72] and New York (USA), where all parks were supposed to be free WLAN zones [cf. 73].

Miniaturization, on the other hand, is part of a larger set of technological developments making the process of convergence more likely and easier, but this convergence was beginning to speed up already at a time when 'mobile phones' would still fill the trunk of a car. Miniaturization definitely plays a central role when it comes to the performance of microprocessors as well as the possibility to assemble devices with multiple functionalities. The calculation power of any microprocessor-containing device, from dish washers and cars to telephones and personal computers is still exponentially rising. And accordingly are the possibilities for computer-aided applications [cf. 2, p. 2223]. Modern more luxurious cars, for example, use them for a plethora of functionalities from engine control to navigation, adaptive cruise controls, blind spot detection or even automatic parking. Moreover, miniaturization of all components has enabled devices such as smartphones with their bundle of functions. And with these technological developments, also production technologies and material properties became more and more equal [cf. 30, p. 355].

After a ground was thus laid for a blurring of boundaries by technological change, another important driver was the deregulation of the telecommunications markets [cf. 38, p. 1; 53, p. 318]. In the USA, e.g., the "Telecommunications Act of 1996" aimed at letting anyone compete in any market with any communications business [cf. 74]. This act was the first amendment to the Communications Act of 1934 and opens with the pledge "To promote competition and reduce regulation in order to secure lower prices and higher quality services for American telecommunications consumers and encourage the rapid deployment of new telecommunications technologies" [75, 110 STAT. 56]. Another good example is Europe, where the "Commission Directive 90/388/EEC on competition in the markets for telecommunications services" was introduced in 1990 to progressively introduce competition to the telecommunications sector [cf. 76, p. 10], followed by further legislation in the years afterwards [cf. 12, p. 33]. 12 years later a series of directives (including 2002/77/EC) was trying to introduce a common regulatory framework for "all electronic communications services and/or networks which are concerned with the conveyance of signals by wire, radio, optical or other electromagnetic means (i.e. fixed, wireless, cable television, satellite networks) [77, p. 21]. It even mentions a convergence phenomenon having "shaped the information technology, media and telecommunications industries over recent years" [77, p. 21]. Similar developments have taken place in most countries, opening the hitherto monopolized telecommunications sector (and related sectors) to competition and thus furthering convergence. Bierly and Chakrabarti argue that deregulation has fuelled industry fusion in the ICT sector and forecast companies from the different telecommunication (long-distance and local), the computer and the computer software industries to be competing against each other in the future [cf. 19, p. 36].

Next to technological (especially digitization) and regulatory aspects [cf. 2, p. 2223], a third trend has propelled developments in this industry—the trend for one-stop shopping or multi-purpose devices See, e.g., [2, p. 2228; 26, p. 9]. In this context, Kaluza et al. highlight increased pressure from substitutive products during the process of convergence [cf. 2, p. 2241]. This can be well observed in the domain of internet service providers (ISPs), cable and satellite television providers and telephone providers. Each of them has faced competition from other providers in his area since deregulation. But today, all of them are directly competing with each other. In the German market, for example, telecommunication companies like the Deutsche Telekom offer broadband, telephone and television access, just like cable providers such as Unitymedia or ISPs like 1&1. Thus, convergence leads to intensified competition and results in decreasing prices for these services.[45]

[45] The prospect of diminishing prices is actually one of the reasons for the deregulation activities of the past two decades see, e.g., [75, 110 STAT. 56].

Firms from the initial ICT sectors were among the first and most obvious to react to the blurring or reorganization of industry boundaries.[46] According to an estimate by the European Commission in 1997, 15 % of the total value of worldwide M&A activity of US$ 1 trillion was rooted in ICT [cf. 15, p. 6]. One of the most prominent examples is probably the merger of AOL and Time Warner in 2000,[47] bringing together internet and more traditional media competencies [cf. 78]. At the beginning their merger was dressed in very optimistic wording, aiming at creating the "[...] World's First Internet-Age Media and Communications Company" [78].[48] They acclaimed the uniqueness of this all-stock combination valued at US$ 350 billion. According to the board's statements, they had expected to be able to build on a range of growth opportunities, from music and entertainment to broadband and telephony. Although supposedly a merger between equals, many saw it as rather a takeover of Time Warner by the (revenue wise) much smaller AOL. This was reflected in the name of the new company (AOL Time Warner), in the distribution of shares (55 % of AOL Time Warner was held by former AOL shareholders) and the fact that AOL's chairman was also chairman of the combined board. And the basic idea behind it was preparing for convergence [cf. 78–80]. But somehow this merger, which has even been called "the worst deal of the century" [81] and "one of the largest blunders in corporate history" [82], never really delivered on its promises. Most likely, this was more due to the outdated AOL business model of dial-up internet access combined with the provision of own content than to a serious misjudgment of the convergence phenomenon. Today AOL's business spans dial-up and VoIP telephony, but increasingly more important—advertisement (for that reason the majority of services is now offered to customers for free).[49] Ironically, a company that set out to spearhead convergence-related industry changes has been immensely affected by the increase in broadband-access, a prerequisite for convergence in other services. Whatever the reasons were, the stock market value of the combined company fell by about 80 % since the merger [cf. 84]. In 2003, "AOL" was dropped from the name [cf. 85] and finally in 2009 Time Warner spun off AOL [cf. 83].

But even long before them, other firms tried to harden themselves against increased competition from new fields of business [cf. 2, p. 2225]. In 1984, IBM wanted to strengthen its position through acquisition of telecommunication equipment maker Rolm [cf. 30, pp. 356f.], while AT&T purchased, e.g., 25 % of

[46] Actually, some attribute convergence as a major impact to M&A activities, especially in ICT see, e.g., [68, pp. 227f.].

[47] The deal was announced on January 10th, 2000. The merger was then completed after gaining approval from the Federal Trade Commission and the Federal Communications Commission on January 11th, 2001. Hence parts of the literature list the merger as happening in 2000 or 2001, respectively.

[48] Capitalization is according to original text.

[49] The shift in focus is also reflected in revenues. The share of subscription income in total revenues is down from 74 % in 2006 to 46 % in 2008, including an increase of 11 % in advertising revenues and a decrease of 67 % in subscription revenues [cf. 83, pp. 42ff.].

computer maker Olivetti in 1983 [cf. 50, p. 449; 86] and in 1991 acquired NCR (National Cash Register). Rolm was a small Silicon Valley company specialized in PBX (private branch exchange, i.e. routing calls within offices) [cf. 50, 87, p. 449]. Olivetti and NCR were supposed to open up the personal computer market to AT&T as well as a specific distribution channel into the financial services market (in the case of NCR). But all three maneuvers were later reversed [cf. 1, p. 1082; 19, p. 36; 28, p. 145; 50, pp. 448ff.], just as Matsushita Electric's takeover of MCA (Music Corporation of America) and Universal Pictures in 1990. IBM sold Rolm to Siemens after yearly losses of more than US$ 100 million [cf. 14, 87, p. 4]. AT&T sold their stake in Olivetti in 1989 [cf. 88, p. 9] as well as spinning off NCR in 1996 [cf. 89]. Markoff refers to experts who had by the late 1980s changed their opinion on ICT convergence. Initially they had awaited a blending of data and voice streams but had later acknowledged that the massive integration of both services would not be due for a longer time [cf. 87].[50] Additionally, others have criticized the way and the propositions under which many of these early years' convergence-driven alliances and M&As were conducted. Despite possible economies of scope, many of these strategic actions lacked the necessary inner fit. Large companies, incumbents in their own sectors, often underestimated the differences between selling personal computers and providing telecommunications services, for example. Disillusioned by their experiences, this has led some of the players to question the underlying logic in general. IBM is said to have realized "that convergence between computer and communications technologies was difficult to achieve" [50, p. 450, based on Electronic Business, June 15th, 1988, p. 63] and AT&T's top management that "the concept of converging communications and computer markets, which drove the NCR deal, is an illusion" [cf. 50, pp. 448f., based on Business Week, October 2nd, 1995, p. 28]. One of the rather few who stayed with their enlarged portfolio is Sony with its stake in the media market [cf. 16, p. 7].[51]

Whilst most scholars report M&A activities caused by convergence (or expect it for the future),[52] only few try to analyze the ex or even the results of it. Bower introduces a conceptual framework for five different types of M&A activities and their strategic triggers. One of those is a scenario of industry convergence, which in his analysis of 1997–1999 deals in the USA above US$ 500 million is the reason for 4 % of M&As [cf. 90]. Typical examples would include the 2000 deal of AOL/Time Warner or 2001 Allianz/Dresdner Bank takeover [cf. 91, pp. 36ff.].[53] Gambardella and Torrisi add to the discussion by questioning the performance

---

[50] Thus it seems to be of critical importance to have the proper timing. When firms act to hastily, this can result in costly adventures. On the other hand, waiting for too long may prove to be as detrimental [cf. 12, p. 237].

[51] See also the more thorough discussion of Sony's activities around convergence in ICT later in this chapter.

[52] For authors discussing M&A activities as a reaction to convergence cf., e.g., [38, 50, 90, 91]. For an overview and short discussion of prior literature see also [12, pp. 41f.].

[53] See also Sect. 2.3.5 for a brief discussion of convergence between insurance and banking.

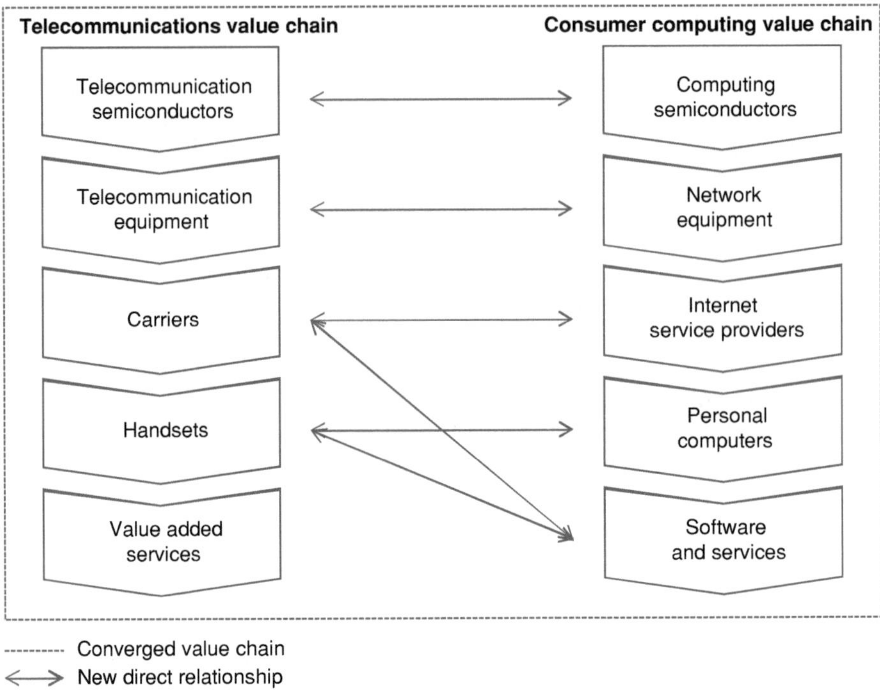

Fig. 2.4 Converging value chains in ICT (*Source* Adapted from [11, pp. 85ff.])

level after such a strategic change. In their analysis of 32 companies from Europe and the USA (all ICT companies listed in 1990s Fortune 500), they find that firms with stronger focus actually performed better [cf. 50, pp. 450ff.].

The blurring boundaries in ICT have also led to fundamental changes in value chains [cf. 12, pp. 48f.]. Figure 2.4 depicts these changes resulting in heavily increased competition between companies active in telecommunications and those in consumer computing.[54] Whereas originally both sectors had separate value chains with separate players for each of their steps, connections increased and lead to faded boundaries between the two. As a result, firms active as semiconductor or equipment providers, are now able to offer their goods and services also in the respective other area, as their technological and customer bases are merging into one larger, unified area. On the downside, this also means competition from their counterpart, of course. But these connections are not only established on similar levels of the value chain. Accordingly, telecommunication handset manufacturers are also offering software or services from the domain of computing.[55]

As can be seen when looking at the example of Apple, it is of little importance today, in which sector a firm was originally active. Apple is a good example for a

[54] [Cf. 11, pp. 84ff.].

[55] See Footnote 54.

company starting with computer devices and software offerings. In 1976 it was founded as "Apple Computer, Inc.", stressing its focus of building and selling computers. For the next decades it concentrated on the development of computers, peripherals and specialized software. Specifically, its inclusion of a graphical user interface (GUI) had a great influence on other operating systems' providers. It was initiated by an agreement between Xerox (who had invented the GUI with their system "PARC") and Apple,[56] allowing Apple to study Xerox's GUI and Xerox to invest US$ 1 million in Apple.

The first real move away from the computer industry value chain can be seen in the production (and licensing) of the Apple Bandai "Pippin" multimedia platform in 1995. It was supposed to be an inexpensive network computer with computer games functionality, but mainly perceived as a games console, inferior to its competitors like Sega Saturn or Sony PlayStation [cf. 92].[57] The final move into the consumer electronics market was the introduction of the portable digital audio player "iPod" in 2001, complemented by the 2003 launch of the strongly intertwined "iTunes Store" (offering online music downloads) [cf. 93, 94]. On January 9th, 2007, a Form 8-K was filed with the United States Securities and Exchange Commission (SEC) announcing its renaming to "Apple Inc." [cf. 95]. The name change was supposed to reflect Apple's strategic move away from just personal computers and software to the addition of media offerings (iTunes Store), consumer electronics (iPod range and Apple TV) and mobile phones (iPhone range) [cf. 96, p. 1]. This could be interpreted as the desire to have its name also reflect its claiming convergence ground in ICT. On the same day Apple's first smartphone "iPhone" and its digital media receiver "Apple TV" were presented to the public [cf. 97], marking the inclusion of a further foreign value chain element—the production of telecommunication handhelds. Being a smartphone from the very beginning, the latest iPhone model also contains a mobile navigation system.

But Apple is not the only company incorporating the ICT convergence in its fields of activities. Sony Corporation is another very prominent example. Founded in 1946, it (including subsidiaries and joint ventures) nowadays sells a wide array of products, from audio and television devices, cameras, batteries and semiconductors to personal computers, music, motion pictures and telephones.[58] Originally it developed audio devices, like magnetic tape recorders and transistor radios. Later adding television sets and semiconductors, Sony teamed with CBS in 1968 to form the 50:50 joint venture CBS/Sony Records, which is nowadays the completely owned Sony Music Entertainment. So, already more than 40 years ago, the company moved away from just hardware manufacturing to producing content as well. Next to selling further different consumer electronics (like the portable

---

[56] This was in advance of Apple's initial public offering (IPO) in 1980.

[57] Game consoles themselves are today seen as convergence products by some researchers. They can be used as gaming devices, DVD (or Blu-ray Disc) players or even to access the internet [cf. 49, pp. 202ff.].

[58] Today through Sony Ericsson Mobile Communications, a 50:50 joint venture between Sony Corporation and Telefonaktiebolaget LM Ericsson [cf. 98–100].

music playing device “Walkman” and photo cameras), Sony began to sell personal computers in 1982 and phones in 2001 [cf. 101]. Hence, it is another suitable example of a company trying to mirror convergence trends in the whole industry within its own organization. In contrast to Apple, Sony has not only broadened its in-house capabilities, but also reacted to developments in ICT by collaborating with other players in the form of joint ventures (e.g. initially CBS Records, Prudential Insurance or Ericsson) or by acquiring them (e.g. Columbia Pictures Entertainment or AIWA) [cf. 101]. Strikingly, the company is not only reacting to this phenomenon, but also explicitly addressing convergence. In 1998, Sony and Microsoft issued a joint press release announcing plans to create convergence of personal computers and consumer audio-visual electronics by several collaboration activities [cf. 102]. Today, they describe themselves as being focused on electronics, game, entertainment and financial services. Amounting to a “globally unique” setup, Sony wants to “aggressively [carry] out [their] convergence strategy” [103].

In contrast to AOL/Time Warner, Apple and Sony can be viewed as convergence examples grounded on a broader foundation. Their development is not only pushed by outside events such as technological change, demand shifts or deregulation, but also decisively by managerial triggers and drivers (see Table 2.1). Similarly, Nyström discerns between the four drivers organizations, technology, deregulation and demand in the ICT domain. To the common factors of technological change in general and digitization in particular, she also emphasizes the role of a transition to IP (Internet Protocol, a protocol for the transfer of packet-switched data) technology [cf. 12, pp. 29ff.]. Due to this technology’s indifference to the type of transferred data, it can and does serve as a platform for all different types, mainly texts, photos, sounds and video. Nyström also emphasizes the necessity of a combination of different drivers for convergence to happen. According to her line of argument, ICT boundaries would not have blurred if it was not for technological **and** regulatory developments [cf. 12, p. 35].

Albeit all the decisive changes in ICT related industries, it is still being debated how much convergence actually has happened in this area. Lind stresses that while telecom, computer, media and consumer electronics industries were substantially transformed by the events of the last decades, they are still around as largely separate industry sectors. He sees manifestation of changes in their value propositions and business models, but thus far no challenging of their actual existence. Furthermore, in his assessment the number of industry sectors or at least markets has increased, with entry barriers in these markets as high as they had been in traditional settings [cf. 4, pp. 13f.]. Regarding convergence on the media edge of ICT, Gordon summarizes the status quo in the year 2002 to be far from finalized, but with convergence’s outlines taking shape. The vision had been for all media content to be stored and distributed digitally and via the same distribution lines to multi-purpose devices. But while he rightly analyzes that this is not the case in his book chapter published in 2003, many of those have changed in the meantime. Gordon identifies major obstacles in limited transmission speeds and differing formats [cf. 7, pp. 60ff.]. Although there is still a broad selection of file formats, compatibility has probably improved as much as connection speeds. In their 1998 study on strategic alliances and technological competencies in a convergence setting in the ICT industry, Duysters and Hagedoorn find that despite technological convergence firms appear to sustain their original technological base. Analyzing strategic alliances and patent data of "leading" computer and telecommunication equipment manufacturers, they discuss this concentration on their traditional respective technology areas as probably the most efficient approach. This is accompanied by a tendency of (at least initial) inertia to change. The authors still expect this single focus approach to alter in the future [cf. 30, pp. 359ff.].

Another reason for a much slower realization of convergence in ICT may lie in the fact that most developments in the earlier phase were focused on just the technological or the regulatory side and only to a lesser extent meant to address the end-customer. Tarjanne argues that people need to be the focus of attention if convergence is supposed to be successful [cf. 12, pp. 43f.; 104, pp. 41ff.]. Similarly, Katz sees drivers of convergence in ICT to be twofold—supply-side digitization drove convergence at the beginning, before demand-side effects took over. In his opinion this is due to a re-thinking of corporate customers [cf. 1, p. 1084].[59] But this insinuation of each driver happening sequentially over a course of time is probably over-simplified. It appears to be not so much a question of one driver necessarily being the off-setter before another driver takes over, but rather the co-existence of different drivers at the same time. In contrast to smartphones[60] or triple play[61] services which provide benefits for the consumer and address his tendency towards one-stop shopping, earlier developments evolved without a clear added benefit. For instance, a shift from analog to digital telephony may be scientifically fascinating or resulting in cost reduction possibilities for the providing firms. But only if this leads to cheaper connection charges, better sound quality or added functionalities, it is of interest to consumers as well. Hence, it could be expected that full industry convergence will only happen if companies are successful in transforming technology convergence[62] into customers' perception of improved products or services.

---

[59] In this approach to emphasize the influence of the customer base Katz also predicts that "within a decade, many people will take multimedia 'phones' for granted", driven by the wish to engage in one-stop shopping [cf. 1, p. 1086].

[60] Nyström relates to smartphones (or in her wording mobile phones) as the point of convergence, with influences and inputs from the areas of communication (e.g., speech, messaging, e-mail), broadcast media (e.g., TV, radio) and mobile data (e.g., internet, mobile services) [cf. 12, pp. 221f.].

[61] 'Triple play' is a term used in marketing and referring to the simultaneous provision of data, voice and television streams. Sometimes it is even extended to 'quadruple play', including mobile phone access as well. In most cases it is administered through the networks of either broadband service providers or cable providers see, e.g., [49, p. 214].

[62] This study views convergence as happening sequentially, based on different levels. See Sect. 2.4 for a detailed discussion of different phases of convergence.

Furthermore, convergence may also lead to divergence. Kaumanns et al. describe changes in customers' perception and usage of products and services to be marked by a growing diversity of offerings, media consumption patterns, target groups and consequently business models [cf. 105, p. 29]. Based on their representative survey of 1,000 German consumers between 14 and 69 years of age, they discern between different 'usage modes', depending on the type of activity. Most people consume more traditional audio-visual media in a 'lean back-mode', whereas they largely switch to a 'lean forward-mode' when accessing the internet. Recent developments enable consumers to turn into 'ProSumers' and actively influence or even produce media content [cf. 105, p. 2]. In their assessment those companies are profiting from convergence that combine devices with services and content, such as Apple's iPod [cf. 105, p. 30].

This convergence/divergence dichotomy (or at least its perception) changed from the time of the first discussion of convergence in ICT. In the 1980s companies appeared to be able to use their core technologies to produce a broadening range of products, whereas in the 1990s they were expected to include more and more different technologies into one product. Today companies need to do both and balance convergence in technologies/products with divergence in the respective other [cf. 28, pp. 126ff.]. Only companies aware of changing market dynamics and with the capability to address those will be able to use a convergence setting to their advantage. In many cases this will not be the more established and path-dependent large companies, but rather small and more flexible ones.[63]

Furthermore, when assessing developments in ICT different levels and patterns of innovative activity need to be considered. Patent growth rates, technological pervasiveness or share of self-citations may vary greatly even within the ICT sector [cf. 106].[64] Because of all these characteristics in ICT, Bröring concludes "[...] there is no total substitution of hitherto separate sectors, but rather a fragmentation into different sub-segments and value chains" [60, p. 280].[65] Accordingly, the International Engineering Consortium (IEC) is expecting a realignment of the five traditional information industries photography, publishing, telecommunications, computing and entertainment into the three new sectors information content, information appliances and information highways. In contrast to the traditional boundaries along the different forms of information (images, text, voice, data and audio/video), the new sectors are expected to be based on the functions of information (create/collect, display/store and process/distribute). Governmental

[63] See, e.g., [60, pp. 280ff.] for a discussion of efforts undertaken by AT&T, Palm and Research in Motion (RIM) developing the market segment of PDAs and later smartphones. While AT&T withdrew quickly because of their failure to successfully market their PDA 'EO 440', Palm and RIM were highly successful in transporting the added benefits of their handhelds/BlackBerrys to the end-customer.

[64] Corrocher, Malerba & Montobbio discern between different "Schumpeterian patterns of innovative activity in the ICT field". Section. 3.3 will discuss differences in, e.g., the propensity to patent and how this may influence patent analyses in greater detail.

[65] See also, e.g., [16, p. 11] for a similar assessment.

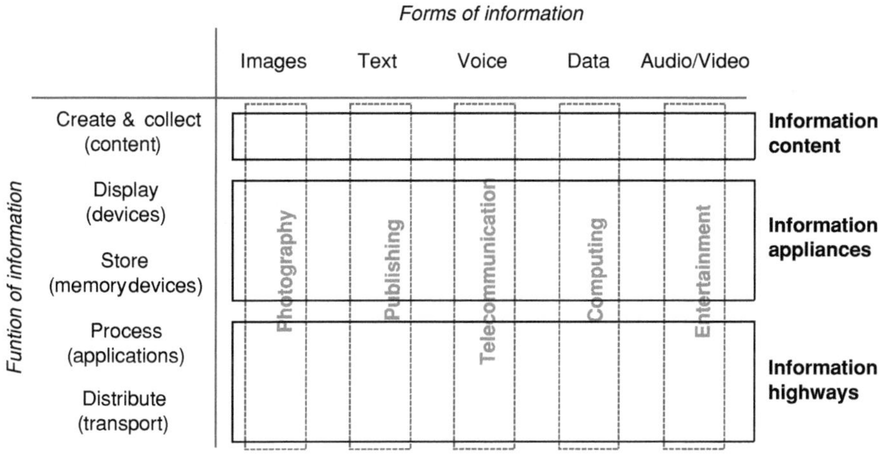

Fig. 2.5 Convergence in ICT based on a shift of focus from forms of information to functions of information (*Source* Adapted from [107, pp. 7ff.])

deregulation of barriers between the different forms on the one hand and the technological ability to carry out any of the function irrespective of the form of data to be used on the other hand, are believed to be leading to this reshaping of the information industry (Fig. 2.5) [cf. 107, pp. 3ff.]. Initially, the IEC has proposed this fading of boundaries in 1999 and anticipated it for ten to fifteen years later [cf. 107, p. 3; 108, p. 305]. Table 2.2 depicts this change in industry structure and the related core competencies. In spite of companies that are trying to bridge the gap between these three functional sectors (like Apple or Sony), recent developments have indeed led to a concentration based on functions rather than forms of information. Section 5.2.6 will also briefly address manifestations of this new structure in patent literature.

Table 2.2 Convergence in ICT with new industry sectors and their relation to traditional sectors

New industry sector	Examples of needed core competencies/ expertise	Traditional industry sectors linked to it
Information content	• Fostering and managing creativity • Information gathering • Programming skills	• Publishing • Entertainment
Information appliances	• Global presence (economies of scale) • Strategic sourcing • Design and manufacturing expertise • Concurrent engineering and marketing	• Consumer electronics • Personal computers
Information transport	• Network management • Interactive communications • Billing systems • IT platforms	• Telecommunication • Cable TV • Broadcast media • Private networks

Source Adapted from [107, pp. 18f.]

2.3.2 The IT and Electronics Industry with Other Industries

Constituting a largely generic technology with broad applications in virtually any field of modern life, IT has been discussed to be converging with many different other fields as well. An often employed example is the area of "intelligent buildings" [cf. 11, p. 3] or "smart homes" [cf. 109, pp. 158ff.]. Here building technologies, especially regarding temperature control and intrusion and fire safety, are expected to merge with information technology [cf. 11, p. 3]. The employment of information technology to control household appliances as well as the abovementioned building technologies can be accepted as a given fact. In contrast to this, more far-fetching visions of houses completely managed by a home computer system that would signal a finished turn of the washing machine on the television set and would consist of completely intertwined and 'intelligent' systems, have thus far proven to be overoptimistic. Notwithstanding these drawbacks, the promoters of smart homes appear to be just increasing the level of 'smartness' in homes since the first discussions in the 1980s. Today they are still talking about fridges connected to the internet and automatically restocking their content as well as the possibility to access all the control systems from wherever the homeowner would be in the world [cf. 109, pp. 158ff.]. Certainly, many of these functions are technically possible today, but still most homes are built rather 'unintelligent'. Independent of the degree of realization of these predictions, smart homes can hardly be interpreted as an area of converging industries. As much as architects, builders and producers of household appliances are increasingly using information technology, as little are their industry sectors likely to converge into one common sector. Even more, following the definition of convergence within this study they would not even been characterized as converging at all (not even on the technology level) but rather the one technology using the other.

Furthermore, the IT industry is part of a reported convergence with the security industry. As security services are largely shifting away from analog technologies to digital technologies, IT companies enter security industry's market [cf. 55, pp. 1ff.; 56, pp. 1ff.].

The intersection of the packaging and paper industry on the one hand and the chemical industry on the other hand is another increasingly intertwined area largely based on the development of nanotechnology competencies. Printable electronics are a product range that is part of this possible convergence area. They include OLEDs,[66] RFID[67] tags and other (semi-)conducting modules that can be printed onto substrates and serve a broad array of purposes. Their functionalities include displays, sensors and indicators [cf. 11, p. 3; 60, pp. 282ff.; 110, pp. 396ff.]. Whereas the products are competing with other products from the electronics and the packaging and paper industry, they are to a high degree only an example of technological convergence and not of market convergence. Especially the chemical

[66] Organic light emitting diodes.

[67] Radio frequency identification.

industry collaborates with players from the electronics domain to develop printable electronics solutions consisting of the chemical ingredients and the machinery equipment needed to actually print and sell these products [cf. 60, p. 282].

2.3.3 The Chemical Industry

The chemical industry, in the middle of a paradigm shift, is full of further examples of blurring boundaries. First, most firms have closed down their huge central laboratories and redeployed their researchers to smaller units managed by the individual business unit. This set up is aiming at less basic research and more customer-oriented development, with customers altering their demands and buying behavior substantially. Second, this traditionally technology-driven industry has realized that most impulses for innovations have in the recent past come from its neighboring disciplines. While their substantial R&D budgets for the assessment of inventions used to be spent on experimental trial and error, the so-called 'innovation scissors' (the difference between the increasing development time and decreasing commercial exploitation time), became a serious problem since the 1960s [cf. 111, p. 57]. Most companies tried to battle the fact that they were developing fewer really new molecules that often only showed incremental changes in properties, compared to existing materials, by increasing the speed of their trial and error approach. These ultra high throughput screening methods[68] are still industry standard today, but they are supported by methods developed in neighboring fields, especially informatics (leading to computational chemistry and molecular modeling) and biotechnology. Working with informatics allows for fewer physical experiments and thus increased turnover and methods developed by researchers in biotechnology make a directed approach possible. Through understanding of physiological mechanisms within a plant or the human body, researchers can start to look for a specific key to a lock much more target-oriented. Consequently, (partial) convergence has happened particularly between chemistry and biotechnology in science and technology. But lacking a real industry of its own, there is no real biotechnological industry that could converge with the chemical industry. Biotechnological companies are characterized as such due to their technological base and not because of independent markets or customers. Their products serve as substitutes for either chemically derived products or pharmaceutical products.

Further drivers of convergence tendencies within the chemical industry structure include the alteration of the raw materials base. For about half a century almost all chemicals for the different kinds of applications were synthesized from

[68] UHTS (ultra high throughput screening) is based on fully automated testing of tens of thousands of substances from a company's substance library on single targets, especially in the development of new pharmaceutical drugs or chemicals to be used in crop protection.

oil. But while the demand for oil has surged for decades, the world's reserves are definitely limited. Additionally, environmental awareness has increased in most countries and led to an increasing interest in products derived from renewable raw materials. This has led to the blurring of boundaries between the incumbent materials manufacturers and new competitors with access to renewable raw materials. Hence, chemical companies like Dow or Bayer are today competing with agricultural companies like Cargill or Archer Daniels Midland. This could lead to an industry convergence of parts of the chemical industry and parts of the agricultural industry [cf. 112, pp. 1ff.].

2.3.4 The Pharmaceutical Industry

In contrast to this competition between similarly sized firms, the pharmaceutical industry has in the past mainly taken over considerably smaller biotechnological firms to fill its product pipeline or to broaden its technological knowledge base [cf. 18 pp. 8ff.]. This has to be seen in the light of a setting, where on the one hand one active ingredient alone may account for over US$ 1 billion p.a. in sales. However, on the other hand, companies have to deal with extremely high failure rates of lead compounds tested in the discovery phase: the estimated success rates are about 0.01–0.02 % [cf. 113, p. 20]. Hence, many firms experiencing backlashes with predicted blockbusters[69] have turned to biotechnological firms to fill their drying product pipelines. While this often results in acquisition of the biotech firm (e.g. Roche buying Genentech, Merck KGaA acquiring Serono, or AstraZeneca's US$ 15 billion takeover of MedImmune), building alliances with others outside the firm's boundaries may often be preferable in regard to costs, flexibility and broadness of cooperation possibilities.

In the pharmaceutical industry's relation with biotech firms, often not convergence but mere utilization of technologies or even whole companies plays the decisive role. But at the borders of this segment two clear examples of industry convergence have earned some considerable attention—the emerging sectors of NFF and of Cosmeceuticals. The NFF sector has been emerging since the early 1990s at the intersection of pharmaceutical and food industry sectors cf., e.g., [20–22], starting in Japan, which is still a very strong market for NFF [cf. 114]. The NFF sector is marked by input-side convergence as well as output-side convergence. Food technology competencies are converging with LDL-cholesterol lowering technologies. Regulations regarding pharmaceutical products are increasingly transferred also to food products with enhanced characteristics. And finally, customers are seeking to purchase foods with added health-benefits in an attempt of one-stop shopping [cf. 20, p. 83]. Thus firms do not only need knowledge and technological

69 Blockbusters are pharmaceutical products that account for at least US$ 1 billion in annual sales.

skills formerly unimportant to them, but also find themselves competing with one another in one unified market of nutrition and disease prevention [cf. 115, pp. 11ff.].

A similar setting can be found in the sector of Cosmeceuticals [cf. 11, p. 42], a term coined by Klingman in the 1970s [cf. 116]. It is reported to be the fastest growing segment of the "natural personal care industry" with cosmetic-pharmaceutical hybrids aiming at enhancing both "beauty" and health. [cf. 117, p. 155] While health benefits appear to be more disputed than for most products in the NFF sector,[70] some experts are calling for a harmonized regulation through bodies like the FDA (US Food and Drug Administration) [cf. 117, p. 158]. The broader the health claims marketed with the Cosmeceuticals, the more likely they will have to undergo scrutiny by drug administration bodies, leading to a convergence of regulations and standards. As examples from the area of Cosmeceuticals and especially NFF constitute the prime empirical base for the analyses conducted in this study, they will be analyzed in greater detail in Sect. 4.3.

With "many of the innovations that have transformed health care [having] come from the combination of progress in seemingly unrelated areas" [118, p. 34], the pharmaceutical industry is also affected by developments involving nanotechnology, biotechnology, information technology and cognitive science (NBIC), which are aiming at the "improvement of human performance" [cf. 11, pp. 187ff.; 119, pp. 281ff.; 120, pp. 35f.]. With possibly far-reaching implications, Roco and Bainbridge speak of the world standing "at the threshold of a New Renaissance in science and technology, based on a comprehensive understanding of the structure and behavior of matter from the nanoscale to the most complex system yet discovered, the human brain" [119, p. 281]. Similarly, Hacklin argues that nanotechnology will act as a kind of catalyst or platform for a range of existing scientific disciplines, including mathematics, physics, chemistry, biology and engineering sciences [cf. 11, pp. 188f.]. The combination of these different disciplines and technological fields[71] will lead to products like biosensors, nanosensors or biological nanoscale chips which could be employed in drug delivery or other areas to overcome human illnesses and 'limitations'. Roco & Bainbridge report on six visionary topics in combination with NBIC:

- Overall potential of converging technologies
- Expanding human cognition and communication
- Improving human health and physical capabilities
- Enhancing group and societal outcomes
- National security
- Unifying science and education.

[70] See, e.g., [116] who also cites Dr. Nicholas Perricone, founder of NV Perricone MD Cosmeceuticals, opposing accusations of Cosmeceuticals' low credibility: "Cosmeceuticals are science-based rather than marketing-driven, giving the customer results and value for their investment."

[71] Other authors describe similar developments but limit them to only nanotechnology and biotechnology cf., e.g., [121, pp. 105ff.].

These topics are largely related to anticipated alterations in basically any function of the human mind and body, including communication, thinking, sensory and physical strength. Through a massive increase in human performance, they expect not only enhanced national productivity but also an improvement of societal outcomes and quality of life [cf. 119, pp. 281ff.]. While these developments could certainly lead to massive changes in the underlying technologies and the way products and processes are constructed, industry convergence would need distinct industry sectors, which could then begin to converge. As discussed above, it remains arguable whether these are existent in connection with the respective scientific disciplines.

2.3.5 Service Sector and Other Industries

Certainly, convergence is mainly a phenomenon discussed in connection with research-intensive industries. Following the line of argument in most studies (including the present one), this is also rightly so as it is mainly driven by substantial changes in the underlying technology and knowledge bases. In contrast to this, convergence may also be driven by political or societal developments.[72]

One of the largest sectors affected by convergence trends is the health care system. Health insurers are said to engage more in health care management and health information management, aiming at reducing their expenses. Hospitals and physicians spend more effort in delivering non-directly health-related services, such as wellness, nutritional and fitness consulting [cf. 122, pp. xxiiff.]. Some even discuss a convergence of health care and the internet, with more and more information searched on the internet and sometimes even online discussion of possible treatments [cf. 123, pp.8ff.]. Despite several further (and probably to a certain extent unknown) parameters influencing the path of the health care system, Pierce concludes: "Convergence will be gradual and the impact may not be immediately obvious, but the long term outcome will dramatically reshape the industry" [122, p. xxx].

Part of this reshaping is a blurring of boundaries between traditional health insurers and banking institutions. Especially in the USA new legislations are expected to lead to a convergence of both sectors [cf. 124, pp. 3ff.]. The predicted merger of banking and insurance institutions (termed 'bank assurance') is not necessarily limited to health insurers. Furthermore, banks are experiencing more turmoil from other angles. For example, car manufacturers are offering customers more and more banking services. While potential buyers used to ask their local bank for a loan, they nowadays often turn to the banking affiliate of their car maker. And those banks do not only offer services in direct connection with the

[72] See Sect. 2.2.3 for an overview of different triggers and drivers of convergence and their possible classification.

purchase of cars, in the meantime they have established quite a vast array of regular banking services [cf. 125, p. 348].

These examples share a market perspective rather than a technological perspective driving their convergence. This is different with the final example. Rosenberg's technological convergence is focusing on developments in all "machinery and metal-using sectors" [8, p. 423]. Due to the employment of a common set of technologies, previously unrelated industries (because of the characteristics and the applications of the products) converge technologically. He uses the examples of firearms, sewing machines and bicycles [8, pp. 423ff.]. Undoubtedly, these industries witnessed technological convergence. As markets and companies did not really begin to merge, industry convergence (as defined earlier) did not occur on the basis of metal-using technologies.

2.4 Process of Convergence

The examples presented in Sect. 2.3 exemplify the variety of convergence phenomena in different industrial settings. This is in part caused by different triggers and drivers and, of course, varying strategic responses by different firms. To another part this is due to diverse parameters and characteristics of the respective industries. But the underlying logic is in most cases similar. For the anticipation of convergence, it is of particular relevance to consider the level at which convergence is supposed to be measured. Thus, this chapter will analyze the sequence of events and the modules on which convergence is usually build upon. A few of the aforementioned authors have tried to categorize convergence on separate levels, on the basis of different points in time or different breadths of actions. Prior to the introduction of an idealized sequence of levels of convergence, three approaches to categorization are briefly discussed.

As well the European Commission as Tarjanne discriminate three levels from each other: (1) technology, (2) industry and (3) services/markets. Following their delineation, convergence on the *technology* level addresses different products/services fulfilling similar needs [cf. 12; 15; 33, p. 37]. So, e.g., NFF would (for a part of the customer base) have medical effects similar to those caused by pharmaceutical drugs. On *industry* level, convergence leads to a higher degree of direct interaction between distinct firms, resulting in the formation of alliances, joint ventures or even M&A activities. Convergence on the level of *services* tries to classify the establishment of new services at the intersection of the formerly disjoint industries. In their opinion, convergence on one of these levels does not necessarily transpire to similar events on the other two levels [cf. 15, pp. IVff.].

Fransman employs a rather comprehensive distinction between five areas[73]: (1) networks (representing infrastructure), (2) industries/markets, (3) products/

[73] Although Fransman uses the term 'areas', it is strongly connected to the discussed 'levels'.

services, (4) firms and (5) technologies. An overview of these five areas, their explanation and examples are provided in Table 2.3. Basically, convergence in *network infrastructure* represents the offering of similar access to one point. One example would be the possibility to connect to the internet via dial-up services over telephone landlines, access through mobile phone networks, broadband, or cable services. Internet telephony is his example for convergence of *industries* and *markets*, with separate industries beginning to compete with each other on the basis of similar products/services. Accordingly, *products/services* convergence reflects the merger of different functionalities afterwards combined into one single product, as can be seen in smartphones (in comparison to cellular phones, digital cameras and portable audio players etc. as separate devices). As a fourth area *firms* are introduced, describing any form of strategic partnerships between companies from different industries. Finally, convergence in *technologies* is sub-divided into three further cases. First, generic technologies for different applications, such as digitization in ICT are mentioned. Second, technologies as a sort of umbrella technology, such as the internet protocols are listed. Third, Fransman identifies competition between distinct technologies providing similar benefits as a sub-division, such as the technologies behind mobile telephony competing with those enabling fixed telephony [cf. 33, pp. 35ff.].

The probably most detailed taxonomy is discussed by Hacklin. He employs four stages of convergence in his "evolutionary and sequential perspective": (1) knowledge convergence, (2) technological convergence, (3) applicational convergence and finally (4) industrial convergence [cf. 11, pp. 56ff.]. According to his definition the first is marked by (to an individual firm) serendipitous *knowledge* spillovers between previously disjoint knowledge bases and leading to a new combination of knowledge sets, initiated by a large scale erosion of isolated industry specific-knowledge [cf. 25, p. 3]. Accordingly, the ability to enrich conventional food with substances that reach the desired area in the human body and can thus trigger a physical reaction, could be viewed as such a combination of distinct knowledge bases. In Hacklin's line of argument this knowledge would not only relate to the technological knowledge sets of food processing and of drug delivery, but also to the related market knowledge. [cf. 11, pp. 56f.] Consequently, he describes *technological* convergence as a "transition of knowledge convergence into a potential for technological innovation" [11, p. 60]. Again, it is not viewed as a conscious-driven act, but rather as a partly serendipitous evolution, based on the combination of different knowledge sets. Digitization would be an example of a technological convergence that in his reasoning would not necessarily lead to convergence in the higher hierarchies of convergence. It could also be seen as just a creation of some sort of widely used, generic technology.

On the next step, the *applicational* convergence, managers' will becomes a decisive factor. Here they need to spot opportunities for a realignment of their business activities, going further than just the creation of new products/services. Such realignment can be seen in the food industry, where many companies have realized and utilized the potential of the new knowledge and technological capabilities and transformed those into new fields of activity. The final step,

Table 2.3 Fransman's classification of convergence areas and their examples

Interrelated convergence areas	Kind of convergence	Examples of convergence
Networks	• Interconnection/interoperability between networks	• Telecom[a] (PSTN,[b] ISDN,[c] fixed and mobile) and data and broadcasting (cable, terrestrial and satellite) networks
Industries/ Markets	• Previously separate industries based on separate markets now offering similar products/services to the same customers	• Telephony, fax transmission, video conferencing and internet access offered by information processing and software companies and telecom companies
Products/ Services	• Products/services incorporating some of the functionalities of other products/services previously sold in disjoint markets	• Television (TV) sets becoming Personal Computer (PC)-like • PCs becoming TV-like • Hybrid PCTVs • Computer-telephony integration • VoIP
Firms	• Firms, previously in separate industries/markets, converging in terms of markets, products/services and technologies	• Joint ventures/strategic alliances between computer hardware and software, telecom, broadcasting companies
Technologies	(a) Digital technology providing a 'common currency' (b) Technologies integrating other technologies (c) Technological competition as a result of and response to convergence	(a) Computing, switching, transmission and broadcasting now all digital (b) TCP/IP[d] Internet protocol or Platform-independent Java programming language (c) Fixed versus mobile telephony or VoIP versus telecom networks or broadband versus cable modems versus satellite

Source Adapted from [33, p. 37]

[a] Telecommunication

[b] Public switched telephone network (*PSTN*), originally network of fixed-line analog telephone systems, today mostly digital

[c] Integrated services digital network (*ISDN*), telephone network system that can transmit voice, video and data streams at the same time over PSTN

[d] Transmission control protocol/internet protocol (*TCP/IP*), two central parts of the Internet Protocol Suite that are often used synonymously for the whole set of communication protocols necessary to communicate through local area networks (*LANs*) or the internet

termed '*industrial* convergence' denotes a degree of convergence, where firms from previously distinct industries are increasingly competing with each other as the industries' boundaries blur. This is set off by changing business models and value chains and will in many cases lead to increased strategic alliances and M&A activity [cf. 11, p. 61f.]. The ICT setting serves well as an example for industry convergence. Undoubtedly, different knowledge sets had to be combined to be able to stretch technological developments such as digitization across industries as diverse as camera manufacturers, state-owned telecommunication monopolists and personal computer manufacturers. New business models have been developed by many companies in ICT and several firms have tried to conquer new markets by siding with firms from other industry sectors.[74] Summing up, Hacklin views convergence as an "evolutionary process" [11, p. 65], with each of the steps triggering the next sequential convergence. In his perception these are preceded by distinct knowledge bases, unassociated technologies, intersected technologies and integrated technologies, respectively [cf. 11, p. 103].

According to the short explanations of the three levels by the European Commission and Tarjanne, it remains unclear what is exactly included on each of the levels and where their borders run. Furthermore, their approach does not entail an explanation of how convergence—in general—evolves. As comprehensive as Fransman's categorization is, it poses the challenge to clearly distinguish between the individual areas. For instance, technologies and networks would appear to be closely related, as competition of different networks could also be viewed as competition between underlying technologies. Similarly, the delineation of industries/markets, firms and products/services seems to be rather arbitrary. It is not clear why products/services should not be part of either the firm perspective (as they are offering them) or the market perspective (as that is where they will be competing against each other). Similarly, although Hacklin's distinction between four sequential stages of convergence is rather convincing, it bears the problem of differentiating between knowledge/technological convergence as well as between applicational/industrial convergence. This is also reflected in his own use of the terms. Several tables in his book, discerning between the four stages, list common results for each of the two pairs.[75] Following his delineation, it is particularly difficult to grasp the concept of knowledge convergence and to decide where knowledge convergence would end and technological convergence would begin. The term 'knowledge' itself is, despite its ubiquitous use, very complicated to define.[76] Hacklin's examples for a merging of knowledge sets tend to be rather intertwined with technological developments [cf. 11, pp. 56ff.]. Thus, a differentiation between (patentable) technological knowledge and technologies themselves appears to be rather complex. Especially when assessing the degree of convergence on different levels, it is important to select distinct proxies for each of the convergence steps. Hence, the present study needs to build its concept on sharply distinguishable sequences and their related indicators.

Convergence has been discussed mainly in connection with industries that are to a high degree science and technology driven. In the ICT as well as the chemical sector and their neighboring industries shortened lifecycles and up to two-digit R&D spending have resulted from pivotal developments in the respective scientific fields. Thus, in most cases a convergence of technologies or even whole industries

---

[74] For a thorough discussion of convergence in ICT see Sect. 2.3.1.

[75] See, e.g., [11] or Table 3.11.

[76] For a thorough definition of knowledge and its delineation from terms such as 'data', 'information' and 'wisdom' see a comprehensive discussion in, e.g., [126, pp. 52ff.].

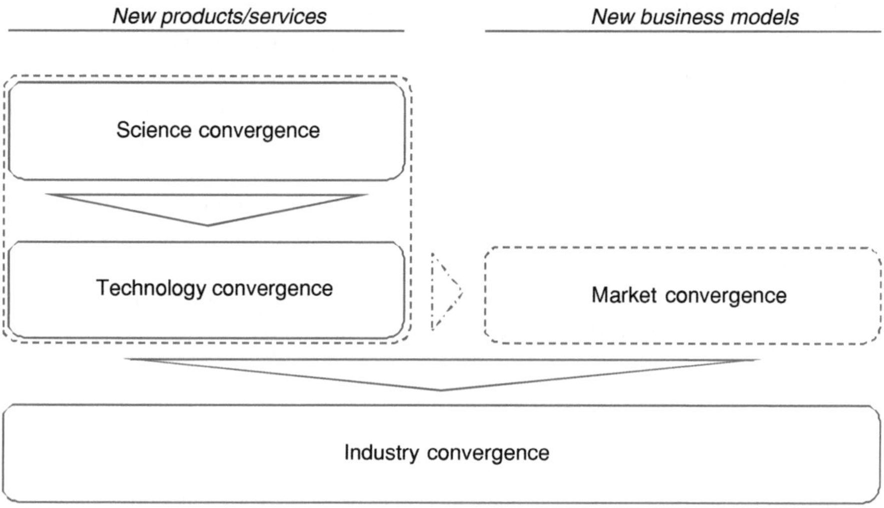

Fig. 2.6 Sequential process of industry convergence. *Dashed lines* indicate facultative dependencies, whereas *solid lines* represent obligatory connections between individual sequences (*Source* Adapted from [127], p. 259)

will succeed a growing overlap of scientific disciplines. The hierarchical nature of these interdependencies is the starting point for the sequential convergence process, which is depicted in Fig. 2.6.

When looking at further cases of convergence, there is mostly a similar pattern of events. At first, different scientific disciplines realize that their research is of interest to or affected by other disciplines. Once they begin to use more and more research results of one another [cf. 47], a scientific convergence will start with cross-disciplinary citations and eventually develop further into closer research collaborations. After the distance between basic science areas has been decreasing for some time, applied science and technology development should follow [cf. 128, 129], leading to technology convergence. This can trigger market convergence, with new product-market combinations. Finally, firms begin to merge with each other, completing the convergence process with industry convergence. Whilst in most cases convergence will thus build up over each of the steps, on some occasions industry convergence may evolve as a direct result of just one of both developments—either new business models or new products/services. For example, convergence of the banking and the insurance sectors has not been driven by technological developments, but rather by new business models. Hence, the idealized time series of events introduced earlier [cf. 130, pp. 386f.; 131, pp. 640f.], has been developed further to also include such phenomena. Also, convergence does not necessarily imply the development of a new industry. In some cases industry areas or, e.g., scientific disciplines may create a new area of collaboration, without changes in the anteceding areas and even nothing more than loosely coupled to them. This may mean a movement from both areas, but as well only one of the fields may be evolving their expertise further in such a direction.

Being aware that this is a simplified and idealized process, this sequential model is still deemed useful to discuss different means for anticipating and (later in the time series) realizing occurrences of convergence. Independent of which of the parts of this process will be part of any case of convergence, it can still be used as a theoretical basis. Naturally, caution has to be exerted when deriving conclusions from such measurements, as it has to be borne in mind that not all steps need to show a tendency towards convergence. Chapters 3 (concept) and 6 (discussion) will discuss this study's approach in anticipating convergence and limitations to the interpretation of accordingly derived results in greater detail.

References

1. Katz, M.L.: Remarks on the economic implications of convergence. Ind. Corp. Change **5**(4), 1079–1095 (1996)
2. Kaluza, B., Blecker, T., Bischof, C.: Implications of digital convergence on strategic management. In: Dahiya, S.B. (ed.) The Current State of Economic Science, vol. 4, pp. 2223–2249. Spellbound Publications, Rohtak, India (1999)
3. Stieglitz, N.: Strategie und Wettbewerb in konvergierenden Märkten. Deutscher Universitäts-Verlag, Wiesbaden, Germany (2004)
4. Lind, J.: Ubiquitous convergence: market redefinitions generated by technological change and the industry life cycle. Paper presented at the DRUID Academy Winter 2005 PhD Conference, Aalborg, Denmark, 27–29 Jan 2005
5. OECD: Telecommunications and broadcasting: convergence or collision? No. 29. In: OECD Digital Economy Papers No. 5. OECD Publishing (1992)
6. Oxford Dictionary of English: In. Oxford University Press, Oxford, UK (2006)
7. Gordon, R.: The meanings and implications of convergence. In: Kawamoto, K. (ed.) Digital Journalism: Emerging Media and the Changing Horizons of Journalism, pp. 57–73. Rowman & Littlefield, Lanham (MD), USA (2003)
8. Rosenberg, N.: Technological change in the machine tool industry, 1840–1910. J. Econ. Hist. **23**(4), 414–443 (1963)
9. Oettinger, A.G.: Compunications in the national decision-making process. In: Greenberger, M. (ed.) Computers, Communications, and the Public Interest. Johns Hopkins Press, Baltimore (MD), USA (1971)
10. Farber, D., Baran, P.: The convergence of computing and telecommunications systems. Science **195**(4283), 1166–1170 (1977)
11. Hacklin, F.: Management of convergence in innovation—Strategies and capabilities for value creation beyond Blurring Industry boundaries. Physica-Verlag, Heidelberg, Germany (2008). Contributions to Management Science
12. Nyström, A.-G.: Understanding change processes in business networks—A study of convergence in Finnish telecommunications 1985–2005. Åbo Akademi University Press, Turku, Finland (2008)
13. Curran, C.-S., Leker, J.: When two worlds collide—Measuring industry convergence. In: 2009 Academy of Management Annual Meeting, Chicago (IL), USA, 07–13 Aug 2009
14. Bally, N.: Deriving managerial implications from technological convergence along the innovation process: a case study on the telecommunications industry. In: MEMO-STENCIL, Preliminära Forskningsrapporter, vol. 223. Åbo Akademi School of Business, Åbo, Finland (2005)

15. European Commission: Green paper on the convergence of the telecommunications, media and information technology sectors, and the implications for regulation—Towards an Information Society Approach. In., vol. COM(97)623. Brussels, Belgium (1997)
16. Lind, J.: Convergence: history of term usage and lessons for firm strategists. In: 15th Biennial ITS Conference, Berlin, Germany, 2004
17. Balthasar, A., Bättig, C., Thierstein, A., Wilhelm, B.: "Developers": key actors of the innovation process: types of developers and their contacts to institutions involved in research and development, continuing education and training, and the transfer of technology. Technovation **20**, 523–538 (2000)
18. Bierly, P.E., Chakrabarti, A.K.: Managing through industry fusion. In: Brockhoff, K., Chakrabarti, A.K., Hauschildt, J. (eds.) The dynamics of innovation: strategic and managerial implications, pp. 7–26. Springer, Berlin, Heidelberg, New York (1999)
19. Bierly, P.E., Chakrabarti, A.K.: Dynamic knowledge strategies and industry fusion. Int. J. Manuf. Technol. Manage. **3**(1/2), 31–48 (2001)
20. Bröring, S.: The front end of innovation in converging industries: The case of Nutraceuticals and functional foods. DUV, Wiesbaden, Germany (2005)
21. Bröring, S., Cloutier, L.M., Leker, J.: The front end of innovation in an era of industry convergence: evidence from Nutraceuticals and functional foods. R&D Manage. **36**(5), 487–498 (2006)
22. Bröring, S., Leker, J.: Industry convergence and its implications for the front end of innovation: a problem of absorptive capacity. Creativity Innov. Manage. **16**(2), 165–175 (2007)
23. Fai, F., von Tunzelmann, N.: Industry-specific competencies and converging technological systems: evidence from patents. Struct. Change Econ. Dyn. **12**(2), 141–170 (2001)
24. Malhorta, A., Gupta, A.K.: An investigation of firm's strategic response to industry convergence. In: Academy of Management Proceedings, 2001, Best Paper Series: G1 (2001)
25. Pennings, J.M., Puranam, P.: Market convergence & firm strategy: new directions for theory and research. In: ECIS Conference, The Future of Innovation Studies, Eindhoven, Netherlands, 20–23 Sept 2001
26. Weaver, B.: Industry convergence: driving forces, factors and consequences. Paper presented at the 19th NFF conference, Bergen, Norway, 9–11 Aug 2007
27. Müller, K.: Strategieprozess und Marktkonvergenz—Der Zusammenhang zwischen unternehmensstrategischem Handeln, Unternehmensstrategie und konvergierenden Märkten aus der Perspektive der Strukturationstheorie—Mit einem Beispiel aus dem Bereich Mobile TV. (2008)
28. von Tunzelmann, N.: "Convergence" and corporate change in the electronics industry. In: Gambardella, A., Malerba, F. (eds.) The organization of economic innovation in Europe, pp. 125–157. Cambridge University Press, Cambridge, UK (1999)
29. Kern, T.: Convergence makes a comeback. Broadcasting & Cable (2005)
30. Duysters, G., Hagedoorn, J.: Technological convergence in the IT industry: the role of strategic technology alliances and technological competencies. Int. J. Econ. Bus. **5**(3), 355–368 (1998)
31. Rikkiev, A., Mäkinen, S.: Success factors for technology integration convergence collaboration: empirical assessment. In: XXXX (ed.) PICMET '09 Conference: "Technology Management in the Age of Fundamental Change", pp. 374–382, Portland (OR), USA 2009
32. Myring, J.: Industry convergence is good news for users. NMA 24 Aug 2006, p. 7 (2006)
33. Fransman, M.: Convergence, the internet and multimedia: implications for the evolution of industries and technologies. In: Bohlin, E., Brodin, K., Lundgren, A., Thorngren, B. (eds.) Convergence in Communications and Beyond. North Holland, Amsterdam, The Netherlands (2000)
34. Gallouj, F.: Innovation in the Service Economy: The New Wealth of Nations. Edward Elgar Publishing, Cheltenham, UK (2002)

35. Tidd, J., Hull, F.M.: Managing service innovation: variations of best practice. In: Hull, J.T.F.M. (ed.) Service Innovation: Organizational Responses to Technological Opportunities & Market Imperatives. Series on Technology Management, vol. 9, pp. 3–34. Imperial College Press, London, UK (2003)
36. Quinn, J.B., Guile, B.R.: Managing innovation in services. In: Quinn, B.R.G.J.B. (ed.) Managing Innovation: Cases from the Services Industries. Series on Technology and Social Priorities, pp. 1–8. National Academy Press, Washington (DC), USA (1988)
37. Central Intelligence Agency: The CIA World Factbook—GDP Composition by Sector. https://www.cia.gov/library/publications/the-world-factbook/fields/2012.html?countryName=United%20Kingdom&countryCode=uk&regionCode=eu&#uk (2009). Accessed 17 Nov 2009
38. Borés, C., Saurina, C., Torres, R.: Technological convergence: a strategic perspective. Technovation **23**(1), 1–13 (2003)
39. Yoffie, D.B.: CHESS and competing in the age of digital convergence. In: Yoffie, D.B. (ed.) Competing in the Age of Digital Convergence, pp. 1–36. Harvard Business School Press, Boston (MA), USA (1997)
40. Gill, T.: Convergent products: what functionalities add more value to the base? J. Mark. **72**(2), 46–62 (2008)
41. Choi, D., Valikangas, L.: Patterns of strategy innovation. Eur. Manage. J. **19**(4), 424–429 (2001)
42. Guilhon, B.: The emergence of the quasi-markets for knowledge. In: Guilhon, B. (ed.) Technology and Markets for Knowledge: Knowledge Creation, Diffusion and Exchange within a Growing Economy, pp. 21–40. Kluwer Academic Publishers, Norwell (MA), USA (2001)
43. Adner, R., Levinthal, D.A.: Technology speciation and the path of emerging technologies. In: Day, G.S., Schoemaker, P.J.H., Gunther, R.E. (eds.) Wharton on managing emerging technologies, pp. 57–74. Wiley, New York (NY), USA (2000)
44. Kim, J.: Will technology fusion induce the paradigm change of university education? Int. J. Technol. Manage. **38**(3), 220–234 (2007)
45. Kodama, F.: Technology fusion and the new R&D. Harvard Bus. Rev. **70**(4), 70–78 (1992)
46. Fujimoto, M., Miyazaki, K., Tunzelmann, N.V.: Technological fusion and telemedicine in Japanese companies. Technovation **20**, 169–187 (2000)
47. Bainbridge, W.S.: Transformative concepts in scientific convergence. Ann. N. Y. Acad. Sci. **1093**, 24–45 (2006). doi:10.1196/annals.1382.003
48. Yasunaga, Y., Watanabe, M., Korenga, M.: Application of technology roadmaps to governmental innovation policy for promoting technology convergence. Technol. Forecast. Soc. Chang. **76**, 61–79 (2009)
49. Ip, B.: Technological, content, and market convergence in the games industry. Games Cult. **3**(2), 199–224 (2008)
50. Gambardella, A., Torrisi, S.: Does technological convergence imply convergence in markets? Evidence from the electronics industry. Res. Policy **27**, 445–463 (1998)
51. Nyström, A.-G.: What is convergence? Perceptions from the Finnish telecommunications Sector. Paper presented at the 18th European Regional ITS Conference, Istanbul, Turkey, 02–04 Sept 2007
52. Hacklin, F., Marxt, C., Fahrni, F.: Coevolutionary cycles of convergence: an extrapolation from the ICT industry. Technol. Forecast. Soc. Chang. **76**, 723–736 (2009)
53. Hacklin, F., Raurich, V., Marxt, C.: Implications of technological convergence on innovation trajectories: the case of ICT industry. Int. J. Innov. Technol. Manage. **2**(3), 313–330 (2005)
54. Theilen, F.: Geschäftsmodellbasiertes Konvergenzmanagement auf dem Markt für mobile Financial Services. Diskussionsreihe Bank & Börse. Springer, Wien, Austria (2004)
55. Weaver, B.: 'Convergence' in the security industry—Definitions, drivers and consequences. In: LUSAX Memos., vol. LXM-BW2—Convergence, pp. 1–6 (2008)

56. Weaver, B.: Security industry convergence - bridging the knowledge divide. In: LUSAX Memos., vol. LXM-BW1—Knowledge Gap, pp. 1–5 (2007)
57. Moore, G.E.: Cramming more components onto integrated circuits. Electronics **38**(8) (1965)
58. Intel Corporation: World's First 2-Billion Transistor Microprocessor. http://www.intel.com/technology/architecture-silicon/2billion.htm?id=tech_mooreslaw+rhc_2b (2009). Accessed 03 Nov 2009
59. Hauschildt, J., Salomo, S.: Innovationsmanagement, 4th edn. Vahlen, München, Germany (2007)
60. Bröring, S.: Developing innovation strategies for convergence—Is 'open innovation' imperative? Int. J. Technol. Manage. **49**(1/2/3), 272–294 (2010)
61. Mowery, D.C., Rosenberg, N.: The influence of market demand upon innovation: a critical review of some recent empirical studies. Res. Policy **8**, 102–153 (1979)
62. Howells, J.: Rethinking the market-technology relationship for innovation. Res. Policy **25**, 1209–1219 (1997)
63. Herzog, P.: Open and closed innovation—Different cultures for different strategies. Gabler, Wiesbaden (2008)
64. Nordhoff, S., Höcker, H., Gebhardt, H.: Renewable resources in the chemical industry—Breaking away from oil? Biotechnol. J. **2**(12), 1505–1513 (2007)
65. Narayanan, M.P.: Managerial incentives for short-term results. J. Finance **40**(5), 1469–1484 (1985)
66. Yu, L., Slavin, M.: All natural whole-wheat functional foods for health promotion and disease prevention. In: Functional Food and Health. ACS Symposium Series, pp. 125–142. American Chemical Society, Washington, DC (2008)
67. Gilbey, J.: Valuing the digital economy. Nature **462**(7270), 165 (2009). doi:10.1038/462165a
68. Athreye, S., Keeble, D.: Technological convergence, globalisation and ownership in the UK computer industry. Technovation **20**, 227–245 (2000)
69. Basu, I.: Estonia Becomes E-stonia. Government Technology (2008)
70. de Bruxelles, S.: Swindon to become first UK wi-fi town. Times Online (2009)
71. City of Milpitas: Information services—Citywide free public WiFi internet service. http://www.ci.milpitas.ca.gov/government/is/wifi_service.asp (2009). Accessed 22 Nov 2009
72. Reardon, M.: Free San Francisco Wi-Fi project dies. cnet news (2007)
73. Chan, S.: Deal collapses for free Wi-Fi in parks. The New York Times (2009)
74. Federal Communications Commission: Telecommunications Act of 1996. http://www.fcc.gov/telecom.html (2008). Accessed 22 Nov 2009
75. Anonymous: Telecommunications act of 1996. In: 104th US Congress, vol. Public Law No. 104-104, pp. 110 Stat. 156–161. Washington (DC), USA, (1996)
76. Anonymous: Commission directive 90/388/EEC of 28 June 1990 on competition in the markets for telecommunications services. In: The Commission of the European Communities (ed.), vol. 90/388/EEC. Official Journal L 192, 124/107/1990 P. 0010–0016. Brussels, Belgium (1990)
77. Anonymous: Commission Directive 2002/77/EC of 16 September 2002 on competition in the markets for electronic communications networks and services. In: The Commission of the European Communities (ed.), vol. 90/388/EEC. Official Journal L 249, 217.209.2002, pp. 2021–2026. Brussels, Belgium, (2002)
78. AOL Time Warner: America online and time warner will merge to create world's first internet-age media and communications Company. http://www.timewarner.com/corp/newsroom/pr/0,20812,667602,00.html (2000). Accessed 23 Nov 2009
79. McCullagh, D.: AOL, Time Warner to Merge. Wired.com (2000)
80. Glasner, J.: Why AOL wants time warner. Wired.com (2000)
81. Rushe, D.: AOL and time warner split after 'deal of the century' goes bad. Times Online (2008)
82. Musgrove, M., Ahrens, F.: Time Warner, AOL to Split by Year-End. The Washington Post (2009)
83. Bewkes, J.: AOL information statement. In: Time Warner & AOL shareholders (ed.), pp. 1–240. New York (NY), USA (2009)

84. Smith, A.: Time warner to split off AOL. CNNMoney.com (2009)
85. AOL Time Warner: AOL time warner to rename company ‘Time Warner’. http://www.timewarner.com/corp/newsroom/pr/0,20812,670030,00.html (2003). Accessed 25 Nov 2009
86. Spekman, R.E., Bolon, M.: AT&T and Olivetti: Analysis of a Failed Strategic Alliance (2008)
87. Markoff, J.: I.B.M. to Sell Rolm to Siemens. The New York Times, 14 Dec 1988 (1988)
88. Spekman, R.E., Bolon, M.M.: AT&T and Olivetti: Analysis of a Failed Strategic Alliance (1993)
89. NCR Corporation: NCR History Timeline. http://www.ncr.com/about_ncr/company_overview/history.jsp (2009). Accessed 25 Nov 2009
90. Bower, J.L.: Not all M&As are alike—and that matters. Harvard Bus. Rev. **2/2001**, 93–101 (2001)
91. Sodeik, N.: Projektmanagement wertorientierter Mergers & Acquisitions, vol. 124. Planung, Organisation und Unternehmensführung. Josef Eul Verlag, Lohmar, Germany (2009)
92. Snow, B.: The 10 worst-selling consoles of all time. http://www.gamepro.com/article/features/111823/the-10-worst-selling-consoles-of-all-time-page-2-of-2/(2007). Accessed 20 Nov 2009
93. Foljanty, L.: The apple museum. http://www.theapplemuseum.com (2009). Accessed 20 Nov 2009
94. Sanford, G.: Apple history. http://www.apple-history.com (2009). Accessed 20 Nov 2009
95. United States Securities and Exchange Commission: Form 8-K: Apple Inc. In: vol. Commission File Number 000-10030, 09 Jan 2007. Washington (DC), USA, (2007)
96. Yoffie, D.B., Slind, M.: Apple Inc., 2008. In: vol. Case Study No. 9-708-480, pp. 1–32. Harvard Business School (2008)
97. Anonymous: Apple Inc. http://en.wikipedia.org/wiki/Apple_inc (2009). Accessed 20 Nov 2009
98. Sony Corporation: Affiliated Companies. http://www.sony.net/SonyInfo/CorporateInfo/Subsidiaries/index.html (2009). Accessed 22 Nov 2009
99. Sony Corporation: Annual Report 2009. In. Tokyo, Japan (2009)
100. Sony Ericsson Mobile Communications: Sony Ericsson. http://www.sonyericsson.com/cws/corporate/home (2009). Accessed 22 Nov 2009
101. Sony Corporation: Corporate history. http://www.sony.net/SonyInfo/CorporateInfo/History/history.html (2009). Accessed 22 Nov 2009
102. Sony Corporation: Microsoft and Sony to collaborate in creating the convergence of personal computer, digital television and consumer audio-visual platforms. Press Release, 07 Apr 1998 (1998)
103. Sony Corporation: About Sony Group. http://www.sony.net/SonyInfo/(2009). Accessed 22 Nov 2009
104. Tarjanne, P.: Convergence and implications for users, market players and regulators. In: Bohlin, E., Brodin, K., Lundgren, A., Thorngren, B. (eds.) Convergence in Communications and Beyond, pp. 39–43. Elsevier Science, Amsterdam, The Netherlands (2000)
105. Kaumanns, R., Neus, A., Pörschmann, F.C.: Konvergenz oder Divergenz? Erwartungen und Präferenzen der Konsumenten an die Telekommunikations- und Medienangebote von morgen. In: IBM Global Business Services—Strategy & Change. IBM Corporation, Stuttgart, Germany (2007)
106. Corrocher, N., Malerba, F., Montobbio, F.: Schumpeterian patterns of innovative activity in the ICT field. Res. Policy **36**(3), 418–432 (2007)
107. International Engineering Consortium (IEC): The new information industry. White Papers (2007)
108. Rhéaume, L., Bhabra, H.S.: Value creation in information-based industries through convergence: A study of U.S. mergers and acquisitions between 1993 and 2005. Inf. Manage. **45**(5), 304–311 (2008). doi:http://dx.doi.org/10.1016/j.im.2008.03.002

109. Schnaars, S., Thomas, G., Irmak, C.: Predicting the emergence of innovations from technological convergence: lessons from the twentieth century. J. Macromarketing **28**(2), 157–168 (2008)
110. Karvonen, M., Lehtovaara, M., Kässi, T.: Managing technological convergence: evidence from printed intelligence industry. In: Kocaoglu, D.F., Anderson, T.R., Daim, T.U., Jetter, A., Weber, C.M. (eds.) Proceedings of PICMET '09, Technology Management in the Age of Fundamental Changes, pp. 396–408, Portland (OR), USA, 2–6 Aug 2009
111. Gassmann, O., Reepmeyer, G., von Zedtwitz, M.: Leading Pharmaceutical Innovation: Trends and Drivers for Growth in the Pharmaceutical Industry, 2nd edn. Springer, Berlin/Heidelberg, Germany (2008)
112. Preschitschek, N., Curran, C.-S., Leker, J.: Are plastics turning Organic? Antecedents of Convergence in the Chemical Industry. Paper presented at the The XXI ISPIM Conference—The Dynamics of Innovation, Bilbao, Spain, 06–09 June 2010
113. McNamara, L.: Market dynamics in search of double-digit revenue growth: Can big pharma hit its numbers? Int. J. Med. Mark. **4**, 18–26 (2004)
114. Cole, M.: When food meets medicine. Int. Rev. **June/July**, 15–17 (1991)
115. Donnelly, W.J., Fenelon, M.A., Giblin, L., Stanton, C.: Obesity. The food research agenda. Int. J. Dairy Technol. **61**(1), 11–15 (2008)
116. Crompton, S.: Hype or beautiful science? Times Online (2007)
117. Dureja, H., Kaushik, D., Gupta, M., Kumar, V., Lather, V.: Cosmeceuticals: An emerging concept. Indian J. Pharmacol. **37**(3), 155–159 (2005)
118. Wilkinson, J.M.: Cultivate technology convergence for product innovation. Med. Device Technol. **17**(3), 34–36 (2006)
119. Roco, M.C., Bainbridge, W.S.: Converging technologies for improving human performance: integrating from the nanoscale. J. Nanopart. Res. **4**(4), 281–295 (2002)
120. Loveridge, D., Dewick, P., Randles, S.: Converging technologies at the nanoscale: the making of a new world? Technol. Anal. Strateg. Manage. **20**(1), 29–43 (2008). doi:10.1080/09537320701726544
121. Thomas, T.C., Acuna-Narvaez, R.: The convergence of biotechnology and nanotechnology: Why here, why now? J. Commer. Biotechnol. **12**(2), 105–110 (2006)
122. Pierce, M.E.: Convergence of the health industry. Leadersh. Health Serv. **18**(1), xxii–xxxi (2005). doi:10.1108/13660750510578402
123. Frank, S.R.: Digital health care—The convergence of health care and the internet. J. Ambul. Care Manage. **23**(2), 8–17 (2000)
124. Weinstock, P.G., Kalahurka, S.E.: Health savings accounts and the convergence of healthcare and banking. Bank. Law J. **125**(1), 3–48 (2008)
125. Fahrni, F.: Konvergenz als Herausforderung für neue Wertschöpfungsketten und Geschäftsmodelle. In: Klumpp, D., Kubicek, H., Roßnagel, A., Schulz, W. (eds.) Informationelles Vertrauen für die Informationsgesellschaft, pp. 347–356. Springer, Berlin/Heidelberg, Germany (2008)
126. Niedergassel, B.: Knowledge sharing in collaborative R&D Projects. University of Münster (2009)
127. Curran, C.-S., Leker, J.: Patent indicators for monitoring convergence—examples from NFF and ICT. Technol. Forecast. Soc. Chang. **78**(2), 256–273 (2011). doi:10.1016/j.techfore.2010.06.021
128. Meyer, M.: Does science push technology? Patents citing scientific literature. Res. Policy **29**(3), 409–434 (2000)
129. Murray, F.: Innovation as co-evolution of scientific and technological networks: exploring tissue engineering. Res. Policy **31**(8–9), 1389–1403 (2002)
130. Curran, C.-S., Bröring, S., Leker, J.: Anticipating converging industries using publicly available data. Technol. Forecast. Soc. Chang. **77**(3), 385–395 (2010)
131. Curran, C.-S., Leker, J.: Employing STN AnaVist to forecast converging industries. Int. J. Innov. Manage. **13**(4), 637–664 (2009). doi:10.1142/S1363919609002455

Chapter 3
Anticipating Convergence

Scrutinizing cases of industry convergence yield at least one central cognition: the blurring of boundaries often leads to substantial changes in industry dynamics and structure. Being aware of these changes can enable firms to prepare for opportunities while being forearmed against threats.[1] Consequently, a way to reliably anticipate them is of high practical importance. Chapter 2 introduced definitions and delineations of those convergence aspects important for this study's concept of anticipating converging industries. Based on the sequential process model discussed in Sect. 2.4, the following chapter will outline the above-mentioned concept and highlights potential fields of analysis for individual steps. It is not intended to provide complete overviews of technology foresight and forecasting, scientific publications and bibliometrics, or patents and patent analyses. However, it will briefly introduce the respective concepts and then focus on their applicability within this study's anticipation approach.

3.1 Anticipating Converging Industries

3.1.1 Goals and Possible Outcomes of Anticipation

Fading science, technology, market, or industry boundaries can affect far more than only the directly connected stakeholders. Some convergence phenomena have the power to not only change the shape of industries but also the way in which we lead our everyday lives.[2] In analogy to the classification of triggers and drivers depicted in Table 2.1, effects of convergence can basically be found in any of the five domains from technology through society. First, the technologies being developed have to stay abreast of developments resulting from convergence.

[1] See also Sect. 6.4.2 for a discussion of the relevance of anticipating convergence for different interest groups, including managers, politicians and regulators and academic researchers.

[2] See especially Sect. 2.3.1 for a more thorough discussion of convergence events in ICT.

C.-S. Curran, *The Anticipation of Converging Industries*,
DOI: 10.1007/978-1-4471-5170-8_3, © Springer-Verlag London 2013

Accordingly, the existence of smartphones makes it necessary to increase calculation power and interface quality to be able to switch smoothly among desktop computers, notebooks, and smartphones. But to achieve such results, academic scientists as well as company researchers and engineers need to work jointly on common projects. Second, managers have to decide on how to react to convergence. They have to discern between cases of industry convergence and only partial convergence phenomena, in order to align internal activities and structures to a changing environment. A long list of alliances and M&As as well as reorganizations and divestitures can be traced back to reactions to convergence phenomena, being the result of managerial action while at the same time changing industry characteristics. Third, these industry characteristics are also mirrored in the composition of industry sectors.[3] Fourth, political decisions may trigger or even drive convergence. But at the same time authorities might find themselves in a position where they will have to adapt their regulations, incentives, and sometimes even their jurisprudence to changes in the environment.[4] Finally, the omnipresence of mobile communication, for example, has led to changes in the way we communicate. Managers are expected to be answering to e-mails as well as phone calls basically anytime and anywhere, leading to much faster response times to any incident around the globe. In their private lives, people tend to less frequently agree on fixed appointments and to rather turn to a lower level of commitment through postponing decisions to a later point in time. This 'let's text'-attitude leads to a more flexible and ad hoc lifestyle in peoples' free time—with all the positive and negative sides to it.

For all of the stakeholders, it is of great importance to be able to judge on the degree of complete industry convergence (or the presence of convergence at all) in their field of expertise. Bearing resemblance to Heisenberg's uncertainty principle,[5] a compromise has to be accepted in respect to the accuracy of a convergence assessment and its anticipation at an early moment. The earlier forecast is supposed to be made, the less accurate it will be. And the later and thus more accurate

[3] This can be observed well in the pharmaceutical industry, where companies have moved away from a mere utilization of biotechnological techniques and are shifting their business models to an increasingly biotechnological approach. This happens partly due to alterations to their R&D activities and strategic partnerships and partly due to acquisitions of biotechnological companies of varying size. [cf., e.g., 1, p. 131; 2, p. 103].

[4] Both is the case in ICT. Deregulation regarding telecommunication fueled convergence, while many countries nowadays struggle to adapt their jurisprudence to the blurring boundaries between the different forms of data or the outdated distinction between data and physical objects (e.g., when it comes to theft or the handling of stolen goods/data).

[5] In his groundbreaking paper, Heisenberg deduces this uncertainty principle on the basis of the position and the momentum of an electron. He concludes that the more precisely one of these two factors is determined, the less precisely the other factor can be determined. In physics, this principle has also been attributed to other pairs of properties and is in most cases a result of the way such properties can be measured. E.g., in the case of the electron an exact measurement of the position (through light waves and microscopes) leads to an increase in the momentum of it (resulting in an imprecise measurement of the momentum) [cf. 3, pp. 174ff].

it will be made, the less valuable it will be. Once convergence is already happening, knowing about it and reacting to it may lead to costly adjustments to ones activities. This is particularly undesirable where access to limited resources is involved. Polymer production on the basis of either oil or renewable feedstocks is a good example for such a situation.[6] Materials' manufacturers did not want to shift away from oil too early, as the interest in renewable feedstocks could as well have turned out to be a short-lived trend. Also, first movers had to engage in cost-intensive R&D, process development, and marketing efforts. At the same time waiting for too long could prove to be detrimental, especially because of the limited access to renewable feedstocks (bearing in mind also their relevance as a source of human and animal nutrition). Similarly, Day and Schoemaker present examples of effects of emerging technologies[7] on incumbent firms as well as new entrants. They describe emerging technologies' as often having "traumatic effects" on incumbents and stress the importance of anticipating and mastering the challenges posed by them.

In the light of mostly failed predictions of convergence [5, pp. 144f.], the task of deriving accurate and reliable forecasts from the vast amount of data sources and the myriads of pieces of information appears to be unaccomplishable. While this study does not claim to derive a full-scale solution applicable to all areas and in every case, it does claim to develop a theoretically sound concept reinforced with empirical proof. Consequently, this leads to two possible approaches. Ideally, this study would yield the anticipation of areas of future convergence, which could then be verified or falsified in the future. The concept's reliability could then be assessed depending on the degree of verification/falsification. Acknowledging the fact that most occurrences of convergence have taken decades to gather full speed, the whole concept and all future alterations could be evaluated only in the more distant future. Hence, many scholars conducting research in the area of forecasting have shifted their attention to testing hypotheses on the basis of historical data [cf. 5, p. 158]. Accordingly, the anticipation approach will be tested against different sets of empirical data from mainly the areas of NFF and partly ICT, where applicable. While both sectors still experience decisive evolution, their longstanding history also makes the collection of historical data possible.[8] The final judgment on the appropriateness of this approach should then be made at some point in the future, based on the findings of this study. Until then, the extrapolation of trends from historical data is the only possibility for an empirical proof of concept.

Before introducing this study's concept and its levels of analysis, the next paragraph will briefly cover forecasting and related concepts.

---

[6] See also Sect. 2.3.3.

[7] Convergence does not have to be based on emerging technologies, of course. But the potentially massive effects of both bear a strong resemblance. Accordingly, the findings of Day and Schoemaker can be clearly related to the situation of companies dealing with emerging technologies [cf. 4, p. 8].

[8] See Sect. 5.2 for results from ICT and mainly NFF and Chap. 6 for a discussion of the significance and relevance of findings.

3.1.2 Predicting the Future

In contrast to convergence, 'foresight/forecasting' and its related terms have been the objects of numerous research projects [cf. 6, p. 44; 7, p. 449]. Consequently, they are not only widely discussed in a vast amount of scientific articles[9] and books, but have even led to the creation of a few mostly well-reputed specialized journals.[10] These include, e.g., Foresight, International Journal of Forecasting, International Journal of Foresight and Innovation Policy, Journal of Business Forecasting, Journal of Forecasting, Journal of Foresight, Technological Forecasting and Social Change, and The Journal of Financial Forecasting. Obviously, foresight/forecasting techniques are applied in various fields, driven by the desire to know 'what will be coming' or at least 'what might be coming' [cf. 10, p. 93; 11, p. 442]. This is the case in people's everyday life in the form of weather forecasts, prognoses on election results or GDP growth rates. While sometimes even complete certainty about the future would not help the ones knowing about it (as they might not be in the position to—realistically—change anything about it, as in the case of GDP growth rates), it regularly allows for preparations to better cope with the opportunities and challenges to be faced (as in the case of weather forecasts). Thus, foresight/forecasting principles and techniques have been vastly applied in management settings as well as when debating about policy foci [cf. 12, p. 47]. Some researchers criticize the lack of a systematic development of a conceptual framework for the field as a whole [cf. 10, p. 93]. They instead identify "isolated but uncoordinated research on improving methods, selection of methods, or integration of analysis and stakeholder engagement" [13, p. 288]. This chapter will thus give a short introduction into the mainstream delineation of different processes and techniques, without trying to provide a full overview of different streams of the literature.

Often the terms 'forecast' and 'foresight' are used interchangeably or at least in similar settings, when their meaning in fact is substantially different.[11] Semantically, they refer to:

---

[9] This study will use the term 'scientific article' when speaking of publications in scientific journals (see Sect. 3.2.1 for a more precise definition), despite the fact that 'scientific paper' appears to be disseminated more broadly [cf. 8, p. 3]. This is indicated by, e.g., the number of hits received in a Google search (176,000 for 'scientific article'/663,000 for 'scientific articles' in comparison to 481,000 for 'scientific paper'/1,330,000 for 'scientific papers'). Nevertheless, this study is based on a distinction between 'article' and 'paper', with articles entailing papers in a narrower sense, but also reviews, commentaries, letters etc. An even broader term is 'scientific publication', which further includes scientific books, presentations or other forms of scientific communication. Analyses are carried out on the basis of all forms of scientific publications that are dealing with original research (directly or indirectly), if they appear in scientific journals. While this is a deviation from narrower definitions of scientific articles [See, e.g., 9, pp. 19ff.], an early warning system should refrain from ruling out sources of information too eagerly.

[10] The question of reputation of scientific articles and journals and approaches to its quantification will be discussed in Sect. 3.2.

[11] Historically both were used synonymously, while now a real difference has largely been agreed upon. Additionally, the term 'foresight' is now often favored over 'forecast', as the latter

- **Forecast:** "A conjectural estimate, based on present indications, of some thing in the future, esp. of coming weather; a prediction" [14, p. 1014].
- **Foresight:** "The action or faculty of foreseeing (being aware of beforehand; predict); prescience. [...] Care or provision for the future; providence, prudence" [14, p. 1017].

Thus, already the dictionary's definitions point out a decisive difference between both terms. While for *forecast* it puts stronger emphasis on the estimate on the basis of given facts, it stresses the general ability and occupation to foresee in combination with preparing for the (possible) future for *foresight*. This is one central difference that is also commonly reflected in scholars' theorizing about foresight's meaning and its delineation from forecasting. Martin, for instance, underlines "the belief that there are many possible futures" [15, p. 140]. In his opinion, foresight is not forecasting, as it is less about the valid prediction of the *most likely future*, but rather the discussion of a set of *possible futures*. They should be systematically analyzed to allow for informed decisions to be made, again highlighting the involvement of a "consciously 'active' attitude towards the future, recognizing that the choices made today can shape or even create the future" [15; see also 16, pp. 57ff.]. Similarly, van der Meulen even proclaims that the very idea of foresight is to not try to forecast. It should rather produce "visions of the future" [17, p. 8], and thus motivate actors deriving science and technology policies to actively develop the future. Then a discussion of the different possible futures and their implications could take place, trying to avoid the one and to make the other happen. This reasoning is supported by others, proclaiming "the real purpose of foresight is not to predict how a driver of change will evolve, but to change the mental models that decision makers carry in their heads".[12] Both mention a second important difference. According to their opinion, forecasting is more of a technique (or a set of techniques) attempting to identify the most likely path of developments. Contrastingly, foresight is a process, involving iterative steps with consultation of relevant actors on the different forms of the future [cf. 15, p. 140; 17, p. 8]. Martin defines foresight as "the process involved in systematically attempting to look into the longer-term future of science, technology, the economy and society with the aim of identifying the areas of strategic research and the emerging generic technologies likely to yield the greatest economic and social benefits".

Rohrbeck et al. go further and describe forecasting as one part of foresight (in building an anticipation of the future on historical data) on as well the company level as the regional, national, or supranational level [cf. 19, p. 3]. In their view, foresight includes the use of data gathered to make decisions. Deviating from

(Footnote 11 continued)
is of lower reputation with decision makers in the public sector due to "wrong" forecasts in the past [cf. 10, p. 95].

[12] This quotation is attributed to a former head of the scenario planning unit at Shell [cf. 18, p. 6].

Martin and van der Meulen, they call these investigations *futures analysis*[13] if organizations, processes, and actors are to be included. This term has also been used by, e.g., the Technology Futures Analysis Methods Working Group,[14] who phrases the comprehensive analysis of future-related investigations accordingly. This includes individual forms like technology intelligence, forecasting, roadmapping, assessment, and foresight [cf. 13, pp. 287ff.].[15] This collectively authored paper adds a definition of *technology futures analysis* (TFA) to represent "any systematic process to produce judgments about emerging technology characteristics, development pathways and potential impacts of a technology in the future" [13, pp. 288f.]. They discern between the public sector's technology foresight studies focusing on future technological developments and their (undesired or desired) interactions with society at large[16] on the one hand and industry's technological forecasting on the other hand. The latter are aiming more at identifying the emergence and performance of new technologies at a certain point in the future [cf. 13, pp. 288f.]. Table 3.1 provides a summary of the main differences between forecasting and foresight.

Building on this notion, foresight can be divided further into different forms.[17] With foresight being the overarching process including all divisions, *technology foresight* is the systematic monitoring of new and changing technologies. *Strategic foresight*,[18] in turn, can be defined as an "integrated approach to early recognition of new technologies, upcoming competitors, new markets, changing customer

[13] Rohrbeck et al. actually call it "future analysis", probably a typing error [cf. 19, p. 3].

[14] The *Technology Futures Analysis Methods Working Group* is formed by 19 well known scholars in this field: Alan L. Porter, W. Bradford Ashton, Günter Clar, Joseph F. Coates, Kerstin Cuhls, Scott W. Cunningham, Ken Ducatel, Patrick van der Duin, Luke Georgehiou, Theodore Gordon, Harold Linstone, Vincent Marchau, Gilda Massari, Ian Miles, Mary Mogee, Ahti Salo, Fabiana Scapolo, Ruud Smits and Wil Thissen [cf. 13, p. 287].

[15] *(Technology) roadmapping* refers to widely employed processes enabling decision making on science and technology developments in connection with an entity's strategy. It can be compared to a map, with the entity (mostly a company) moving like a car through the streets of possible future roads. Ideally it includes the intersections (at which important decisions have to be made) and potential outcomes of these decisions, based on qualitative and quantitative data. For further classifications and applications of roadmaps see, e.g., [20–31].

[16] This view is also supported by others. See, e.g., van der Meulen, who argues that "in technology foresight the process is focused on identification of (critical, generic) technologies that are likely to have a large impact on economic, ecological and social developments" [17, p. 8]; or Martin, who highlights the critical role of technological foresight to help make decisions on which areas of science and technology to support—a move perceived necessary even in "rich" nations [15, p. 162].

[17] Naturally, as with most of the literature (especially on futures analysis), different types of foresight are used synonymously. This is particularly the case for 'foresight', 'strategic foresight', 'corporate foresight' and 'technology foresight'. In spite of these fuzzy delineations, this study will strive to be as consistent as possible.

[18] See also Rohrbeck, Arnold and Heuer [19, pp. 3f.], who subdivide *Strategic Foresight* into the four dimensions *Technology Intelligence*, *Competitive Intelligence*, *Political Environment Foresight* and *Consumer Foresight*.

Table 3.1 Main differences between forecasting and foresight

Forecasting	Foresight
Basic points, topics, and research questions have to be clarified in advance	Basic points, needs, and research questions are still open and looked for as part of the foresight process
More quantitative than qualitative	More qualitative than quantitative
Questions what the future in the selected area might look like	Looks for 'information' about the future for priority setting
More result oriented, can also be performed by individual people or in single studies (depends on methodology)	Brings people together for discussions about the future and for networking, makes use of the distributed intelligence
Not necessarily assessments, different options and choices or the preparation for decisions	Criteria for assessments and preparation for decisions
Describes future options, results more important than the communication aspects	Communication about the future as an objective
Long-, medium-, and short-term orientations as well as the path into the future are the major points	Long-, medium-, and short-term orientations with implications for today
No information about consensus necessary	Finds out if there is consensus on themes
Mainly 'experts' and/or strict methodologies, less dependent on opinions	'Experts' and other participants, very dependent on opinions

Source Adapted from [10, p. 100]

requirements and changing societal phenomena and the systematic integration of early recognition results into strategic planning" [32, pp. 274f.]. Finally, *corporate foresight* is the application of strategic foresight within the boundaries of or in respect to a corporation. While the methods used in foresight are plentiful,[19] they can be grouped into four main classes: [cf. 32, pp. 283f.][20]

- **Heuristic methods** (semi-random approaches: e.g., screening of websites or publications)

[19] [cf. 13, pp. 290f], for an extensive overview of 51 TFA methods and their variations as well as their classification regarding methodological "family", qualitative/quantitative and exploratory/normative characters. See also Martin [15, pp. 156ff], for a further typology of foresight methodologies in respect to organizations, specificity, functions, orientation, 'intrinsic tensions', time horizons as well as foresight processes' main elements and stages; or Rohrbeck, Arnold and Heuer [19, p. 7], distinguishing *market-oriented methods* (e.g., customer scenarios, competitor analysis or lead-user analysis), *integrating methods* (e.g., roadmapping or scenario technique) and *technology-oriented methods* (e.g., bibliometrics, patent analyses or simulations); finally Linstone [33, p. 3] stating that "In the thirty years of publication of Technological Forecasting and Social Change, more papers have appeared in it on Delphi than on any other forecasting technique".

[20] Interestingly, in their survey of 30 small- and medium sized biotechnology firms, they find very different approaches to these methods. While some are rather concentrating on heuristic or interactive methods, others are purely set on using science-driven foresight methods, thus focusing on bibliometric analysis of publications and patents, reasoning that "scientists want to do science".

- **Questionings** (semi-quantifiable: e.g., surveys sent to suppliers, customers, or other informants)
- **Interactive methods** (expert-based methods, such as workshops, collective SWOT analyses, or scenario analysis)
- **Quantitative methods** (quantifiable data sources: e.g., data from customer relationship management, market research, or bibliometric analyses).

As these methods are to a large extent expert based, two main challenges have to be met. First, van der Meulen highlights the importance of a method mixture, counterbalancing the three constitutive elements: expertise, interaction, and creativity of a foresight method triangle. He deems futility of efforts to arrive at a sensible forecast, the logical result of an unbalanced method selection, as each of the methods alone is unbalanced in this triangle [cf. 17, p. 19]. Second, any expert-based method has to be interpreted with caution due to an "optimism bias" found by Tichy. In his study, the degree of optimism correlates positively with the degree of self-rated knowledge of the involved experts [cf. 34, pp. 341ff.].[21] Notwithstanding various qualitative and quantitative methods, McMaster stresses the fact that "foresight is a matter of interpretation rather than information" [36, p. 151], as foresight could not be viewed as a mere extension of what we have learned from the past. On a similar note, Martin warns that foresight seldom yields proper results when first carried out. Like in other scientific areas, advances are often based on 'trial and error', which is attributed to the combination of many (often conflicting) factors. Accordingly, he reasons that foresight needs to be employed on a continuous basis to minimize the risk of interpreting the results incorrectly.

Johnston (one of the forefathers of forecasting) already in 1970 highlighted the fact that "[...] it is sadly evident that neither the reasoned judgment of an insightful observer nor the refined techniques of the objective technician who works with masses of quantitative data can guarantee results which will meet the test of actual historical developments" [37, p. 174].[22] Actually, Goldstein and Gigerenzer report on a substantial amount of studies from different areas testing varying degrees of complex forecasting methods in competitions against each other [cf. 38, pp. 760ff.]. Nearly, all of these studies arrive at the conclusion that while the more complex models are often better at fitting data in hindsight, they suffer from overfitting in forecasting future events. Consequently, in contrast to the widespread belief that 'complex problems call for complex solutions' simple

[21] Cf. Tichy, who provides a detailed overview of the role of expert-status as well as, e.g., affiliation on the quality and tendency of assessments [See also 35, p. 160].

[22] Johnston also describes three different types of 'outlook statements'. He discerns between *predictions* (based on the application of a "covering law"), *projections* (conditional predictions, not necessarily based on causal relationships) and finally *forecasts* (representing the most likely projection).

models are mostly turning out to be the most reliable ones.[23] This conclusion is also supported by an extensive examination of 111 forecasting approaches, conducted by Makridakis et al. in 1979, stating that "[...] under certain circumstances, simpler methods do as well or better than sophisticated ones" [39, p. 116]. Strikingly, Armstrong points out that in a survey of econometrics experts more than two-thirds agreed (wrongly) with the above-mentioned belief [cf. 40, p. 549].

In addition to their potential higher reliability, simpler methods also offer the opportunity to be based on easily understood assumptions, which cannot as easily be (unconscious or conscious) misused by the investigator [cf. 41, p. 75]. Generally, there is a demand for better forecasting methods [cf. 42, p. 49] and adjustments to specific settings. Linstone, for example, mentions the peculiar setting of convergence with boundaries fading in the sciences (physics, chemistry, and biology) as well as between technologies ('infotech' and 'biotech') [cf. 43, p. 187]. This challenge of foresight/forecasting methods is also voiced by Porter et al. concentrating on a convergence of information and molecular technologies creating new capabilities, but at the same time also leading to important structural changes in industries [cf. 13, pp. 297ff].

Finally, unlike the briefly mentioned studies on forecasting and foresight, this study takes a slightly different approach. As the title indicates, it focuses on an **anticipation** of converging industries. While the concept is focusing on an *early identification of signals* for possible convergence, it also carries a decisive *monitoring* flavor.[24] The term 'anticipate' already describes quite well the promises and pitfalls of this approach. When looking at its definition in the dictionary [14, p. 92], it covers five different sentiments:

- "Take into consideration or mention before the due time", but also
- "Consider something too soon; raise a matter too soon" and
- "Observe or practice in advance of the due time; cause (a future event) to be a reality beforehand; cause to happen earlier, accelerate", as well as
- "Seize or take possession of beforehand", and finally
- "Take action in advance regarding; forestall".

Depending on its context of use it may thus have a positive or a negative connotation. Taking something into consideration at an early point as well as taking action in advance may be considered to be per se positive, while considering something *too soon* definitely points in a negative direction. Causing something to happen earlier, or accelerating it, may be perceived to be either positive or negative—depending on one's standpoint and the nature of *it*. Two facets are making it particularly appropriate for this study. First, the definition, as

[23] In their article they report on complex problems such as predicting outcomes of Championships and tournaments, estimating the best customers in the future or locating a serial criminal in geographical profiling.

[24] While a degree of monitoring effort is part of most forecasting/foresight approaches, it is more prominent in Ayres or Watts and Porter [cf. 42, p. 51; 44, p. 28].

well as the term's common use, refers to the future orientation not so much in the form of an exact mathematical formula with a calculated probability and margin of error, but rather simpler and more intuitive methods. Second, more like foresight (in contrast to forecasting) there is an immanent active and preparatory twist to it. The aim is not so much to engage in very detailed academic discourses, but rather to use an approach integrating thinking in terms of possible concrete next steps as well as indicators without the need of complex calculations. Irrespective of the degree of robustness and validity caused by simpler models, one very important asset is the transparency achievable through them [cf. 38, p. 770]. This should enable a further refinement of the approach by other researchers and an easy adaption to different settings, e.g., different industries or different levels of convergence.

3.1.3 Where to Anticipate

When aiming at anticipating convergence, the question arises: "Where to look for convergence indicators?" Building on the (idealized) sequential process of industry convergence (Fig. 2.6), this question should be answered taking into consideration the different levels. As a consequence, this would not only enable investigators to assess on what level convergence was happening (or had happened), but also to gain insights into the driving forces behind the respective convergence. For example, if their findings would indicate science and technology convergence without market convergence, this would limit possible triggers and drivers largely to nonmarket related causes.

The complexity level of this investigation is determined to a large extent by the breadth of the search strategy. Naturally, at earlier points in time (when knowledge about possibly converging areas is still weak) it would have to be broader than at later points. This leads to two implications. First, the more areas to be included, the less likely it would be that any investigator could have expert knowledge in all of them. Second, more areas would also mean more data leading to more complex analyses and an increase in costs. Thus, this concept places special emphasis on using quantitative and easily accessible data. While qualitative data, like expert interviews and case studies, are of special importance in understanding developments and getting a "feeling" for new developments [cf. 45, 46], it often proves difficult, time consuming and expensive to find the right sources at the right time. But time and money are resources, most firms do not have to spare. Hence, this study is focusing on using publicly available data.

Table 3.2 Possible sources for a comprehensive anticipation concept

Convergence of	Main sources
Science areas	Scientific publications
Technologies	Patents
Markets	General firm and market information
Industry segments	General firm and business information

Source Adapted from [50, p. 644]

To limit errors caused by a weakness of convergence signals, more than one indicator should be used at each level, whenever possible.[25] Table 3.2 lists the four levels of convergence discussed in Sect. 2.4 and possible sources for quantitative data to assess the degree of convergence. **Science convergence** is marked by a blurring of boundaries between hitherto disjoint scientific areas or even disciplines. This can be observed through changes in the way research is conducted in any given area. If affected by convergence, researchers in such an idealized area[26] will start to read (and eventually cite) the literature from the other area, use terms and combinations of words common to the other area and publish in the other area's journals. They will co-author articles together and eventually their affiliations might more often change to institutions of the other area or of the converged field. Therefore, scientific publications appear to be a valid source for calculating the degree of convergence in science. For some indicators the full publication would need to be accessed, while for others databases containing bibliographical information could be sufficient.[27] Besides scientific publications, also other proxies could be analyzed. For instance, the relocation of leading researchers, the share of scientists from different disciplines or the creation of interdisciplinary collaboration projects or institutions could also be outcomes of science convergence. In addition to the difficulty of attributing such developments (beyond doubt) to convergence, it poses two further challenges. First, such events will be less common and only happen when convergence has already gained considerable momentum. Second, monitoring these changes would most likely necessitate a time-consuming surveillance of a large number of websites, in absence of an already aggregated information interface.

---

[25] The use of multiple indicators has been demanded for some time, as the large amount of studies on firms' innovative performances has thus far not led to the one single and broadly accepted indicator or at least a universal set of indicators. [See, e.g., 47, pp. 1365ff.] Similarly, see also Daim et al. [48, p. 1009], who use bibliometrics and patent analysis to forecast emerging technologies, reasoning that multiple methodologies can improve the results of technology forecasting; or Thomas [49, pp. 110ff.], employing multiple methods to forecast new product market potential based on the assumption that a combination of independent methods leads to greater accuracy than a single method. At the same time he reports on the very different results obtained by other scholars on the question *how many* indicators should be employed.

[26] It should be borne in mind that, especially in the light of blurring scientific boundaries, it may be extremely challenging and thus a decisive simplification to actually define these boundaries. .

[27] See also Sects. 3.2.2 and 3.2.3.

Technology convergence is resulting from the application of technologies associated with one of the areas being employed in the other one. Additionally, it may be accelerated by the development of technologies to be used in both areas. If these technologies are nonconfidential and of interest to a greater audience, they may be mentioned in scientific publications or even newspaper articles and alike. But in industries with a considerable propensity to patent,[28] the easiest way to monitor these advances will be based on patent analyses. Measures as discussed for scientific publications will be complemented by patent–specific ones, such as the co-classification of IPCs[29] or the relationship of IPCs and company affiliations. An added benefit of patents is their high degree of formalization, their customary compilation in databases, and their accessibility free of charge at most national patent offices.

Scrutiny of the two remaining levels is a two-edged sword. On the one hand, convergence in markets or industries should be much more obvious than on the previous two levels. Most people will have realized that things have changed in regard to products offered and industry restructured. They might be unlikely to attach the 'convergence' name tag to these observations, but an informed researcher would be expected to identify the underlying patterns in such cases. On the other hand, operationalization and quantification are extremely challenging. Especially the question whether, for instance, a given merger between two companies is the result of blurring industry boundaries or of different business logic, is very difficult to answer. **Market convergence** appears to be the most difficult to assess on the basis of automated and large-scale monitoring concepts. It would be identifiable through an overview of actual products or services in the market, or more generally by identifying general customer trends. For a proof of concept this could be done by individually collecting data on companies' product portfolios (especially the introduction of new products in possibly converging areas) and through expert interviews, but in managerial practice this would probably remain a weakness.

Finally, for a thorough assessment of complete **industry convergence** compared to just technological or market-side convergence, additional analysis of collaboration projects [cf. 51], press releases, or general business media could be employed [cf. 52, 53, pp. 2ff.]. Data on strategic alliances, shifts in industry's product portfolios, and on mergers and acquisitions should be employed to judge the degree of complete industry convergence. Databases like Factiva or archives of newspapers concentrating on business news could be a valid source for such kind of information. Once firms active in the field are identified, dedicated M&A databases such as Thomson Reuters' SDC Platinum™, Bureau van Dijk's Zephyr or Standard and Poor's Capital IQ Platform would present other viable sources for a thorough investigation of industry changes.

The relevance of an analysis on each of the above-mentioned levels is depending on the underlying intentions. In order to be able to fully answer the question whether

[28] See Sect. 3.3.1 for a more detailed explanation of differing propensities to patent in different countries and industries.

[29] International Patent Classification, for a more detailed explanation see Sects. 3.3.1 and 3.3.2.

two industries are converging, especially the industry level would have to be scrutinized. When debating possible triggers and drivers, all four levels should be analyzed. For anyone trying to anticipate industry convergence, the earliest phase would be the most appropriate one. Due to the high time sensitivity of innovation processes, it is of strategic importance, particularly to firms, to realize at the earliest possible moment that convergence might be affecting them. But convergence in scientific disciplines may often not translate into connected developments in markets and industries. An assessment of technology convergence on the basis of patents could partly heal this defect, as they are (in larger quantities) an acceptable proxy for commercial interest in a given invention.[30] All in all, when interested in employing ways of anticipating convergence, a **robust early warning system** would be most interesting. As the word ‘early’ implies, such a concept should focus more on the early parts of converging industries than on the latter ones. Once markets begin to converge, strategic actions would already be mostly too late. While this would especially be the case for so-called ‘first movers’, other companies could also be affected where, for instance, resources are scarce or entry barriers particularly high. Furthermore, ‘robust’ implies the need for more than one column to base strategic decisions upon.

During the assessment of (especially) science and technology convergence, a further decision has to be made regarding measures and measurement strategies. On a general level, it should be discerned between measures that could be of academic interest and those more easily applicable within a governmental or industrial setting[31] (where the ones evaluating degrees of convergence will most likely not be forecasting experts). As this study is aiming at the development of more a practical concept for anticipating than of a highly complex approach for monitoring convergence, the following paragraphs will report on possible measures and indicators in brief reviews of the respective literature streams, but focus on their expected practical appropriateness for deriving a robust forecast.

3.2 Scientific Publications

3.2.1 Particularities of Scientific Publications

Ideally, scientific publications are the very essence of science. From today’s viewpoint, they are supposed to introduce, describe, and discuss research conducted by one group of scholars and share it with all other researchers. Hence, they serve three central purposes:

---

[30] See Sect. 3.3 for a discussion of possible conclusions from patent analyses.

[31] Following the reasoning of Martin [15, p. 163], that approaches to anticipate the future benefit as much as innovations from “having a specific ’customer’ with well-defined needs which are taken into account at an early stage”, this study will focus on easy applicability by non-foresight-experts.

- Broaden the world's knowledge by describing research, its results, and conclusions.
- Spark further research in proximate and distant disciplines.
- Validate the correctness of the results reported through intensive discussions with reviewers, editors, and the readership.

Additionally, they are employed as an external assessments of the quality of research scholars are conducting [cf. 54, p. 5]. The more they will be able to have their articles published in well-respected journals without initiating further dispute, the more reputation they will gain. More generally, they are regarded as an output indicator of scientific activity, productivity [cf. 55, pp. 179f.], and sometimes even quality [cf. 56, pp. 4ff.]. Despite the fact that the importance of communication in nonscientific articles[32] is being debated, many scholars assume that all important research findings are *eventually* published in scientific journals [cf. 55, p. 181].

The core of scientific publications is marked by primary journal publications, for which the Council of Biology Editors (CBE, now the Council of Science Editors, CSE) proposed the following definition: "An acceptable **primary scientific publication** must be the first disclosure containing sufficient information to enable peers to assess observation, to report experiments and to evaluate intellectual processes; moreover, it must be susceptible to sensory perception, essentially permanent, available to the scientific community without restriction and available for regular screening by one or more of major recognized secondary services (e.g., currently *Biological Abstracts*, *Chemical Abstracts*, *Index Medicus*, *Excerpta Medica*, *Bibliography of Agriculture*, etc., in the United States and similar services in other countries)" [57, pp. P1f.].[33] Through the emphasis on 'enabling peers' a further refinement is made. Scientific articles, *per definitionem*, refer to articles published in peer-reviewed publications [cf. 9, p. 20],[34] consisting mainly of journals and conference proceedings. As well, primary scientific articles may take different forms, largely depending on journals' preferences. They consist of lengthier 'full papers', mostly following the traditional IMRAD (introduction, methods, results, and discussion) order [cf. 9, pp. 20f.][35] and of other forms, such as shorter 'letters' or 'notes'.

Other scientific publications include review papers, conference abstracts, and scientific books. Although review papers may report new insights through the combination and condensation of primary articles' results, they are not considered to be primary scientific papers. Conference contributions may be delivered in the form of abstracts, presentations, or short papers and may contain primary research results as well as secondary reviews or outlines of future research. If they are not published in conference proceedings, they are hardly accessible for the largest part

[32] E.g., personal communication, presentations, working papers.

[33] Cited after [58, p. 1,190] and [9, p. 19]; accentuation not in original text.

[34] This chapter will later shortly discuss the peer-review system.

[35] One very rare exception is the well-renowned journal Cell, with its IRADM structure.

of the scientific community. Even if part of published proceedings, their contribution, originality, and quality is perceived to be on average weaker than that of scientific papers [cf. 9, pp. 22ff.]. Books themselves can take different forms. They may be based on dissertations or other primary research, but in many cases they will be either textbooks, dictionaries, or edited books with broader individual sections, thus not based on original research [cf. 59, p. 13]. Further means of communication are, for instance, (government) reports, grant proposals, working papers, and websites. With the exception of working papers, these normally do not include original research, but rather the presentation of more general findings and their practical implications.

Historically, scientific communication started, in line with its central motivations, in the form of individually "published" letters or books. For example, the Babylonians used clay tablets in the sixth century BC and the Persian scientist Alhazen wrote a seven volume treatise on optics in the early eleventh century AD. Later, scholars like Galileo Galilei or Tycho Brahe shared their seminal contributions in the form of letters at the end of the sixteenth century [cf. 60, pp. 1f.]. Finally, the seventeenth century saw the foundation of the first scientific journals, the *Journal des Sçavans*, first published in Paris by Denis de Sallo on January 5, 1665 and shortly afterwards the *Philosophical Transactions of the Royal Society*, which were first published on March 6, 1665 in London [cf. 61, pp. 3f.]. In spite of these different means of communication (and their being mostly used up to the present day), scientific articles have been the one preferred central medium for the dissemination of scholarly contributions since then [cf. 62, p. 154; 63, pp. 1ff.]. Interestingly, while they are the primary means of information to stay abreast of the latest developments in basically any discipline, the importance of other means varies greatly between disciplines. The German Science Foundation (DFG, Deutsche Forschungsgemeinschaft), for example, found in a large scale survey of more than 1,000 scholars in Germany that conference proceedings are very important sources in the engineering domain. Furthermore, scholars in the social sciences and especially the humanities rely much more often on books and edited books [cf. 64, pp. 22f.]. Table 3.3 exhibits these considerable differences. In general, the natural sciences and medicine focus much more on scientific journals than mathematics, engineering, and the humanities/social sciences. The broadest distribution of important sources can be found in the humanities, where more than half of the respondents use any publication type—other than grey literature.

Notwithstanding these differences in respect to disciplines, scientific articles pose for authors a possibility to claim ownership of a certain theory and communicate with a larger audience, independent of the need to acquire the funds necessary for a publication in book form [cf. 65, p. 367ff.]. This has led to "an explosion in the number of scientific studies" and thus the amount of new knowledge generated [60, p. 2]. For readers it meant being able to keep track of recent developments in all the fields they were interested in. From the very beginning, reproducibility was a major

Table 3.3 Frequently used types of publication to obtain current information in scholar's own discipline (in percent)

	Humanities	Social and behavioral sciences	Biology	Medicine	Agricultural sciences, forestry, veterinary medicine	Chemistry	Physics	Mathematics	Geoscience	Engineering sciences
Scientific journals	93.5	93.3	100.0	95.7	100.0	98.1	96.0	87.9	95.1	90.7
Edited books	87.7	58.3	41.6	34.5	50.0	39.4	33.7	34.4	57.4	40.2
Books	85.4	52.9	21.3	19.3	23.1	23.8	30.2	50.0	39.7	36.1
Conference proceedings	60.0	24.3	28.6	23.5	28.6	24.0	40.8	33.3	32.8	83.9
Recensions	55.5	15.0	9.2	4.5	7.1	3.9	2.2	9.7	1.8	6.2
Grey literature[a]	8.3	22.2	4.5	2.7	7.7	4.0	9.0	29.6	12.1	9.7

Source Adapted from [64, pp. 22f.] and [69, p. 31]

[a] "Grey literature" refers to a body of materials that is not distributed through the established publication and distribution channels. It includes means of communication like reports, white papers, or working papers

concern and served as a central indicator in quality control. Until today, research findings are supposed to be presented in a manner that any informed reader would be able to repeat experiments and calculations and thus arrive at the same results.[36] With the evolution of industrial printing technologies and later electronic media, journals could be produced at lower prices, fueling an increase in the distribution of the included articles [cf. 61, p. 9f.].

Journal publishers entail mainly two groups. The first of the two large groups to publish scientific journals were learned societies like the *Royal Society* in the UK, the French *Académie des Sciences* or the *St. Petersburg Academy of Science* in Russia. They consisted mainly of individuals who were researchers themselves or highly interested in the generation of knowledge. The second group consists of the commercial publishers, who try to generate revenues through publishing in some cases a vast array of scientific journals [cf. 70, p. 371; 61, p. 3; 60, p. 2]. Elsevier, for instance, the "world's largest global academic journal publisher" produces over 200,000 new research articles in about 1,100 journals every year [71, p. 33; 72, p. 25]. This branch of Reed Elsevier also offers an abstract and citation database (Scopus) and information resources for medical professionals, generating total revenues of £ 1.7 billion and an adjusted operating profit of £ 568 million (both 2008) [cf. 71, p. 13ff.].[37] Elsevier's journal portfolio spans virtually all science disciplines from Astrophysics and Economics to Forensics and the Social Sciences [cf. 72, p. 36; 73].

Unsurprisingly, in the light of such operating profits, there has been a long-standing discussion about "proliferating scientific literature" and whether the steady increase in the number of scientific publications is due to a "scientific pull" or more a "publishing push" [70, p. 369ff.; cf. also, e.g., 74 p. 529ff.]. The rise in numbers of scientific articles can be largely attributed to an increase in generated knowledge, which in turn is caused by more researchers employing more advanced and automated tools and methods. While a reliable assessment of the number of articles is difficult, estimates speak of 1.35 million in 2006 [cf. 75]. In addition to the number of articles, also the number of journals has increased decisively [cf. 70, p. 369ff.]. Reported reasons for the establishing of new journals are diverse: predominantly, the lack of a dedicated journal for a certain topic, followed by a felt need for more rapid publication and commercial interests cf., e.g., [70, p. 371; 76,

---

[36] As there have been several occasions where it was later proven that authors had presented false (or even falsified) results, an increasing number of journals is demanding original and raw data to be submitted alongside the manuscript. This is especially important in the light of some authors' claims that a large portion of published research results are actually false. Cf., e.g., Ioannidis [66, pp. 696ff.] and Kleiner [67]. They discuss published, but supposedly wrong findings in medical studies, caused by poor study design, researcher bias and selective publication of research results. Strikingly, Kleiner quotes Solomon Snyder (associate editor of the *Proceedings of the National Academy of Sciences* and neuroscientist) in response with "[...] I'm not reading it to find proof like a textbook. I'm reading it to get ideas [...]" [67]. See also Niedergassel for a short discussion of reproducibility in scientific publications. [68, pp. 82f.].

[37] As a reference, the European Commission estimated the global core science, technology and medicine publishing market at between 7 and 11 billion USD. [72, p. 5].

p. 36ff.]. Although there are several scientific areas with more than one (dedicated) journal, the nearly 24,000 journals[38] cover ever more specialized scientific subfields. As a result, the number of journals has grown at rates between 3.3 and 4.7 % p.a. throughout the twentieth century. In addition to scientific journals, there were another estimated 160,000 nonacademic journals at the beginning of the new millennium [62, p. 154ff.].[39] Altogether, the amount of scientific publications is said to have doubled about every 15 years in the past 300 years [cf. 77, p. 12].

According to Mabe and Amin, an average journal will publish about 100 papers p.a. On the assumption of between 0.8 and 1.1, unique authors per publication and year and allowing for a rejection rate of about one third, this would mean that a scientific community of 150 potential authors should be sufficient for a new journal.[40] Indeed, Ziman criticizes the fact that the majority of papers rejected by the "core scientific journals" are later published in journals of lesser importance (as perceived by a large part of authors and readers). According to him, this is caused by the creation of journals aiming at relieving the pressure on the renowned journals and thus 'harmlessly filing away relatively trivial papers' [70, p. 371]. This perceived decline in quality of the scientific literature is not without objection [70, p. 369ff.]. Braun and Zsindely [74, p. 529] have named this 'damnation' of 'pollution' with the more and more scientific literature of limited value the "Barnaby Rich Effect". Rich, who himself has published at least 26 books in the early seventeenth century, is quoted complaining about "the disease [...] of the multiplicity of books". Consequently, they question the appropriateness of some authors' negative stance toward the number of publications—who in turn were proven to have authored scientific publications "fairly above the productivity of average authors in their respective fields" [74, p. 530].

Besides the question whether people righteously criticize the increase in journals and articles[41] or not, two problems are directly linked to this phenomenon.

---

[38] 2006 figures [cf. 75].

[39] In their study, Mabe and Amin distinguish between three periods of growth in the number of scientific journals. 1900–1944 at a rather constant annual growth rate of 3.3 %, where developments are mainly driven by scientific societies. The next period, until 1978, was marked by the belief that science had won World War II and could thus also win the Cold War, if governments provided sufficient funding. In that period, growth rates reached 4.7 % p.a., increasingly driven by commercial publishers. Finally, driven by the oil crisis, increased ecological awareness and the peace movement, the last period exhibited considerable skepticism towards science and its achievements. In their argument, this resulted in a slower growth rate of again 3.3 % p.a.

[40] Deviating from their calculation are, Björk, Roos and Lauri who report an average 111 articles p.a. per journal indexed in Thomson Reuters Institute for Scientific Information's (ISI) bibliographic databases. Contrastingly, journals not indexed publish only 26 articles p.a. [cf. 75].

[41] In principle it does not matter much whether the number of journals or just the number of articles (in each journal) increases. It is about the growing number of scientific publications. This is already system-immanent, as even at a constant rate of articles p.a. the body of literature increases from year to year. In addition, because of the growing number of yearly articles (at least partially driven by the growth of world population), this collection is growing even faster. Thus,

First, the more journals being published, the more difficult it gets to know the whole literature of any discipline and especially to stay abreast of recent developments. This is in part caused by the physical limitations of any human being to read, understand, and store information. Even more importantly, especially for smaller research institutions and developing countries, the sum of subscription prices of different journals or sets of journals may often be too expensive to read all the desired journals [cf. 60].[42] Second, following Ziman's line of argument, lesser journals would possibly also lead to a more intense competition for publishing slots. Consequently, the establishment of new journals in many cases leads to little-known journals seeking authors willing to publish in them, which in turn may lead to a fear of driving away possible authors by too harsh quality criteria.[43]

Quality of journals is traditionally measured in two ways by the scientific community [cf. 79, p. 181][44]: either survey based [See, e.g., 81; 82] or citation based [See, e.g., 83–86]. A prominent example of the earlier is the *VHB-JOURQUAL* of the German Academic Association for Business Research (Verband der Hochschullehrer für Betriebswirtschaftslehre, VHB), which ranks German and international journals of the various business research disciplines based on its members' assessment. To arrive at this ranking, all the VHB members (representing more than 90 % of all researchers with at least a PhD (Doctor of Philosophy) in business administration in the German speaking countries) are invited to take part in an online survey and to assess the quality of journals based on the quality of published articles and the quality of review processes

(Footnote 41 continued)
'more journals' implies more articles per issue, more issues per volume or any other cause for an increase in the number of scientific publications per year.

[42] There are different approaches to solve this problem. Some authors suggest the heavier usage of so called 'self-archiving', where authors make their working papers and even the prepublication version of their articles openly available to readers over the internet. They can be stored in central archives, enabling more advanced searching functionalities [cf. 78]. Another approach is the establishment of open access journals, which are offering their articles free of charge to readers. They are either sponsored by the publishing institution or operate on the basis of a submission fee to be paid by authors wishing to publish in the respective periodical. But while Harnad [78] thus envisions the end of the "papyrocentric" dissemination of scholarly knowledge, others like Mabe and Amin [62, pp. 160f.] believe in the perseverance of the more traditional journals. This is notwithstanding the fact that today nearly all journals offer online access to their articles (to their subscribers that is, for the non-open access journals) in addition to their print issues.

[43] Naturally, this does not have to be the case. Sometimes new journals will be introduced as a more specialized sub-journal of highly regarded journals (e.g., Nature Publishing Group's journals 'Nature', one of the highest-quality journals in the natural and life sciences and other journals 'Nature Biotechnology', 'Nature Materials' or 'Nature Physics'), or for new topics being dealt with by established learned societies or alike.

[44] For a good first overview of literature on this topic in economics see also [80].

[cf. 79, pp. 180ff.]. While this procedure is contended even within the same organization,[45] it is still widely used as a basis for the assessment of individuals' and institutions' publication performance [cf. 79, p. 202].

The latter approach is probably the most widely applied; it is based on different calculations aiming at the reliable reflection of a journal's scientific impact. Although building on earlier works, it is mostly connected with Garfield, who in his 1955 seminal paper coined the term "**impact factor**" [89, p. 109]. Later he founded the *Institute for Scientific Information* (*ISI*) and with Sher introduced the *Science Citation Index* (*SCI*) in the early 1960s. They created the journal impact factor as a simple method for selecting journals for the SCI, regardless of their size [cf. 90, p. 979]. ISI, now owned by Thomson Reuters, publishes *Journal Citation Reports* (*JCR*) for over 6,400 selected journals in its 'Science Edition' and more than 1,800 journals in its 'Social Sciences Edition' [cf. 91, p. 2].[46] The impact factor employs a simple calculation, the numerator (number of citations in all indexed journal articles of the current year to all articles in assessed journal in the two previous years) divided by the denominator (number of "substantive articles"[47] in the same 2 years) [cf. 90, p. 279]. As this approach is being widely criticized for not taking into account discipline and journal specifics enough, Schutte and Švec started to use the editorial of *Folia Phoniatrica et Logopaedica's* to boost its impact factor from 0.655 [cf. 93, p. 281] in 2006 to about 1.4 in 2008 [cf. 94, p. 274] by citing every single article published in their journal in the two preceding years. They justified their course of action as follows: "Our journal has taken an active approach with regard to impact factor measures and we intend to keep this initiative to protect the field of phoniatrics and logopedics against blindly applied general criteria of scientific measures handicapping under-researched interdisciplinary fields [...]" [cf. 94, p. 274]. Apparently, this tradition perceived by them as an act of 'self-defense' has been discontinued in subsequent volumes of the journal, maybe due to their title being suppressed in JCR due to "excessive self citations" [95].

---

[45] For instance the Commission on Technology and Innovation Management (Kommission für Technologie- und Innovationsmanagement, TIM-Kommission) within the VHB presents its own ranking of journals in the innovation management domain, based on a meta-analysis of 31 international rankings [cf. 87, pp. 185ff.; 88, pp. 1ff.].

[46] While the SCI covers science journals, social sciences journals are covered by the *Social Sciences Citation Index* (*SSCI*).

[47] Alone the selection of 'substantive articles' is the source of widespread criticism. While it is generally acknowledged that they should entail all original research (no matter whether reported in full papers, short papers or letters), the enclosure of reviews and editorials is discussed very controversially. What is included and what not is a matter of individual agreement between ISI and each journal, depending on the content of, for instance, an average 'letter' in their journal. Reviews are included, which leads to higher impact factors for journals publishing more review articles, as they are on average cited more often than papers [cf. 92]. Interestingly, even those parts of a journal that are not taken into account for the number of 'substantive articles', are considered when counting citations to the journal.

Even more important than the criticism regarding metrics and databases is the criticism in respect to scope. In contrast to its initial conceptualization of being a measure to be cautiously used in the evaluation and comparison of journals in restricted disciplines, it is now a nearly universal measure of quality in any scientific context [cf. 96, p. 1,107]. This is not restricted to journals or disciplines, but it is also applied in judging researchers, institutions or even whole countries, thus moving away from a fair assessment of the entity's own work to a (less labor intensive) probability-based indiscriminate surrogate [cf. 90, p. 980; 96, p. 1,107].

Notwithstanding its impressive dissemination, the impact factor (and its derivatives) is the object of very controversial discussions leading Garfield to draw a surprising comparison: "Like nuclear energy, the impact factor has become a mixed blessing" [90, p. 979]. As such, longstanding discussion appears to be an integral part of any evaluation approach, this study will not delve deeper into it but rather emphasizes the great caution that should be exerted when dealing with such systems.

In the absence of any unquestioned quantitative quality proxy, the scientific system largely depends on peer review for the assessment of an article's or project's quality. Especially when trying to evaluate quality *ex ante*, measures such as the number of citations or the impact factor of the publishing journal are inapplicable. The expert approach in turn is based on the ability of a knowledgeable peer group to righteously judge the quality and contribution of any piece of work they are supposed to review. This is in principle just like any other grading system. Accordingly, teachers in school are expected to grade the performance of their pupils in regard to expectations and the distribution of performances. But as teachers' grading in school, peer review is not without discussion [cf. 61, p. 15]. No matter how much editors try to ensure a (double) blind review process,[48] it will often be undermined by clear indications for authors' identities. This may result from the way authors cite their prior research and publications, from openly advertising articles currently under review, or from reviewers searching for scholars active in the respective field. Irrespective of the question whether this undermining is done on purpose, it leads to a tendency of the review system to favor submissions from prestigious institutions [cf. 97, pp. 77ff.; 61, p. 15].

Even more importantly, some reviewers may be too evaluative instead of taking a developmental approach [cf. 98, p. 376]. While their task is to evaluate the quality of a submission, it is not only about finding flaws in a manuscript but also about helping it to be developed into a better article. Just like editors, who have to be reminded that they "are in the business of publishing, not rejecting, manuscripts" [99, p. 489], some reviewers seem to forget to not only be authors' prosecuting but also their defense attorneys [cf. 100, pp. 192ff.]. This may lead to rejection of manuscripts with potential large contributions to the body of

[48] In a blind review process the reviewer will not know the identity of the author(s). Most scientific journals follow a double blind approach, where neither reviewers nor authors know the other party's identity. This is supposed to ensure a fair review process independent of favors, or bias in respect to authors' standings or the reputation of their affiliations.

knowledge [cf. 61, p. 15; 98, p. 376]. While this may be caused by overzealous reviewers, their rationale may also be less honorable. Especially in narrow fields of science reviewers and authors are often direct competitors for publication slots and grant money. Thus, some reviewers may be tempted to minimize the quality of another scholar's contribution [cf. 61, p. 15]. Finally, on some occasions reviewers will not understand the groundbreaking content of papers. For instance, Rosalyn S. Yalow's seminal 1956 paper on insulin transporting antibodies was rejected by *Science* and initially rejected by the *Journal of Clinical Investigation.* As her research's results (for which she received the Nobel Prize in Physiology or Medicine in 1977 [cf. 101]) were apparently beyond the apprehension of her reviewers, she had to adjust parts of her co-authored paper according to their remarks [cf. 102, p. 1,237].

Consequently, questions of quality[49] and possible distortions in the scientific literature are caveats when using scientific articles as a source of analysis. Two central measures in bibliometrics[50] are citation and authorship patterns. It has to be borne in mind that citations are not only a result of an article's quality,[51] but also its type and topic [cf. 79, p. 181]. Furthermore, they may be subject to the influences of authors' "citation cartel memberships" [79, p. 181], where authors cite each other as an act of courtesy or in order to improve institutions' or journals' citation counts (and thus reputation).[52] Many authors believe that the chance for a submission's acceptance is increased by overly citing articles from the respective journal and its related publications. Sometimes this practice is even actively encouraged by reviewers and editors.

Most researchers agree—in theory—with the International Committee of Medical Journal Editors (ICMJE) that "Authorship credit should be based on [:]

---

[49] While varying definitions exist for quality, Martin provides a convincing delineation of quality, importance and impact of publications. According to his line of argument, *quality* describes "how well the research has been done, whether it is free from obvious 'error', how [...] original the conclusions are and so on." *Importance* is defined as a publication's "*potential* influence on surrounding research activities [...] if there were perfect communication in science". Finally, *impact* in his view is "its *actual* influence on surrounding research activities at a given time". Strikingly, he adds that "While this will depend partly on its importance, it may also be affected by such factors as the location of the author, and the prestige, language and availability of the publishing journal" [103, pp. 348f.].

[50] See Sect. 3.3.2 for a brief introduction to bibliometrics, their definition and possible applications.

[51] For a lively debate on citations as a proxy of quality see, e.g., a series of articles in *Scientometrics* [104–117].

[52] Reference lists may be expanded for other questionable reasons, as well. Some researchers view the references as a broadest possible list for further reading and some reviewers put special emphasis on having their own works cited or that of high-profile scholars. For an overview of 15 reasons to cite attributed to Garfield, see Smith [118, p. 84]. These include diverse reasons such as paying homage to pioneers or peers, criticizing previous works or providing background reading. Furthermore, she provides several reasons not to cite other publications, such as the irrelevance to the present work, the unavailability or the potentially citing authors' unawareness of the respective work.

(1) substantial contributions to conception and design, acquisition of data, or analysis and interpretation of data; (2) drafting the article or revising it critically for important intellectual content; and (3) final approval of the version to be published. Authors should meet conditions 1, 2 and 3" [119]. In practice, bodies like the CSE have felt the necessity to counter developments such as *guest, honorary,* and *ghost authorships*[53]—an indication for their existence in the scientific literature. Especially when taking into account, the number and diversity of articles supposedly co-authored by very renowned scholars and it remains questionable, whether this could really always be the result of substantial contributions. The whole issue of authorship appears to be so flawed as to trigger the CSE to install an *Authorship Task Force* "to inform as wide an audience as possible about" guidelines on authorship [120]. They specifically acknowledge the fact that practice concerning which contributions would merit authorship is extremely dependent on disciplines and even individual departments [cf. 120].

3.2.2 Bibliometrics

Conducting bibliometrics, i.e. the statistical analysis of the literature [cf. 121, p. 160], dates back to the 1890s [cf. 122, p. 149]. Pritchard introduced the term 'bibliometrics' in his 1969 documentation note "Statistical Bibliography or Bibliometrics?", to describe the "application of mathematics and statistical methods to books and other media of communication" [cf. 123, p. 349]. This was supposed to avoid any misunderstanding as could be experienced when using the expression 'statistical bibliography'.[54] While researchers seem to agree that bibliometrics somehow "uses quantitative and qualitatively descriptive methods [...] to examine documents" [124, p. 63], various definitions exist ranging from very broad to more narrow ones.[55] Diodato defines it in his *Dictionary of Bibliometrics* as "mathematical and statistical analyses of patterns that arise in the publication and use of documents" [126, pp. viiif.]. It is commonly delineated from 'scientometrics' (quantitative studies of science) and 'informetrics' (dealing with any form of information and thus including the earlier two terms) [cf. 55, p. 180]. There is a very strong link between scientometrics and bibliometrics, as the earlier is often

[53] Guest authorship denotes authorship based on the expectation that inclusion of this particular author will improve likelihood of acceptance for publication. Honorary authorship typically describes authorship based on an individual's position, e.g., as head of a department. Finally, ghost authorship relates to *de facto* authors not listed as such, because of conflicts of interest or an attempt to overemphasize the remaining author(s)' contribution [cf. 56, p. 18].

[54] According to Pritchard, this was experienced by most researchers, as 'statistical bibliography' could be interpreted as statistics on bibliographies as well as bibliographies on statistics, thus it had only been used four times in the 46 years prior to Pritchard's note [cf. 123, p. 348], p. 348.

[55] For a more detailed overview of various definitions see, e.g., [125, pp. 373ff.].

relying on the latter for an unbiased[56] evaluation of scientific activity [cf. 55, p. 181].

In using bibliometrics, Tijssen discerns between items (pieces of information in or about scientific publications) and entities (the publications themselves or sets of publications) [cf. 77, pp. 13f.; 55, p. 181]. He further distinguishes between three main parameters of scientific publications [cf. 55, pp. 181f.]:

- **Product** (scientific activities, which can be measured in numbers of publications)
- **Process** (transfer of knowledge, dissemination of knowledge measured through citations[57])
- **Structure** (social and cognitive networks, to be seen in co-authorship, affiliations, and reference lists).

Based on these parameters, various measures have been used in bibliometric analyses. They range from very simple measures such as the *number of publications* by one author,[58] institution, or country to more advanced measures such as the *author proportiometric index*[59] or the *h-index*.[60] When dealing with bibliometrics, it has to be taken into account that deriving reliable results may be hampered by issues of incomplete data and only partial coverage of databases. While one missing publication alone may be detrimental to, e.g., the h-index of an author, even leading databases may only cover as little as 26 % of a scientific field (as is the case for ISI's SCI, and the field of mathematics) [cf. 55, pp. 186f.]. Consequently, any analysis depending on full coverage of author's or even a

[56] As briefly discussed in Sect. 3.2.1, this can only be 'relatively unbiased' or less biased especially by the respective scientists themselves.

[57] Although many authors use the terms 'citation' and 'reference' synonymously, the first is (in its original sense) the quotation a publication is receiving from an earlier publication. The latter is the quotation of an earlier publication in a later publication. At the same time, 'citation' refers to the actual linkage between two documents, irrespective of the direction. As such it is also used within the present study.

[58] Strikingly, Martin finds that (in his analysis of empirical papers published in *Scientometrics*) "scientometric analysts would seem to rely rather heavily on the simplest of indicators, despite their well known limitations". Indeed, publication counts were used in 60 % of all papers, followed by citation counts and citations per paper/impact factors (31 and 26 %, respectively). Furthermore, mostly only one or two indicators were used and not a set of multiple indicators [cf. 103, pp. 353f.].

[59] The author proportiometric index is a measure for the (un-)weighted number of pages per author in multi-authored articles. This is derived by simple division of the number of pages per number of authors. If weighted, it is corrected based on whether the respective author is the first, second or so forth author [cf. 126, p. 5].

[60] The h-index, or Hirsch-index, was proposed by physicist Hirsch, defining it as the number of papers with citation numbers $\geq$h. He introduced it as a "useful index to characterize the scientific output of a researcher". In contrast to other 'impact'-related measures, it accounts for the importance of a researcher's output without unjustly favoring co-authors of highly cited articles (esp. review articles), penalizing high productivity or arbitrarily picking thresholds [cf. 127, pp. 16569ff.]. For an application see, e.g., [84, 86, 128].

discipline's work will be made obsolete. In addition to general errors in respect to misspellings and authors with identical names, bibliometric studies need to control for differences over time as well as varying practices in scientific fields. Finally, they should be cautious when reporting statistical findings, as evidently significant findings may be caused by "accidental effects" leading to citations [cf. 129, p. 142].

Notwithstanding all limitations [cf. 103, p. 348],[61] widely accepted measures often employ citation-based calculations. They include *straight counts*, *normal counts*, and *adjusted counts*.[62] In most cases, self-citations[63] are either excluded from these analyses or at least clearly stated. These citation-based measures include [cf. 55, pp. 199ff.][64]:

- Number of publications (in SCI) (P)[65]
- Total number of citations to one document, researcher, or institution (C)[66]
- Average number of citations per document (CPP)[67]
- Share of self-citations (%Self-Cites)[68]
- Average Journal Citation Score (JCS)[69]
- Field Citation Score (FCS)[70]
- Journal-to-Field Impact Score (JFIS)
- Combinations of the above[71]
- Attractivity Index (AAI), as a measure for the relative impact of a country's publications.

In general, many measures are based on an attempt to subtract differences between compared entities. This normalization can be done by, for instance, limiting time spans, considering group sizes, and available funds or by relating to mean values such as a FCS_m, a JCS_m, or a $JFIS_m$.

---

[61] For a thorough discussion of different bibliometric measures as well as especially their limitations, [cf. 55].

[62] Straight counts refer to a calculation based on 'the first author takes it all', with all citations allocated to the first author only. In contrast, normal counts allocate full citation counts and adjusted counts the respective fraction (one divided by number of authors) of the citation count to each of the authors.

[63] This may refer to either authors citing their own work or the work of their direct colleagues at the same institution. While this may be done as a reference to earlier works and underlying concepts, it is often perceived to be exercised too abundantly.

[64] For a detailed discussion see, e.g., [130–132].

[65] For an application see, e.g., [53, 128–139].

[66] For an application see, e.g., [128, 129, 131, 133, 136, 138, 140].

[67] For an application see, e.g., [92, 128, 129, 138, 139].

[68] For an application see, e.g., [133].

[69] For an application see, e.g., [83, 128].

[70] For an application see, e.g., [128, 131, 133].

[71] For an application see, e.g., [87, 128, 130, 133, 141].

The *Revealed Literature Advantage* (RLA) of a country constitutes a measure to assess a country's strength in a particular subfield. Its calculation is done by dividing the share of a country's publications in the respective subfield (number of publications by country X in subfield a/number of all publications in subfield a) by its share in publications in the whole field (number of publications by country X in whole field A/number of all publications in whole field A) [cf. 55, p. 202].

Verbeek et al. differentiate the aforementioned measures from approaches to 'map' the underlying structure and evolution of different scientific (sub-)fields [cf. 55, pp. 203ff.]. Whereas most measures are used to assess a specific entity's (individual scholars, departments, universities, journals, or even countries) strength and impact in a given area, maps are designed to use relational techniques to unravel the connections between each of these entities. Thus, they are mostly based on tracking co-citations,[72] co-word usage or a combination of both. Further, possible data points include keywords, classification codes, or journal topics. In contrast to many of the easily quantifiable proxies mentioned earlier, it is the researcher compiling them who is the most decisive factor in producing rich and insightful maps [cf. 142, p. 210]. Consequently, their handling is much more challenging—but at the same time a lot more promising [cf. 143, p. 581], when it comes to unveiling hidden patterns cloaked under large amounts of data.

Maps can be grouped into two different types. First, when based on similarities in journal topics (and, e.g., the co-citation/bibliographic coupling of articles in different journals) or classification codes they should be largely reliable, as the classification (or acceptance to the journal) will in most cases be executed by a smaller group of experts with a high consistency in their decisions. Second, when constructed from co-citation or co-word patterns they could be expected to be less reliable, as the decision to cite another document (or use certain words) is made by individual authors, who may have different reasons for their choices and may thus have less consistent patterns. Wherever keywords are used as the input data of mapping activities, it has to be clarified whether they are 'uncontrolled terms' (provided by the authors) or 'controlled terms' (selected by editors or indexers of database providers). In the case of controlled terms, these maps would belong to the first group, otherwise to the latter one. While in principle co-word analysis should yield the most detailed picture of a given domain, it is probably also the most challenging analysis. Not only is it necessary to obtain the full text of all

[72] While citation analysis focuses on the impact a certain article (or set of articles) has, co-citation analysis is used to uncover relations between different streams of research. Two articles of author A and B are said to be co-cited if they are quoted in the same article later by author C. Additionally, bibliographic coupling refers to author C and D both using an identical (sub)set of references in their publications. Caution should be exerted when using such a form of analysis, as the fact that two papers are co-cited or biographically coupled could indicate different connections. The rationale for it could potentially be their similarity, their coverage of a similar field/topic or the use of a common method. At the same time there could be no connection at all, as different later articles might be referring to completely different parts in the same article (so could one document be referring to a field of analysis and the other to a specific method employed). For application see, e.g., [136].

documents and to employ more complex search algorithms (depending on whether the word pairs are to be automatically selected[73] or provided by an investigator). Also, the analysis is being complicated by the fact that the use of words is anything but unambiguous and to a high degree context dependent.

3.2.3 *How will Scientific Publications be Used in Anticipating Converging Industries?*

After introducing basic principles of bibliometrics and the most common indicators, the following section will discuss their applicability in anticipating converging industries. Their selection has to be based on the exceptional settings in converging scientific fields and on the target to develop a concept which is both valid and robust. Thus, this study will not engage further in the discussion of the general value of each of the indicators,[74] but rather focus on the special considerations for the proposed application.

Measuring convergence in scientific disciplines can be grounded on two aspects. First, if two disciplines begin to merge or to form a common subsegment, collaboration could be expected to intensify and each area to work more with results from the other area. Collaboration could be assessed on the basis of co-authorships of articles or on precise information on collaborative projects, changing researcher affiliations and broadened research agendas. Whilst information on each of the items is available, the only one that can be easily and comprehensively accessed is co-authorship. As this information is incorporated in all of the databases, it is well structured and controlled by database providers. After obtaining authorship information, the task would be to assign authors to their respective scientific field. Where affiliation names indicate a clear connection to one domain, this could be done from the full-text documents. Otherwise it would have to be done manually through the collection of further information for each single author. Literature databases like Elsevier's Scopus® inherently provide author affiliation as well, but it proves to be often outdated or inaccurate and its incorporation in search functions is rather limited.[75] Consequently, to avoid the tedious task of collecting information on authors one by one, a database providing reliable data would be necessary for a larger scale use of this indicator.

---

[73] For an introduction to different ways to automatically identify terms see, e.g., [143].

[74] A short reflection of the most prominent points of criticism has been provided in Sect. 3.2.2.

[75] Like patents (see Sect. 3.3), scientific articles entail, due to requirements of editors and publishers, clear identification of, e.g., authors, references and fields of science. Furthermore, extensive databases have gathered comprehensive collections of scientific articles, added information on subject area and keywords and made them full-text searchable. Still caution has to be exerted when relying on databases as information will regularly be either outdated or flawed (or both).

Second, when research is becoming mutually interesting, authors will begin to cite journals and scholars from the other area and use (key-)words that used to be part of the other area. Thus, both citation analysis and co-citation analysis, based on individual authors where possible and on journal topics or keywords where necessary, should be employed. Sigmund and Wallin express a caveat in respect to citation analysis: "A valid citation analysis is time consuming; it requires insight and, therefore, needs to be performed by experts" [144, p. 11].[76] Analyzing co-words would theoretically be one of the best means, but this could only work if disciplines use a set of precisely discernable words that could be undoubtedly allocated. Hence, this approach does not appear to be easily generalizable. More importantly, even if these sets of words were precisely discernable, how would investigators know which words to look for? In practice, investigators from institutions focusing on chemistry and the chemical industry would have to monitor all scientific publications related to chemistry and then decide which discernable words representative of other disciplines to look for in all the documents. This undertaking would be even more difficult because of the low application orientation of many scientific articles. When thinking of an example like smartphones in ICT, it is very unlikely that early (scientific) articles have used such an expression when describing new combinations of technologies or services.

As a result of this discussion, simpler counts of publications in different journals and co-authored by distinct groups of scholars as a development over time appear to be the only indicators which could be of interest to academics and practitioners alike for an anticipation concept.[77] Two further aspects limit the value of science convergence indicators within the present study. On the one hand, the relevance of a blurring of boundaries between scientific disciplines would in many cases seem to be difficult to assess. Undoubtedly, the boundaries between, e.g., chemistry and biology are a lot less prominent then they used to be. With a rising importance of fields such as biochemistry or biotechnology (both highly dependent on knowledge from chemistry and biology alike), dedicated journals were established just like new departments and university degrees. Furthermore, scholars from both well-established disciplines will now move more freely between faculties and collaborate closely with other researchers of either field (or the new sub-disciplines). Despite this clear impact this has not necessarily led to a unified biology/chemistry discipline. Quite contrarily, both disciplines tend to rather specialize more into smaller areas cf., e.g., [68, pp. 134ff.]. Hence, while it would be difficult to identify the importance of such developments for industries, their relevance to universities and education policy makers is much clearer. On the

---

[76] While this notion is quite convincing and in line with earlier assessments, it is in stark contrast to this study's aim to develop a method which could be applied by at most semi-experts and without too much requirements regarding time, effort and money.

[77] Contrastingly, these indicators could hardly be sufficient for a comprehensive monitoring concept on all stages. The future research opportunities in Sect. 6.4 will thus include a discussion of indicators to be possibly included in future evolutions of this study.

other hand, earlier studies have found substantial time lags between developments in science and their translation into technological or even applicational changes. Adams noted already in 1990 that "a lag effect of roughly 20 years is found between the appearance of research in the academic community and its effect on productivity in the form of knowledge absorbed by an industry" [145, p. 673]. For inter-industry spillovers he finds even a lag of about 30 years, whereas lags are shorter (only 10 years) for computer science and engineering [cf. 145, p. 676].[78] Considering most stakeholders' time horizons, taking into account lags of up to 30 years makes a focus on the scientific literature rather pointless. So even if all the aforementioned drawbacks and challenges could be overcome, the warning signal would come too early for most investigators.

Consequently, the analysis of the scientific literature will only be an added piece of information for the more thorough scrutiny of potential areas of convergence, especially for examinations of triggers and drivers. At the same time, it can serve as an additional indicator in the attempt to spot weak signals.

3.3 Patent Documents

3.3.1 Particularities of Patent Documents

A myriad of excellent books and articles cover the topic of patents[79] and nearly every aspect related to it in great depth. They especially include coverage of underlying theories, history, differences in national systems, differences in propensities to patent, favorable and unfavorable effects of patenting on innovation, and the particularly challenging task of valuating patents. It is thus not the aim of this section to provide yet another lengthy treatise on patents from every angle. Instead, it will focus on a brief discussion of patents in general and the specific relevance of patent analysis in respect to the anticipation of converging technologies and industries.

[78] See also Tansey and Stembridge [146, p. 220], who find a time lag between scientific articles and patents of approximately ten years for the area of monoclonal antibodies. They further predict a similar lag for the area of nanotechnology.

[79] Patents or patent publications encompass all of the available patent literature, i.e. not only granted patents but also published patent applications. Not all patent applications lead to granted patents (either the applicant does not pursue patenting any more or even withdraws his application, or the patent authorities do not grant a valid patent due to legal constraints), but still represent a valid proxy for R&D activities. Furthermore, due to the complex legal obligations and necessary steps granted patents will only be published several years after their initial filing. Thus patent applications will be included in the analyses of this study. In order to increase legibility of it, 'patents' will be used as a synonym for both applications and granted patents throughout, if not stated differently.

Generally speaking, patents are a monopoly granted (for a limited time) by a governmental body to an inventor in exchange for his disclosure of new knowledge. More precisely, Griliches' provides the comprehensive and widely accepted definition: "A patent is a document, issued by an authorized governmental agency, granting the right to exclude anyone else [in a limited territorial area] from the production or use of a specific new device, apparatus, or process for a stated number of years [...]. The grant is issued to the inventor of this device or process after an examination that focuses on both the novelty of the claimed item and its potential utility. The right embedded in the patent can be assigned by the inventor to somebody else, usually to his employer, a corporation and/or sold to or licensed for use by somebody else. This right can be enforced only by the potential threat of or an actual suit in the courts for infringement damages. The stated purpose of the patent system is to encourage invention and technical progress both by providing a temporary monopoly for the inventor and by forcing the early disclosure of the information necessary for the production of this item or the operation of the new process" [147, pp. 1662f.].

This form of intellectual property right (IPR) is a means to bar others from imitating the patent holder's technological knowledge. Thus, it should not be mistaken as a 'right to use', when it is in fact only a 'right to forbid'.[80] Even after a patent has been granted, it may be contested in a court or another appropriate authority, such as the respective national patent office.[81] The only right obtained by a granted and valid patent is the right to appeal to a court against infringement by another entity and to seek compensation. Consequently, inventors will be likely to only choose patents as a protection strategy, if the trade-off between monopoly and disclosure (also referred to as the **patent bargain**) will likely be yielding higher rewards than trade secrets or publications would. If, for instance, competitors can easily 'invent around' a patent but are expected to encounter

[80] The granting of a patent does indeed not indicate legality to produce the subject of the patent (or to execute the process). Albeit, patent authorities carry out their examinations in respect to the requirements of the respective patent laws and not necessarily regarding other (inter)national laws or ethical considerations, inventions which would be challenging morality or public order, are excluded from patentability under regulations such as Article 53a of the European Patent Convention (EPC). But due to examination and granting procedures, an 'immoral patent' may be granted and this may only later be corrected when the respective patent is contested. In respect to illegality, the EPC strikingly explicitly states: "[...] such exploitation shall not be deemed to be so contrary merely because it is prohibited by law or regulation in some or all of the Contracting States" [148, Art. 53a]. For a more detailed discussion of opposition against patents for moral reasons see, e.g., [149–152].

[81] At the EPO (European Patent Office), e.g., an opposition will be decided by the *opposition division* and can potentially be appealed at the EPO's *boards of appeal*. Currently, 24 technical boards of appeal, the *Legal Board of Appeal*, the *Enlarged Board of Appeal* and the *Disciplinary Board of Appeal* handle such appeals. Following the initiative to instigate a European patent, the involved governments are also trying to reach agreement on the installation of a European Patent Court [cf. 153, pp. 3ff.].

difficulties to do so without the information in a patent, then a trade secret will be more appropriate.[82] In cases where *freedom to operate* (FTO)[83] is the main rationale, publications may be another viable alternative. According to Schumpeter, a patent is the promise of monopoly profits as a form of incentive to innovators [cf. 156, p. 305]. As such, it is supposed to further technological progress and eventually increase the wealth of nations by avoiding duplication of inventive activities resulting from the disclosure of technological information. This is contingent upon a certain quality of the invention to be patented.[84]

Historically, patents can be traced back to 700 to 500 BC, where the Greek colony Sybaris granted an exclusive right of use for 1 year to cooks who 'invented' a new dish or recipe. The early Renaissance is said to have seen the first real patent, granted 1421 in Florence for a period of 3 years to Filippo Brunelleschi for his new type of cargo ship. In 1474, Venice introduced the first codified patent law, before most other countries followed over the course of the next centuries [cf. 157, pp. 6f.]. In 1787 the *US Constitution*, the oldest and shortest written constitution in the world [cf. 158], stated: "The Congress shall have power [...] To promote the progress of science and useful arts, by securing for limited times to authors and inventors the exclusive right to their respective writings and discoveries" [159, Art. I Sect. 8]. In 1883, the first 14 nations started to harmonize their patent laws by ratifying the *Paris Convention for the Protection of Industrial Property* (Paris Convention); a number that has increased to 173 to date.[85] The Paris Convention was initiated to help the people of each of the member states to obtain protection for their **industrial property rights**[86] in all other countries. It was followed in 1970 by the *Patent Cooperation Treaty* (PCT), enabling applicants to seek patent protection simultaneously in the member states through means of an "international" patent application.

---

[82] Patents are often perceived to be rather ineffective at protecting process innovations [cf. 154, p. 287]. Also, Brouwer and Kleinknecht find a rather broad mixture of mechanisms to protect processes/products against imitation, including patents but also more importantly (across industries) time leads, keeping qualified people in the firm and the complexity of the process/product design [cf. 155, pp. 617f.].

[83] FTO refers to the question whether a product or process can be commercialized without infringing other entities' IPR. Regularly, it is the task of investigators to ensure FTO. If in turn inventors want to guarantee the possibility of commercializing their inventions, they could file patent applications or scientific publications in order to stop anyone else from obtaining a patent for the same invention. This is particularly interesting, where patents would be too weak or too costly to uphold.

[84] Whether patents really are beneficial to a society is vividly contested. Effects are undoubtedly miscellaneous and highly dependent on the respective national and industrial setting [cf. 157, p. 22].

[85] Thus far the last three nations to ratify the Paris Convention have been Angola, Yemen in 2007 and Thailand in 2008.

[86] In addition to *intellectual property rights*, *industrial property rights* also include, e.g., trademarks and industrial designs.

Granting of a patent is contingent upon three central conditions. The invention has to exhibit **novelty**, an **inventive step** (also referred to as *non-obviousness*) and **industrial applicability**. If substantial information on the invention have already been disclosed (e.g., in the form of personal communication, presentations, or articles) by the inventor or any other person prior to the filing with the patent authorities, it will not be patented. Likewise, building on prior art the inventor will have to add a substantial inventive step, which may not be a 'logical' step to a person having ordinary skill in the art. Finally, there must be a potential industrial application to the invention, although this does not necessarily refer to an implemented application. While some inventions are categorically excluded from patentability[87] by, for instance, the EPC, others—lacking aspects of a technological invention—are not patentable as such: [cf. 148, Art. 52]

- Discoveries, scientific theories, and mathematical methods
- Schemes, rules and methods for performing mental acts, playing games, or doing business
- Presentations of information
- Programs for computers[88]
- Aesthetic creations.

If these conditions are met, the inventor will be able to apply for a patent[89] at the patent office responsible for the respective country. If wishing to obtain a European patent (EP), e.g., the inventor will be required to hand in a request for the granting of an EP, an abstract, a description of the invention, the **claim**(s), and drawings (if necessary). This will be followed by a formalities examination by the EPO, in order to ensure fulfillment of the minimum requirements for according a date of filing. Simultaneously, a European search report is compiled, containing all the pieces of information that might be harmful to novelty or inventive step. After 18 months (based on filing or priority dates,[90] whichever is first), the application is made available to the public. If the inventor decides to pursue the patent procedure further, substantive examination will be carried out before **granting** of the patent. In the case of an EP, it will then have to be validated in each of the designated states. Potential

---

87 Article 53 of the EPC categorically excludes all inventions from patentability, if their "commercial exploitation would be contrary to "ordre public" or "morality"; if they are "plants and animal varieties or essentially biological processes for producing plants or animals (microbiological processes and products thereof are not excluded)" or if they are "methods for treatment of the human or animal body by surgery or therapy and diagnostic methods practiced on the human or animal body." The latter are not excluded if referring to products, substances or compositions for use in such methods [148, Art. 53].

88 For national differences in patentability of software products see [160].

89 For increased legibility, procedures are according to those of the EPO, where not stated otherwise.

90 The priority date is the date of the first submission of a patent application to any patent authority around the world. Within one year of the priority date, applicants may file for additional patents in other countries.

Table 3.4 Content of the eight IPC sections

IPC section	Content
A	Human necessities
B	Performing operations; transporting
C	Chemistry; metallurgy
D	Textiles; paper
E	Fixed constructions
F	Mechanical engineering; lighting; heating; weapons; and blasting
G	Physics
H	Electricity

challengers of the granted patent may file their opposition up to 9 months after publication of the granted patent; this opposition would then be dealt with by an opposition division. Patents not (successfully) opposed could be limited or revoked on behalf of the patent holder.[91] Otherwise, if all fees are paid (and renewal fees as well), the patent will be valid for a maximum of 20 years[92] since filing date.

In order to make search and retrieval of patents as well as handling by the patent authorities easier, different **classification schemes** were introduced. Next to various national classifications one major international classification scheme exists. The *International Patent Classification* (**IPC**) was agreed upon as a universal reference.[93] The IPC structures technology hierarchically into approximately 70,000 subdivisions with alphanumeric codes. They are aggregated into 8 *sections* (Table 3.4) marked with capital letters from A to H (A). Each section is then divided into *classes*, designated by the section letter and a two-digit number (ANN). They are further structured into one or more *subclasses*, denominated by a capital letter (ANNA). Finally, they are completed by *groups* identified by a one- to four-digit number, a forward slash and 00 if a *main group* (ANNA NNNN/00) or at least two numbers other than 00 if a *subgroup* (ANNA NNNN/NNNNNN). Due to the complex classification system, some smaller patent offices experienced problems in classifying correctly according to the 70,000 groups. This difficulty results from the fact that smaller offices also receive less applications and that their examiners will be responsible for much broader technological areas than their counterparts at the larger offices. As a consequence, the IPC was divided into two

---

91 One reason might be the fear of a successful opposition to the patent due to too broad claims.

92 In some territories, such as the EU (European Union) and the USA, assignees may apply for a *supplementary protection certificate*, extending the patent protection for an additional 5 years (25 years in total) for drugs or crop protection products. This regulation was introduced to account for the very long authorization procedures and to permit sufficient commercial exploitation time of the granted monopoly.

93 In comparison to about 70,000 codes according to the *IPC* or 120,000 codes of the *ECLA* (European Classification), the most comprehensive classification is provided by the JPO and its ca. 190,000 items regarding *FI* (File Index) and the roughly 340,000 multi-dimensional viewpoints provided by the *F-terms*. Both ECLA and FI codes are based on a further refinement of IPCs and are only used for the respective European or Japanese patents.

levels of classification, the *core level* (with only 20,000 groups) and the *advanced level*. While the advanced level uses every of the available codes and will be dynamically adjusted to technological changes, the core level only employs a smaller subset down to main groups (and only rarely subgroups) expected to be quite stable [cf. 161, pp. 3ff.].[94]

Although the classification scheme is both application and function oriented, function takes precedence over application [cf. 162, p. 219]. To account for technological changes as well as changes in classification practice, the IPC has been periodically revised since its 1968 first edition. Up to the 7th edition (valid until 2005), it was consecutively numbered, since it is marked with year and month (e.g., IPC-2008.01).[95] This specification is provided in any patent next to the IPCs used. As most inventions do not only fall into one distinct IPC, they are co-classified with one or more other IPCs. Up to the 6th edition of the IPC, this co-classification was done by one main IPC and other IPCs as secondary ones. This practice has not been extended into the later editions of the IPC.[96] As of January 1, 2011 another major reform of the IPC will enter into force, aiming at substantially simplifying its application. As endorsed by the IPC Union Assembly in September 2009, the distinction between advanced and core level will be abolished. Instead, smaller patent offices now using the core level will be using main groups and not the subgroup level. Further amendments include the extension of revision cycles to a yearly pattern (on every January 1st) and the acceleration of efforts to build one unified global patent classification by integrating national or regional schemes (such as ECLA, FI/F-terms and USPC, and US Patent Classification) [cf. 163].

Albeit a detailed discussion of **patent law differences** between countries[97] is beyond the scope of this study, a few particularly relevant differences need to be mentioned, especially within the three hitherto most important patent systems: the

---

[94] An illustrative example is provided in the appendix. The core level will be revised in three-year revision cycles, if necessary. Contrastingly, the advanced level has a planned revision cycle of only three months.

[95] Patent offices are obliged to reclassify backfiles in order to enable searchability with just one version of the IPC. Albeit, a caveat results from possible differences in classification due to changes in precision of the classification system in respect to any given patent.

[96] Due to the major IPC reform all patents published after January 1, 2006, are classified with a set of IPCs printed in italics and bold italics onto the first page of the patent. While the IPCs in bold refer to classification based on 'invention' information, the non-bold symbols refer to classification on the basis of 'additional' information in the patent. Prior to the reform, the classification was conducted on the basis of the claims information only. Now the 'invention' information refers to the claims *and* other inventive aspects, such as drawings and descriptions. For analyses based on classification information this is slightly negative, as a tool is lost to easily assess the main focus.

[97] For a concise overview of the different patenting systems and their effects on practical aspects of managing intellectual property, built on a very comprehensive overview of the literature, see [164]. For further interesting aspects, see, e.g., [165–171].

European, the Japanese, and the US systems.[98] A *patent family*[99] granted by all three of these *Trilateral Offices* is commonly referred to as a *triadic patent*. Most of the differences result from deviations in underlying philosophies. While the main focus of the US system is to protect inventors' rights, the European system is motivated more strongly by the wish to disseminate new technological knowledge and thus yields an increased innovation rate [cf. 162, p. 215]. Based on these motivations, the European and Japanese patent offices (like virtually any other office in the world) are granting patents on a '**first-to-file**' basis. Consequently, in the case of two competing inventors, (s)he will be granted the patent whose application carries the earlier priority date. The USPTO (United States Patent and Trademark Office) operates on a '**first-to-invent**' basis.[100] In this system external sources, such as written protocols or witnesses' testimonies, are used to determine the earlier invention. While this can lead to a high degree of uncertainty for applicants due to lengthy and costly examinations, it is also said to be less biased in favor of larger companies, as they cannot take so much advantage of their resources by quickly filing all potential patent applications [cf. 173]. In order to reduce the risk of competition over researchers' inventions, the USPTO has introduced in 1995 the tool of *provisional patent applications* (PPAs). In contrast to regular applications, a PPA offers the opportunity to obtain an early priority date at lower costs and with less effort. Since a PPA does not need to entail important but time-consuming parts, such as claims and a discussion of prior art, it must be followed by a nonprovisional application within 12 months [cf. 174, pp. 137ff.].[101]

A further decisive difference exists in the number of applications and subsequently of patents granted by the EPO in comparison to the USPTO and the JPO (Japanese Patent Office). This is to a considerable extent caused by two characteristics in the systems. First, US patents do not need to have a technical focus resulting in virtually anything being patentable (including business methods).

---

[98] China is increasingly viewed to be also highly relevant. As the Chinese patent law has been developed in close collaboration with German experts and is very similar to German patent law, it can be expected to largely follow European lines of argumentation [cf. 172, p. 5].

[99] A patent family is a set of patents referring to one invention and with a common priority date, mostly those applied for in different countries by the same inventor. .

[100] Under Sec. 3 of the *Leahy-Smith America Invents Act* (AIA), as signed by President Obama September 16th 2011, the USA will change from its first-to-invent to a first-to-file system as of March 16th 2013. The AIA also entails a wider range of changes to the US patenting system, making it more similar to, e.g., the European system. However, as it is not the purpose of this study to elaborate in detail the peculiarities of the different systems, the AIA and its consequences will not be discussed in greater detail.

[101] Bentwich also suggests the introduction of mandatory *provisional patented paper applications* (PPPAs), i.e., a special form of paper submission to certain 'certified' journals. These should heal the negative effects associated with the patent system, namely the reluctance to disclose new knowledge fully in timely published papers and the so-called *anticommons effect*. This effect refers to the constraints some researchers might experience in conducting further scientific research due to other inventors' patented scientific and technological achievements.

Additionally, inventors may file successor applications to a pending application. Second, inventions in Japan may be split up into 10 different patents, while in the European system they would be handled in one application referring to one common inventive system. As a result, the backlogs at the JPO and the USPTO are decisively larger than at the EPO, leading to average first-action pendency periods for examination of 26, 21, and 9 months, respectively [cf. 173].[102] Furthermore, prior to November 29, 2000, US patent applications were not published, whereas now they will (with a few minor exceptions) be published after 18 months. Also, up to 1995 (after the US had ratified *TRIPS*, the Agreement on Trade-Related Aspects of Intellectual Property Rights) patents were granted for 17 years from the date of issue instead of 20 years from the date of filing.[103] Finally, the USPTO classifies its patents based on its own classification scheme (USPC, which is claims oriented) and translates them into IPCs via a concordance list. This may lead to skewed results when comparing US classified patents with patents granted by other patent offices. One possible approach to circumvent differences in national patent systems when comparing two countries is to use penetration of a third-country market, as entities from both countries will be facing the same situation in this third country [cf. 178, pp. 31ff.].

On the limitation side it has also been argued that, when looking at patents as a proxy for scientific and technological development, the differing **propensity to patent** not only between countries but also individual organizations or whole industries needs to be taken into account.[104] Arundel and Kabla present their findings depicted in an abridged form in Table 3.5.[105] Strikingly, the propensity to

---

[102] To counter this threat to Japan's international competitiveness, especially when it comes to highly contested areas such as biopharmaceuticals development, the JPO introduced three ways to accelerate patent examination. These are *accelerated examination*, *patent prosecution highway* and the *super-accelerated examination*. Basically, all three routes are based on an argument of high importance and high sensitivity. Considering biotechnological innovations, for example, it may often prove detrimental for small firms with a groundbreaking new technology to not be able to gather the necessary funds from other companies or venture capitalists as long as a patent has not been granted. These new measures have also led to the fastest turnover time in Japanese history. A patent on an electrochemical analysis method was granted to Keio University within 17 days in 2008 [cf. 175, pp. 815ff.].

[103] This has in the past led to the existence of so-called 'submarine patents'. They resulted from exceptionally long examinations periods leading to a 'surfacing' of legal monopolies at a time where companies may have already been using certain technologies for years before the patents were granted and thus published. Extreme cases include Phillips Petroleum's polypropylene patent (applied for by Hogan and Banks in 1953 and granted in 1983) and Lamelson and his heirs, who are said to have received up to US$ 1.5 billion in royalties. Using the tweaks of the old US system, inventors could even try to achieve submarine patents by filing additional successor applications to extend the examination phase. As a result, some patents turned valid at a time where they normally would have long been outdated [cf. 176, p. 1; 177, p. 179].

[104] Notwithstanding differences in the propensity to patent, a strong correlation between the number of triadic patents and industry-financed R&D expenditures ($R^2 = 0.99$) has been proven [cf. 179, p. 8].

[105]

Table 3.5 Patent propensity rates for European firms by sector between 1990 and 1992

Sector	N	Sales-weighted patent propensity rate (%) for	
		Product innovations	Process innovations
Food, beverages and tobacco	42	26.1	24.7
Textiles, clothing	9	8.1	8.1
Petroleum refining	17	22.6	29.0
Chemicals	88	57.3	39.0
Pharmaceuticals	32	79.2	45.6
Rubber and plastics products	20	33.7	27.6
Office and computing equipment	8	56.8	20.9
Electrical equipment	26	43.6	21.5
Communication equipment	37	46.6	22.7
Transport and telecom services	23	20.5	12.4
All firms	604	35.9	24.8

Source Abridged from [180, p. 133]

patent[106] varies greatly between different industry sectors. While the investigated European firms are across the board more likely to patent a potential product innovation than a process innovation (35.9 to 24.8 %, respectively), the overall propensity as well as the difference between product and process innovations is highly dependent on the sector. Process innovations are much more rarely patented [cf. 181, p. 226; 182, p. 26], as processes are mostly not sold to other firms. Thus, secrecy is mostly more important in protecting this vital know-how, than patents.[107] If patented, these innovations would have to be disclosed—leading to more cases of imitation and probably also (hardly detectable) patent infringement. The lowest propensity rate can be found in textiles and clothing (8.1 % for product and process innovations), an area traditionally emphasizing fashion and brands

[106] Arundel and Kabla use the *sales-weighted patent propensity rate*, reflecting the share of product/process innovations where the company applied for a patent of all product/process innovations of that same company. The indicator is based on a weighting by sales, as R&D expenditures were not available for all firms in the sample. Even a use of R&D expenditures would have been employed under the assumption that the number of innovations is a direct result of the amount of money spent on R&D (a linear correlation). Due to the lack of data, they used an annual turnover proxy, after testing for applicability against the R&D numbers for a part of the sample (where both numbers were available). Although the propensity rate would have been higher if based on R&D expenditures, the results did not differ structurally. On the other hand, these results may overestimate the propensity rate as the survey did not ask for granted patents but rather patents applied for [cf. 180, pp. 131f.].

[107] Hussinger reports on this hypothesis and on other scholars finding support for it. At the same time, she presents regression analysis results for the German manufacturing industry in 2000. Unsurprisingly, her analyses yield the conclusion that "patents are important to protect intellectual property in the market, whereas secrecy seems to be rather important for early-stage inventions" [183, pp. IIIff.].

over technological capabilities.[108] The highest propensity rate can be found in the pharmaceutical industry, where nearly 80 % of all product innovations and more than 45 % of all process innovations will be filed for patents. Normally, this number could have been expected to be even higher, as the industry is well-known for its emphasis on patent protection. The 20 % of product innovations not connected to a patent application are thus likely to be referring to generics, biosimilars, and products where the granting of a patent would be very unlikely or even impossible.[109]

Two further peculiarities are very interesting when looking at Arundel and Kabla's data. First, the most prominent difference between product and process innovation propensity rates can be found in the office and computing equipment industry (56.8 and 20.9 %, respectively). Most likely, this is rooted in exceptionally complex and expensive production processes with a lot of firm-specific knowledge involved. Second, petroleum refining is the only industry with a higher rate for process (29.0 %) than for product innovations (22.6 %).[110] This is probably more due to the exceptionally low propensity rate in product innovations than because of high likeliness to file for a process patent, which is in turn caused by the very low patentability of this industry's commodity products.

These propensities are also mirrored in the number[111] and especially growth rates of patent filings by technology. Fields of technology (mostly derived from patent classification) are naturally not as stringently bound to a certain industry sector[112] as are the sectors itself, albeit there is mostly a strong correlation to be expected. Accordingly, in fields like telecommunications (116,770), computer technology (144,594), and pharmaceuticals (74,254) more patents are filed and with higher growth rates (4.8, 5.3 and 1.7 %, respectively) than, e.g., in materials/metallurgy (37,705; −1.4 %) or textile and paper machines (51,090; 0.8 %).[113] Similar findings are reported by Brouwer and Kleinknecht [cf. 155, pp. 618ff.] as well as Mansfield [cf. 185, pp. 174ff.], who states that 65 % of innovations in the pharmaceutical industry would not have been introduced and 60 % not even

---

[108] This is true, despite the fact that especially within the last years textiles have seen a lot of 'technization' like stain repellant surfaces, breathability from the inside and waterproofness from the outside for outdoor garments and computerization of machinery equipment involved in the production process.

[109] In the light of extremely high development costs for new drugs, costs (which are one key reason to not file an application in other industries) will only rarely be a decisive reason not to patent in the pharmaceutical industry.

[110] In the unabridged table, only three more industries exhibit slightly higher rates for process innovations: the mining industries, basic metals and fabricated metal products.

[111] For a proper analysis the mere numbers would have to be weighted by sales volumes or workforce numbers. This is not done here, as the reporting of results is largely of informative value.

[112] The acknowledgement of this fact is also the basis of most analyses within this study, where growing cross-industrial use of a technology is taken as a proxy for convergence.

[113] The numbers are for 2005 and the growth rates reflect the average annual growth rate 2001–2005 [cf. 184, p. 41].

developed if patent protection would have been unobtainable. These numbers are decreasing quickly for the chemical industry (30/38 %), fabricated metal products (12/12 %), and textiles (0/0 %).

In respect to other characteristics influencing the propensity to patent, Mansfield finds little support for the hypothesis that firm size would play a crucial role regarding the propensity rate [cf. 185, p. 175]. When looking at Arundel and Kabla's description of contrary findings (indicating a higher propensity in larger firms), it has to be borne in mind that they do not control for the effect of industry affiliation on average sales.[114] Brouwer and Kleinknecht again describe differing results, with larger firms being more likely to seek patent protection. But at the same time, of the firms that have filed patent applications, smaller firms have higher (relative to their firm size) numbers of patents than larger firms [cf. 155, p. 623]. Mansfield's results—in respect to changes over time of firms' likeliness to file patents—support the hypothesis that the more sophisticated a firm's products tend to be, the more likely patent protection will be obtained [cf. 185, pp. 178f.]. Not surprisingly, the propensity to patent is also strongly influenced by the strength of the respective patent law, its restrictions and the resulting fees [cf. 186, pp. 786ff.]. Especially, the last aspect has been widely discussed in connection with the European patents, as not only direct fees tend to be higher than in other countries, but often also translation costs etc., result in higher overall costs to the applicant.[115]

Some have estimated the value of all patents at approximately €1 quadrillion (i.e. €1,000,000 billion) [cf. 187, p. 3], others that of only Japanese patents at US$ 872 billion [cf. 188, p. 24]. The value of individual patents and its estimation is another area of great discussion in the literature. It should be distinguished between the **importance** of a patent, i.e., its technological or societal impact and its **value**, i.e., its economic value to its owner. The first is in most instances nearly impossible to assess quantitatively, if not measured by proxies such as number of citations. Even then, all the caution has to be exerted already discussed in the light

[114] For example, firms from the textile industry exhibit a rather low propensity rate (due to their technological focus). At the same time they are likely to have lower turnovers than firms from the chemical or pharmaceutical industry. Interestingly, this positive relationship between size and propensity is turning negative for very large firms, which may be influenced by petroleum companies. They exhibit, on average, very high turnovers but low propensities to patent (again due to their technological and product focus) [cf. 180, p. 134].

[115] Cf. de Rassenfosse and van Pottelsberghe de la Potterie [cf. 171, pp. 329ff.], using the 3C-index (*costs per claim per capita*) to compare patenting costs to the US and Japan. For a European patent designating 13 countries, the 3C-index is approximately 14 (5) times higher than for US (Japanese) patents, if only the costs prior to granting are considered. If the renewal costs are included for the maximum of 20 years, these the European 3C-index is about 11 (3) times higher than those in the USA (Japan). See also Harhoff et al. [170, pp. 1429ff.], who use 919 country pairs of the years 1995, 1999 and 2003 to analyze major factors affecting the behavior of applicants within the EPC boundaries regarding validation in the member countries. According to their findings, size and wealth of the targeted countries affects this behavior, but less than wealth and size of the applicant's country. Furthermore, validation and renewal fees have a substantial negative impact. Consequently, they recommend the implementation of a *community patent*.

of citation analyses in the scientific literature. Many academics as well as practitioners have tried (and still do) to develop a reliable calculation method to capture a patent's economic value. For both, it is largely accepted that only a small share of all patents is marked by either (or even both) importance or value [cf. 162, p. 221]. Scherer differentiates between four major approaches to assess value/importance: [cf. 189; cited after 162, pp. 221f.][116]

- Patent citation counts
- Number of countries in which a patent is applied for
- Data on patent renewal
- Changes in stock market valuation of firms after granting of patents.

An interesting refinement of citation counts is discussed by Marco, who takes into account several researchers' concern that citations in patents as a value/importance proxy may be biased by a general citation 'inflation'. He suggests the use of a hazard rate based on the inflationary factors, which could then be used to calculate a corrected measure of patent quality. Two examples described are differences between predicted and actual citation lags as well as the ratio of observed citations to predicted citations. In his view, the "advantage of such constructs is that they can be calculated as time-invariant values that are not subject to censoring. Of course, the proposed methods for measuring patent quality are predicated upon a correlation between unobserved heterogeneity in patent citation and patent quality" [190, p. 296]. Sapsalis et al. provide a short summary of the large number of different approaches and conclude with the statement "Indeed, there are almost as many potential methodologies to approximate the value of patent as the number of existing investigations" [191, p. 1632; 147]. Although it would be potentially relevant to include measures of patent value/importance into patent analyses in anticipating convergence, all proposed measures would involve too much complexity in applying this study's concept. Therefore, the topic of value/importance is not further extended here.[117]

3.3.2 Patent Analyses

Patent analysis is one of the most widely applied analyses in the innovation management literature [cf. 162, p. 213] for two main reasons. First, patents can serve as a **proxy for R&D activities** (especially as an early indicator), as technological developments are reflected in patent numbers [cf. 181, p. 237]. Second, patents are probably one of the very few pieces of (historical) information that are not only still 100 % accessible, but also highly structured and reliably fed into

---

[116] For a brief discussion of each of the approaches see also [162, pp. 222f.].

[117] For more information on the assessment of patent value see, e.g., [147, 191–193].

several databases. Many of these databases are offered free of charge,[118] largely depending on the provider and the added benefit they offer. Thus, semi-automatic analyses of large stocks of technological knowledge and identifying patterns in it 'at the touch of a button' become possible, appealing to practitioners and academics alike [cf. 181, pp. 225ff.]. Next to the use of patent analyses in the context of academic research or policy issues, it can serve several different purposes in managerial settings. Ashton and Chen identify five areas of application [cf. 194, pp. 42ff.; 181, p. 226]:

- Technology competition analysis
- New venture evaluation
- Patent portfolio management
- R&D management
- Product area surveillance.

But Ernst also points out the opportunity to use patents' clustering into technological areas as a means to assess technology attractiveness as well as the competitive technological position of a company (or another entity). Accordingly, a company can then direct its expenditures on R&D and collaborative activities in order to strengthen its technological base, especially in respect to its fiercest and possible new competitors [cf. 181, pp. 226]. In summary, as Grupp has pointed out: "No other innovation indicator can be traced back over comparatively long periods of time, may at the same time be disaggregated at a very low level allocable to individual economic units and is also precise and accurate insofar as identification of the timing of the innovation event is concerned" [195, pp. 144f.].

The simplest form of patent analysis is the aggregation of patent-related data in the form of **patent statistics**, as they are published by various scholars, institutions, and especially patent offices. Amongst the most widely disseminated are the

[118] This is of course depending on the type of database an investigator is looking for. Patent authorities around the world generally offer free access to their files through their websites, but with limited functionality regarding search and retrieval operations. The most widely used websites probably being the USPTO's (http://patft.uspto.gov/netahtml/PTO/search-adv.htm for granted patents and for published applications since 2001: http://appft1.uspto.gov/netahtml/PTO/search-adv.html) and the EPO's (http://ep.espacenet.com/advancedSearch?locale=en_EP), with the EPO providing access not only to European documents but also to patent documents from more than 80 countries. In addition to these services, a substantial number of commercial databases exist. Partly they are also offered by the patent authorities (such as EPO's *ESPACE* database range based on the *Mimosa* retrieval software). Others are aggregated or offered by private institutions. STN International (whose services are the basis for most of this study's analyses), for instance, is jointly operated by US based CAS (Chemical Abstracts Service, a division of the American Chemical Society) and German based FIZ Karlsruhe (Forschungsinformationszentrum Karlsruhe, a Leibniz Institute of Information Infrastructure). STN International offers a broad range of databases to be accessed through their interfaces (e.g., *STN Express*® and *STN AnaVist*™), partly aggregated by private institutions, such as the *Derwent World Patents Index*® (DWPI, provided by Thomson Reuters) and partly again by patent authorities, such as EPO's *INPADOC* database. See Chap. 5 for a description of the employed databases and software products.

World Patent Report, published by the World Intellectual Property Organization (WIPO) and the *Compendium of Patent Statistics*, published by the OECD. Together with their online offerings, these reports present a large amount of data on patenting activity at different levels. These include especially the *macro level* (countries, patent offices) and the *meso level* (industries), but also the *micro level* (single firms or research institutes) [cf. 162, pp. 223f.]. In order to lay the foundation for the more detailed description of patent analyses and the discussion of results depicted in this study, a few simple measures will be reported in the following. Next to counts of patents, these reports (like other analyses) also cover data on patent counts in relation to other indicators such as R&D spending.

Notwithstanding differing assessments,[119] the **number of patents** (application as well as granted patents) increases from year to year. The number of patent applications across the world is estimated at 1.76 million (2006), while approximately 727,000 patents were granted in the same year. Interestingly, although growing at an average yearly rate of 5.3 % between 1995 and 2006, the application growth rate is lower than that of other economic indicators (e.g., volume of world trade grew by 7.2 % on average) [cf. 197, pp. 7ff.]. The number of grants has also increased, but is a lot more volatile due to changes in patent laws and amount of patent examiners (thus leading to differing grant rates). As already discussed regarding the core/advanced level of IPCs, the number of applications and patents in force varies greatly between countries. The largest number of applications is regularly filed at the Trilateral Offices, the USPTO, the JPO, and the EPO[120] (about 426,000, 409,000 and 209,000, respectively [cf. 198, pp. 12f.]).

Their relevance is also reflected in the total number of **patents in force**. While the number of patents in force reached 6.1 million in 2006 globally, approximately 1.8 million were in force in the US and 1.1 million in Japan. On the other end of the continuum are much smaller offices, such as Kenya (120 patents in force, 51 patent filings) or Tajikistan (257 and 26, respectively), but also Papua New Guinea or the United Republic of Tanzania (both one patent filing with them as country of origin and no numbers on patents in force) [cf. 197, pp. 64ff.]. As adumbrated earlier, patent documents are available for analysis in databases even long after expiry. Still, especially when estimating the value and importance of a granted patent to the applicant, it is of high importance to calculate its age and validity. Caused by maintenance fees, many patents are not maintained for the full term of 20 years. This leads to an **age distribution**, where more than half of all patents in force in 2006 were younger than 10 years [cf. 197, p. 24].

[119] In 1899 Charles H. Duell, then director of the USPTO is said to have voiced his opinion that „Everything that can be invented has been invented! "[157, p. 1]. Although other authors consider this often used 'quotation' to be a myth. cf., e.g., [196].

[120] In 2006 the *State Intellectual Property Office of the People's Republic of China* (SIPO) has for the first time been the third largest patent filings receiving office, with approximately 211,000 filings. It has to be borne in mind though that numbers for the EPO are smaller due to two different routes applicants may take to be granted an EP. They may file for patent protection either at the EPO directly or at any of the national offices first.

Patents are generally filed by companies. This results in an estimated share of the business sector in overall **patent ownership** of nearly 80 %. While this share reaches nearly 100 % in countries like Finland, Sweden and Japan, it is below 25 % in the Russian Federation, Turkey, and Mexico. The rest of the patents are owned by individuals, but even more importantly by public institutions or the higher education sector.[121] Israel, Australia, and Spain exhibit the highest share of university PCT filings (19.9, 17.5 and 15.2 %, respectively). As surprising as it might seem to not see the USA in this top tier regarding university patenting, this is caused by the even higher propensity to patent of the US industrial sector [cf. 197, p. 37]. The effect of the widely discussed Bayh–Dole Act[122] in the USA can be estimated when looking at a top 20 ranking of PCT applicants from the university sector. And 15 of those are from the USA, with the remaining five from Japan. With 364 applications in 2007 by far the most highly active filer is the *University of California*, followed by the *MIT* and *Columbia University* (175 and 114 applications, respectively). When looking at the business sector, this picture changes decisively. The top 20 applicants are from seven countries and led by *Matsushita Electric Industrial* (Japan; 2,100 applications), *Koninklijke Philips Electronics* (The Netherlands; 2,041 applications), and *Siemens* (Germany; 1,644 applications) [cf. 197, p. 37].

Patent counts and its derivatives have also been used for a plethora of more advanced analyses. To highlight the breadth of these analyses, a few diverse examples are the following approaches to use patents in assessing industries' competencies [cf. 200], using patent trend analysis in forecasting emerging technologies [cf. 48], exploring knowledge transfer [cf. 201], evaluating industry activities [cf. 202], or compiling patent portfolios to enhance market-oriented R&D planning [cf. 203]. A further set of studies is based on factor and regression analysis of different measures (e.g., GDP, number of patents, market capitalization of companies, and patent efficiency) and their effect on patenting behavior cf., e.g., [201, 204–206]. Also, several ways of normalization of counts are being applied. They regularly take into consideration demographic and economic variables, such as population, GDP, or R&D expenditures [162, p. 224].[123]

---

[121] Numbers are based on the applicant's country of residence in PCT filings [cf. 179, p. 24].

[122] The US Bayh-Dole Act adopted in 1980 and sponsored by the two senators Bayh and Dole most importantly includes the permission for universities (as well as small businesses and non-profit organizations) to claim ownership of intellectual property for government-funded inventions. In consequence, US academic institutions are believed to engage more in patenting activities. See [199] for an assessment of effects of the Bayh-Dole Act on academic patent quantity and quality.

[123] When looking at international patent statistics, this normalization yields interesting insights. For instance, resident patent filings per GDP are more than five (four) times higher for the Republic of Korea (Japan) than for the US or Germany. Also, when normalizing by population, Japan and the Republic of Korea report more than 2,500 resident filings per million population, compared to less than 800 for the US. Finally, the US is only ranking 11th in its propensity to patent when it comes to resident patent filings per R&D expenditure, way behind the leading Republic of Korea, the Russian Federation and Japan [cf. 197, pp. 45ff.].

Four further types of analyses can be discerned: specialization indices, impact indicators, patent maps and the more thorough and qualitative content analysis of patents as in text mining analyses, or technology roadmapping.[124] These different streams of analyses are marked by a myriad of indicators trying to quantify the relationship, content, similarity, or quality of patents.[125] First, the most frequently used *specialization index* is the 'Revealed Technological Advantage' (*RTA*) introduced by Soete and Wyatt, measuring the ratio of a country's patenting share in a subdomain divided by the country's share in the main domain [cf. 178, pp. 33ff]. Second, in analogy to the scientific literature, *impact indicators* are based on the analysis of citations to or by a certain patent. Innumerable articles use citations and derived indicators to assess impacts and relations of certain areas of science and technology, or to evaluate patent quality and value.[126] As citations in patents are much more restricted than in scientific publications [cf. 219, p. 413], their use is a lot less debated than in bibliometrics. Third, *patent maps* can be based on content analysis, keywords, classification, or on citation analysis cf., e.g., [220, 221]. They have the focus on a visualization of similarities and connections between each of the items in a map in common.[127] Additionally, they enable an evaluation of the relative position of patents or applicants [cf. 162, p. 227]. Fourth, analyses built on *text mining* have been used, for instance, to evaluate the risk of infringement by calculating and visualizing the similarity of a set of patent documents [cf. 227]. Other more qualitative approaches include keyword-based technology roadmapping [cf. 228] and subject-action-object-format creation to establish inventor profiles for human resource decisions [cf. 229].

---

124 [cf. 162, pp. 225ff] for a brief introduction of the first three types.

125 This is often caused by the application of patent analyses to a very special research question and setting. Because of the richness of patent data, specialized indicators can then be employed. See, e.g., [207, pp. 80ff], using the concentration index CI_i of a company i based on collaborative patent applications CPA_{if} across its technological fields. They further present the relative patent accumulation index RAC_{if} (patent applications of firm in field f compared to all its patent applications) and the relative patent position RP_{if} (firm's patent applications multiplied by overall patent quality divided by the corresponding measures of all competitors) and the rather complex current impact index (*CII*; aggregated impact of all collaborative projects on internal R&D). See also [208, pp. 8ff.], reporting on Euclidian distance matrix (E^d *matrix*), technology centrality index (*TCI*; based on the ties of all patents of a firm), technology cycle time (*TCT*; reflecting age gaps) and density index (*DI*; measuring internal cohesion in networks); or [209, p. 935], using the familiarity of the single component (*FS*) and the familiarity of the cumulated combination (*FC*), both basically mirroring the familiarity of a firm with technological components of a patent.

126 The body of literature on the use of citations in patent analyses is far too large to be adequately represented here. This study will not use citation data for its concept (see Sect. 3.3.3 for a reasoning for this exclusion). For a deeper insight into the possibilities of citation-based patent analyses see, e.g., [133, 190, 210–218].

127 For an overview of possible applications of visualization tools see, e.g., [222–226].

Table 3.6 Advantages and drawbacks of patent analyses

Advantages	Drawbacks
Patents have a close (if not perfect) link to inventions	Not all inventions are patented
Patents cover a broad range of technologies on which there are sometimes few other sources of data (e.g. nanotechnologies)	The propensity to file patent applications differs significantly across technical fields
Each patent document contains detailed information on the inventive process	The value distribution of patents is highly skewed: many patents have no industrial application, whereas a few are of very high value
Patent data are quite readily available (now electronically) from national and regional patent offices	Differences in patent law and practice around the world limit the comparability of patent statistics across countries
The coverage of patent data in terms of space and time is unique (nearly all countries in the world, back to the nineteenth century in most OECD countries)	Changes in patent laws over the years call for caution when analyzing trends over time
	Patent data are complex, as they are generated by complex legal and economic processes

Source Taken from [179, p. 34]

Finally, without going into detail on the dispute for or against the use of patent data,[128] Table 3.6 gives a comprehensive overview of the central advantages and drawbacks of using patent documents as indicators of scientific and technological activities. Although the negative arguments may all hold true to some extent, patent analysis is extensively used in scientific research and commercial applications and is constantly being called upon. This is especially the case when, e.g., dealing with knowledge transfer, strategic management questions, technological foresight, or the performance of different organizations or countries cf., e.g., [47, 139, 140, 181, 201, 206, 228, 231–234]. An additional reason for this popularity of patent analyses lies in the fact that probably more than 80 % of the world's technological knowledge is to be found in patents [cf. 157, p. 1]. Finally, Schmookler has put it rather to the point: "We have a choice of using patent statistics cautiously and learning what we can from them, or not using them and learning nothing about what they alone can teach us" [235, p. 56].

128 For a extensive discussion of the drawbacks and caveats to be considered when analyzing patents, see, e.g., [230]. Like many other criticisms, he focuses on the question whether and to which degree patents may serve as innovation indicators. Furthermore, he reasons that patents will be difficult to compare between different countries due to discrepancies in the respective systems. While both lines of reasoning are undeniably true (to a certain extent), they are not of elevated importance in the setting of this study. Mostly, it is irrelevant whether the patents under scrutiny are mirroring innovative activities or whether they are of high or low value. They always prove an *interest* in a certain domain, even if that interest lies in 'only' securing FTO. The differences between countries can be controlled for by using patent samples from one country only, or by comparing results from different countries.

3.3.3 How will Patents be Used in Anticipating Converging Industries?

A convincing rationale for the employment of patent analyses for technological forecasting has been provided by Fendt: "*Technological changes* threatening the *survival* of business firms are often recognised too *late*, especially by those companies which have been *technological leaders* in a particular area and thus have enjoyed *market success*. Early signals of technological change, with a high potential impact on the *basis of competition* are often very *weak*. If suddenly, *new competitors* capturing strong positions in promising future markets emerge, it is often too late for effective retaliatory responses. However, there exists a relatively *simple* recipe against this scenario. The simple knowledge of the own technological capabilities is to be enhanced by the *continuous monitoring of new technological developments*. It is of substantial importance to detect the slightest technological changes in *key- and pacing technologies as early as possible*. Therefore, the *systematic* evaluation of information obtained from **patent documents** proves to be very valuable" [236, p. 40; cited after and translated from German by 237, p. 365].[129] Despite the different approaches to unveil the hidden patterns in the patent literature, only a few appear to be of higher value to a concept of anticipating converging industries. In analogy to Sect. 3.2.3, this chapter is not aiming at reflecting a comprehensive debate on the pros and cons of the different analysis approaches in general. It will rather focus on easily applicable indicators that can deliver robust results at an early point in time, without the need for either expert knowledge or highly complex investigations and calculations.

Patent analysis has in the past already been employed in the analysis of convergence, especially in the context of technology-driven convergence of ICT [cf. 182, 238, 239]. Fai and von Tunzelmann use the *University of Reading database* for their study of patterns of possible technological convergence at a broadly defined industry sector level. Through their analysis of patenting behavior they find that patterns of technological convergence can be identified when looking at changes within one technological field across all industry sectors. On the other hand, they also report a high degree of path dependency with industrial sectors remaining to engage most heavily in patent activities within their core technological field.[130] Pennings and Puranam argue that based on a validity assumption for classification schemes like IPC or SIC (Standard Industrial Classification) codes, convergence can be found in patent data through growing overlap among

[129] Accentuation according to Ernst.

[130] The University of Reading patent database was compiled by Prof. John Cantwell, based on all patents registered at the USPTO between 1890 and 1995 and grouping 399 technological classes into five broad technological fields while compiling four broad industrial groupings from 875 companies or affiliates [cf. 200, pp. 142ff.].

SIC codes or IPCs and through an increase in patent citations between different classes [cf. 240, p. 12].[131]

More generally, **technology convergence**—in an industry convergence setting—may be marked by either the blending of distinct technologies into one new technology, by the combination of distinct technologies in one product or process setting, or by the employment of 'generic technologies'[132] in distinct industries. The discussed cases of convergence mostly prove to be the result of either a combination of technologies or the increasingly generic character of one technology. When thinking of ICT and its smartphones, for instance, their development is driven (on the technology level) by the omnipresence of digital technologies and smaller electronics parts as well as the combination of the technologies serving each of their subsystems (e.g., phone, camera, mobile Internet, and music player). Both latter cases will lead to patents broader in scope, especially regarding potential claims and applications and filed by entities connected to all the affected industries. This in turn can be observed in shifts in the classification of patents, especially in *co-classifications*. Only a co-classification of patents can serve as a clear sign of mutual interest of or relevance to distinct industries. A further approach is based on the classification of firms. An increasing reference to a technology or set of technologies by firms from different industries will lead to similar conclusions. Accordingly, a co-classification of firms and patents/patent classes will also be a viable instrument. But even if distinct technologies blending into a new one should be observed, this development would result in the co-classification of patents with patent classes referring to each individual technology. In contrast to applicants, who are largely composed of universities, research institutions and firms,[133] inventors will in most cases be much more difficult to relate to their precise affiliation. Hence, inventor analysis should rather be considered after having identified highly active researchers and when trying to gain a deeper understanding of the respective patent area.[134]

Brockhoff mentions a number of caveats when working with IPCs. The classification may be flawed due to: mistakes in filing patents or printing errors, differences in classifying between different patent offices, firms trying to hide their patent in wrong 'classification areas', or lastly new technological developments

---

[131] Ernst has voiced a caveat in connection with IPCs. He commends the opportunity to increase the efficiency of an analysis by using IPCs, but at the same time he reports a misallocation of approximately 15 % of all patents to false technological areas in his sample. The alternative method employed in his case study is the manual allocation of patents, which is inadequate when dealing with large samples [cf. 241, p. 306].

[132] See Sects. 2.2.2 and 2.3.1 for a short introduction into the concept of generic technologies.

[133] This is especially the case when dealing with highly technological patents that imply complex and technologically intensive research and developments. This would apply, for instance, to most patents regarding chemistry or pharmaceuticals.

[134] Such a deep understanding is not the focus of the present study. But it would be highly interesting when looking closer at the analyzed cases of convergence, or for other investigators trying to fully assess the degree of convergence in their area of interest.

not mirrored in the construction of classes [cf. 242, p. 46]. As the present study is largely built on the employment of rather large samples, random errors in filing or printing should not pose any serious problems. As long as numbers are large enough, there is no reason to believe that errors would structurally change the results of a convergence analysis. Differences between classifying offices can be mostly avoided by concentrating on single country samples, or on the other hand a deliberate selection of more than one office in order to minimize potential classification biases. Furthermore, new developments which are not properly representable in patent classes are not an issue when dealing with converging industries, because of their rooting in well-established technological or industrial paths. The only drawback of considerable concern appears to be the case of firms trying to hide their inventions in alien patent classes. If this would be the case to a larger extent, then it could pose a problem. Thus, particular caution should be exerted when analyzing individual firms and tracking their patenting behavior over time. Also, the question remains, whether other allocation procedures would be more feasible. "For the most part, it is too costly and not always technologically feasible to scan the patent description itself in order to decide on its allocation to specific technological fields" [242, p. 46]. Albeit Brockhoff concedes a lack of alternatives, methods to analyze patents based on their content should be seriously considered.

Another approach could focus on content analyses, such as *(co-)word* use or the explicit mentioning of certain applications in the patent sample. This would certainly be easier and more reliable than when dealing with scientific publications, as patents are a sort of *legal-technical hybrid* with a high amount of controlled or at least carefully selected terminology. But just like in bibliometrics, this would involve a lot of effort in the initial preparation of the sample and especially the selection of proper keywords. Furthermore, these tasks would have to be executed for each subfield under investigation. As a consequence, this approach seems less appropriate for this study's targets.

Finally, patents also include *forward and backward citations*[135] to other patents as well as to scientific publications. Due to the legal constraints of the patenting process, these citations are much less determined by individual preferences and unintended (by the system) rationales. Thus, citations would be a viable indicator for convergence. Two patents from different classes or firms could be citing a similar set of earlier documents, indicating a common technology base.[136] Or they could be cited by the same later document. Also, when looking at individual patents, they could each be citing documents related to distinct technologies. Limitations to this methodology lie in the fact that backward citation analysis requires the collection of even more documents than necessary for the compilation of the original sample. The same holds true for forward citations, which pose the

[135] Forward citations are citations received by document X by subsequent documents. Accordingly, backward citations are references in document X to previous documents.

[136] However, it has to be borne in mind that the patents could be referring to completely different parts in the cited documents.

additional challenge of being only useful when the main documents under scrutiny are already of a certain age. These patents can only be cited once they have been published. Most references will be a few years older than the citing patent.

Standard (commercial) databases and database providers generally offer access to all of the aforementioned measures. Hence, each of them and also combinations thereof would be possible. Bearing in mind time, money and effort efficiency, this study will concentrate again on the employment of the more simple indicators and whether they can be successfully used for the proposed concept. This approach follows an argument provided by Ernst, who discusses "three major advantages" of analyzing patent applications without the use of citation data. Besides **accessibility** and **timeliness** of the applications, he stresses their particular **relevance**. Especially, the last is linked to the fact that applications are a direct reflection of a firm's patenting activity (rooted in its R&D activities), without being influenced by external sources (such as through granting or citations) [cf. 237, p. 367].

3.4 Indicators for Industry Convergence

The last two steps and their related measures constitute important proxies for the assessment of market convergence and especially full industry convergence. As this study is aiming at developing an **early warning system** for possible convergence, analyses of many of the measures mentioned in Table 3.8 would be of purely assertive value. Thus, they will be excluded from the remainder of this study.[137] Early warning systems[138] are commonly installed where the effects of incidents or changes are potentially extremely harmful to the *status quo*. Thus, they are used naturally in biological systems, like *meerkat clans* where one or more of the group will stand sentry to watch the entrances to their underground dwellings and warn their clan members of approaching dangers. They also are widely applied in various areas of human lives. Usage stretches from civilian systems, providing advance notice of possible disasters (serving in the prevention of human casualties from, e.g., earthquakes, tsunamis, fires, food emergencies, and epidemic-prone diseases) and military warning systems (such as regarding air raids or ballistic missile attacks), to settings in economic systems (e.g., the International Monetary Fund's *Early Warning Exercise* [cf. 243] or early warning systems in place at the US Federal Reserve or the German Bundesbank [cf. 244, p. 7]).

---

[137] See the discussion in Sect. 6.4 for a discussion of a possible comprehensive monitoring concept for all stages of convergence and the possible role of the aforementioned measures.

[138] This term is used in this study in spite of its negative connotation. Normally, early warning systems are employed to deliver information on a future danger. While convergence may be perceived to be dangerous to some stakeholders, it is a great opportunity to others. As convergence is always connected to the *danger* of missing out on developments and especially changes, 'early warning system' is nevertheless used in this study.

This study's anticipation concept is not intended to automatically flag potential convergence cases and autonomously sound alarm bells. Following the reasoning by Mietzner and Reger, major developments are mostly accompanied by *weak signals*,[139] which will often be information calling for substantial interpretation and analysis effort [cf. 32, p. 274]. Therefore, employing the discussed measures will probably always presuppose a case-by-case interpretation of the gained results by the respective investigator. This is mirrored in this study's approach, which is not built on a mathematical calculation of a comprehensive convergence measure.[140] A corollary of such a method is to try to minimize the investigator's impact in order to objectify it as far as possible. This can be done by combining more than one measure, a procedure demanded anyway in light of weak signals [See, e.g., 12, p. 50]. Moreover, in light of the high uncertainty inherent in weak signals, simpler methods should be preferred over more complex methods, which are prone to mistakes [cf. 246, p. 32].[141]

Albeit structurally deviating from more traditional forecasting approaches, this study will strive to adhere to best practices in the field.[142] As Armstrong points out, it would not be reasonable to try to adhere to all of, for instance, his 139 *principles* in any one case. Instead, their application is conditional on the specific type of forecast and the individual setting. According to his line of reasoning, one should more importantly not ignore "common sense and received wisdom", which forecasters often do. [248, p. 1][143] Bearing in mind that this study does not constitute a pure forecasting approach, Table 3.7 introduces those principles reckoned to be most important for it. Each of the mentioned principles was carefully considered within the study at hand.

In summary of this chapter and before turning to the theoretical foundation and the research setting in the Chap. 4. Table 3.8 presents a brief overview of the most

---

[139] Weak signals as a precursor of strategic surprise and crisis have been thoroughly discussed by Ansoff cf., e.g., [245].

[140] Although this would certainly be welcomed by academics as well as especially practitioners, who are typically changing fields of activity regularly. Consequently, keeping track of recent developments and possessing the knowledge necessary for an 'informed interpretation' will be more difficult for practitioners than for academics. Sect. 6.4.2 will discuss a possible advancement of this approach into a broader early warning system.

[141] See also Hollanders and van Cruysen [247, p. 4], who apply the three principles *simplicity*, *transparency* and *continuity* in their *European Innovation Scoreboard* study for the European Commission.

[142] A comprehensive resource are the 139 principles on forecasting published in "*Principles of Forecasting*", a book edited by Armstrong and involving 40 authors and 123 reviewers. Much of its content and related materials are also provided (and discussed) through the website "forecastingprinciples.com", which is sponsored by the *International Institute of Forecasters* (IIF).

[143] This paper represents a condensed and again carefully reviewed version of the chapters in the book "Principles of Forecasting".

Table 3.7 Principles of forecasting important in this study

Principle	Brief explanation
Consider whether the series of events can be forecasted	Forecasting should only be employed where it can potentially provide more accurate predictions of the future than obtainable without forecasting methods
Decompose time series by causal forces	Decomposition by forces should be done, when there are conflicting forces leading to less accurate overall assessments. E.g., when looking at convergence on the different levels, it should be considered whether there might be a convergence of technologies but divergence in markets, leading to an impression of no convergence in sum
Use diverse sources of data	Selection of sources should be grounded on theory, but at the same time more 'creative' sources may sometimes be necessary. A combination of different sources should reduce the impact of potential biases in each of the sources
Obtain information from similar (analogous) series or cases	Trends obtained from different (but similar) time series may exhibit similar patterns and thus be helpful to better understand and interpret the time series of interest
Use unbiased and systematic procedures to collect data	While this is especially important when using judgmental methods, it should also be considered in situations where, e.g., selection of keywords or thresholds plays an important role
Ensure that the information is valid	As self-evident as this principle might appear, it is still extremely important to always discuss whether the information used can really be translated into the derived conclusions
Obtain all of the important data	Wherever possible, all time periods should be obtained and as little as possible up-front data-processing should be carried out, notwithstanding necessary data cleansing and transformation.
Select simple methods unless empirical evidence calls for a more complex approach	Some complexity may naturally increase the accuracy of the forecast, but is vital to assess the degree of necessary complexity. This is especially the case where knowledge on relationships between data points is scarce. Over-complex methods might negatively affect accuracy through, e.g., unidentified propagating errors
Combine forecasts from approaches that differ	The same reasoning as for the use of different data sources is applicable here

(continued)

Table 3.7 (continued)

Principle	Brief explanation
Provide easy access to the data and full disclosure of methods	This principle should go without saying in a scientific context. Access to data and used methods is imperative for enabling replications and repetitive testing of results
Describe reasons why forecasts might be wrong	This is particularly important, where uncertainty estimates are subjective and where readers will not be able to easily identify likely sources or error
Do not use measures of statistical significance to assess a forecasting method or model	By this principle the selection of invalid methods (and the rejection of valid ones) should be avoided. As different conditions, such as large sample sizes, can lead to significance without relevance, significance tests should only be employed where their benefit has been proven

Sources Based on [249, pp. 89ff.]; [248, pp. 3ff.]; [250]

Table 3.8 Indicators and data sources of interest for anticipating converging industries

Convergence in	Main sources	Indicators	Possible data sources
Science areas	Scientific articles	• *(Co-)Authorships* • *(Co-)citations* • *Journal topics* • *Keywords*	• *CAS* • *SciFinder* • *Scopus* • *STN* • *Web of Knowledge*
Technologies	Patents	• *(Co-)Assignees* • *(Co-)Citations* • *IPCs (co-classification)* • *SIC/NAICS*[a] *co-classification*	• *CAS* • *Patent offices* • *SciFinder* • *STN*
Markets	General firm and market information	• *Customer trends* • *New product launches* • *Product portfolios*	• *Capital IQ* • *Company data (e.g., press releases)* • *Expert interviews*
Industry segments	General firm and business information	• *Mergers and acquisitions* • *Product portfolios* • *Strategic alliances*	• *Capital IQ* • *Company data (e.g., press releases)* • *Factiva* • *LexisNexis* • *Newspaper archives*

Source Own table
Note [a] *NAICS* North American Industry Classification System

suitable indicators and data sources for anticipating convergence. This study will focus on a selection of them (written in italics) for its analyses, combining them into one indicator where appropriate.[144]

References

1. Scott, C.: Never again. Nat. Biotechnol. **28**(2), 131 (2010)
2. Anonymous: PhRMA wants you! Nature Biotechnology **28**(2), 103 (2010)
3. Heisenberg, W.: Über den anschaulichen Inhalt der quantentheoretischen Kinematik und Mechanik. Zeitschrift für Physik **43**(3–4), 172–198 (1927)
4. Day, G.S., Schoemaker, P.J.H.: Avoiding the pitfalls of emerging technologies. In: Day, G.S., Schoemaker, P.J.H., Gunther, R.E. (eds.) Wharton on managing emerging technologies, pp. 24–52. Wiley, New York (2000)
5. Schnaars, S., Thomas, G., Irmak, C.: Predicting the Emergence of Innovations from technological convergence: Lessons from the twentieth century. J. Macromarketing **28**(2), 157–168 (2008)
6. Daim, T.U., Jordan, S.: A foresight based on scientific indicators: a framework drawn from the case of laptop battery alternatives. Foresight **10**(3), 43–54 (2008)
7. Armstrong, J.S.: Research needs in forecasting. Int. J. Forecast. **4**(3), 449–465 (1988)
8. Gross, A.G., Harmon, J.E., Reidy, M.: Communicating Science: The Scientific Article from the 17th Century to the Present. Oxford University Press, New York (2002)
9. Day, R.A., Gastel, B.: How to Write and Publish a Scientific Paper, 6th edn. Cambridge University Press, Cambridge (2006)
10. Cuhls, K.: Form forecasting to foresight processes—new participative foresight activities in Germany. J. Forecast. **22**(2–3), 93–111 (2003)
11. Hand, D.J.: Mining the past to determine the future: Problems and possibilities. Int. J. Forecast. **25**(3), 441–451 (2009)
12. Zweck, A.: Technologiefrüherkennung—ein Instrument zur strategischen Innovationsförderung. In: Industrie- und Handelskammern in Nordrhein-Westfalen, Bergisches Institut für Produktentwicklung und Innovationsmanagement (eds.) Trendforschung—Die Märkte von morgen entdecken, Solingen, Germany, 2006, pp. 47–54, 27 Sept 2006
13. Porter, A.L., Ashton, W.B., Clar, G., Coates, J.F., Cuhls, K., Cunningham, S.W., Ducatel, K., van der Duin, P., Georgehiou, L., Gordon, T., Linstone, H.A., Marchau, V., Massari, G., Miles, I., Mogee, M., Salo, A., Scapolo, F., Smits, R., Thissen, W.: Technology futures analysis: Toward integration of the field and new methods. Technol. Forecast. Soc. Chang. **71**(3), 287–303 (2004)
14. Shorter Oxford English Dictionary: In: Stevenson, A. (ed.) Shorter Oxford English Dictionary, vol. 1. pp. 1–1882. Oxford University Press, Oxford, (2007)
15. Martin, B.R.: Foresight in science and technology. Technol. Anal. Strateg. Manage. **7**(2), 139–168 (1995)
16. Cuhls, K.: Foresight/Vorausschau: Delphi-Studien zur globalen Entwicklung in Wissenschaft und Technik. In: Industrie- und Handelskammern in Nordrhein-Westfalen, Bergisches Institut für Produktentwicklung und Innovationsmanagement (eds.) Trendforschung—Die Märkte von morgen entdecken, Solingen, Germany, 27 Sept 2006, 2006, pp. 57–61

[144] See Chap. 5 for the indicators used and their results, as well as for a description of the employed databases.

17. van der Meulen, B.: The impact of foresight on environmental science and technology policy in the Netherlands. Futures **31**(1), 7–23 (1999)
18. Vecchiato, R., Roveda, C.: Strategic foresight in corporate organizations: Handling the effect and response uncertainty of technology and social drivers of change. Technol. Forecast. Soc. Change. **77**(9), 1527–1539 (2010)
19. Rohrbeck, R., Arnold, H.M., Heuer, J.: Strategic foresight in multinational enterprises—a case study on the Deutsche Telekom Laboratories. Paper presented at the 1st ISPIM-Asia International Conference, New Delhi, vol. 9. 12 Jan 2007
20. Choomon, K., Leeprechanon, N., Laosirihongthong, T.: A review of literature on technology roadmapping: a case study of power line communication (PLC). Int. J. Foresight Innov. Policy **5**(4), 300–313 (2009)
21. Kostoff, R.N., Boylan, R., Simons, G.R.: Disruptive technology roadmaps. Technol. Forecast. Soc. Chang. **71**(1–2), 141–159 (2004)
22. van Lente, H., van Til, J.: A combined roadmapping-cluster approach for emerging technologies. Int. J. Foresight Innov. Policy **3**(2), 121–138 (2007)
23. Möhrle, M.G., Isenmann, R. (eds.): Technologie-Roadmapping: Zukunftsstrategien für Technologieunternehmen, 3rd edn. Springer, Berlin (2008)
24. Gerdsri, N., Kocaoglu, D.F.: Applying the analytical hierarchy process (AHP) to build a strategic framework for technology roadmapping. Math. Comput. Model. **46**(7–8), 1071–1080 (2007)
25. Lee, S., Park, Y.: Customization of technology roadmaps according to roadmapping purposes: Overall process and detailed modules. Technol. Forecast. Soc. Chang. **72**(5), 567–583 (2005)
26. Robinson, D.K.R., Propp, T.: Multi-path mapping for alignment strategies in emerging science and technologies. Technol. Forecast. Soc. Chang. **75**(4), 517–538 (2008)
27. Lizaso, F., Reger, G.: Linking roadmapping and scenarios as an approach for strategic technology planning. Int. J. Technol. Intell. Planning **1**(1), 68–86 (2004)
28. Phaal, R., Farrukh, C.J.P., Probert, D.R.: Technology roadmapping—a planning framework for evolution and revolution. Technol. Forecast. Soc. Chang. **71**(1–2), 5–26 (2004)
29. Phaal, R., Farrukh, C.J.P., Probert, D.R.: Technology management tools: generalization, integration and configuration. Int. J. Innov Technol. Manage. **3**(3), 1–19 (2006)
30. Phaal, R., Muller, G.: An architectural framework for roadmapping: Towards visual strategy. Technol. Forecast. Soc. Chang. **76**(1), 39–49 (2009)
31. Willyard, C.H., McClees, C.W.: Motorola's Technology Roadmap Process. Research Management Sep–Oct, 13–19 (1987)
32. Mietzner, D., Reger, G.: Practices of strategic foresight in biotech companies. Int. J. Innov. Manage. **13**(2), 273–294 (2009)
33. Linstone, H.A.: TFSC: 1969–1999. Technol. Forecast. Soc. Chang. **62**(1–2), 1–8 (1999)
34. Tichy, G.: The over-optimism among experts in assessment and foresight. Technol. Forecast. Soc. Chang. **71**(4), 341–363 (2004)
35. Martin, B.R.: Technology foresight: capturing the benefits from science-related technologies. Res. Eval. **6**(2), 158–168 (1996)
36. McMaster, M.: Foresight: Exploring the structure of the future. Long Range Plan. **29**(2), 149–155 (1996)
37. Johnston, D.F.: Forecasting methods in the social sciences. Technol. Forecast. Soc. Chang. **2**(2), 173–187 (1970)
38. Goldstein, D.G., Gigerenzer, G.: Fast and frugal forecasting. Int. J. Forecast. **25**(4), 760–772 (2009)
39. Makridakis, S., Hibon, M., Moser, C.: Accuracy of forecasting: An empirical investigation. J. Roy. Stat. Soc. A (General) **142**(2), 97–145 (1979)
40. Armstrong, J.S.: Forecasting with econometric models: Folklore versus fact. J. Bus. **51**(4), 549–564 (1978)
41. Fisher, J.C., Pry, R.H.: A simple substitution model of technological change. Technol. Forecast. Soc. Chang. **3**, 75–88 (1971)

42. Ayres, R.U.: The future of technological forecasting. Technol. Forecast. Soc. Chang. **36**(1), 49–60 (1989)
43. Linstone, H.A.: From my perspective: From information age to molecular age. Technol. Forecast. Soc. Chang. **71**(1), 187–196 (2003)
44. Watts, R.J., Porter, A.L.: Innovation forecasting. Technol. Forecast. Soc. Chang. **56**(1), 25–47 (1997)
45. Ernst, H., Teichert, T.: The R and D Marketing interface and single informant bias in NPD research: An illustration of a benchmarking case study. Technovation **18**(12), 721–739 (1998)
46. Moehrle, M.G., Lessing, H.: Profiling technological competencies of companies: A case study based on the theory of inventive problem solving. Creativity Innov. Manage. **13**(4), 231–239 (2004)
47. Hagedoorn, J., Cloodt, M.: Measuring innovative performance: is there an advantage in using multiple indicators? Res. Policy **32**(8), 1365–1379 (2003)
48. Daim, T.U., Rueda, G., Martin, H., Gerdsri, P.: Forecasting emerging technologies: Use of bibliometrics and patent analysis. Technol. Forecast. Soc. Chang. **73**(8), 981–1012 (2006)
49. Thomas, R.J.: Forecasting new product market potential: Combining multiple methods. J. Prod. Innov. Manage. **4**(2), 109–119 (1987)
50. Curran, C.-S., Leker, J.: Employing STN AnaVist to forecast converging industries. Int. J. Innov. Manage. **13**(4), 637–664 (2009). doi:10.1142/S1363919609002455
51. Hagedoorn, J.: Inter-firm R&D partnerships: an overview of major trends and patterns since 1960. Res. Policy **31**(4), 477–492 (2002)
52. Henzinger, M., Lawrence, S.: Extracting knowledge from the World Wide Web. PNAS **101**, 5186–5191 (2004)
53. Järvenpää, H., Mäkinen, S.: Managerial foresight and recognising value creation potential in an uncertain environment: A bibliometric study of distinguishing succesful from unsuccessful technology. In: R&D Management Conference 2007, Risk and Uncertainty in R&D Management, Bremen, Germany, 04–06 July 2007
54. Björk, B.-C.: A model of scientific communication as a global distributed information system. Information Research (2007)
55. Verbeek, A., Debackere, K., Luwel, M., Zimmermann, E.: Measuring progress and evolution in science and technology—I: The multiple uses of bibliometric indicators. Int. J. Manage. Rev. **4**(2), 179–211 (2002)
56. Council of Science Editors: CSE's White Paper on Promoting Integrity in Scientific Journal Publications, 2009 Update. In: Scott-Lichter, D., the Editorial Policy Committee (eds.). Reston (VA), USA, (2009)
57. Council of Biology Editors: Proposed definition of a primary publication. In: vol. November 1968. pp. P1–P2. Philadelphia, (1968)
58. Höpfner, W., Qian, Z., Castaneda, W.R.: Recanalization of thrombosed femoral arteries with a hydraulic catheter. Am. J. Roentgenol. **175**(4), 1190–1191 (2000)
59. Davis, M.: Scientific papers and presentations, 2nd edn. Effective Communication Skills in Science. Academic Press, Elsevier, Burlington (2005)
60. Egger, A.E., Carpi, A.: Scientific Communication: Utilizing the Scientific Literature. Visionlearning (2009)
61. Fjällbrant, N.: Scholarly Communication—Historical Development and New Possibilities. Paper presented at the 18th IATUL Conference—Scholarly Communication in Focus, Trondheim, Norway, 30.06–04.07. 1997
62. Mabe, M., Amin, M.: Growth dynamics of scholarly and scientific journals. Scientometrics **51**(1), 147–162 (2001)
63. Harmon, J.E., Gross, A.G. (eds.): The Scientific Literature—A Guided Tour. The University of Chicago Press, Chicago (2007)
64. Deutsche Forschungsgemeinschaft: Publikationsstrategien im Wandel? Ergebnisse einer Umfrage zum Publikations- und Rezeptionsverhalten unter besonderer Berücksichtigung

von Open Access. In: Over, A., Maiworm, F., Schelewsky, A. (eds.). pp. 1–82. Bonn, Germany, (2005)

65. Kaufer, D.S., Carley, K.M.: Communication at a Distance: The Influence of Print on Sociocultural Organization and Change. Lawrence Erlbaum Associates, Hillsdale (1993)
66. Ioannidis, J.P.A.: Why Most Published Research Findings Are False. PLoS Med **2**(8), 696–701 (2005)
67. Kleiner, K.: Most scientific papers are probably wrong http://www.newscientist.com/article/dn7915 (2005). Accessed 15 Mar 2010
68. Niedergassel, B.: Knowledge Sharing in Collaborative R&D Projects. University of Münster (2009)
69. Deutsche Forschungsgemeinschaft: Publikationsstrategien im Wandel? Tabellenband—Ergebnisse einer Umfrage zum Publikations- und Rezeptionsverhalten unter besonderer Berücksichtigung von Open Access. In: Over, A., Maiworm, F., Schelewsky, A. (eds.). pp. 1–221. Bonn, Germany, (2005)
70. Ziman, J.M.: The proliferation of scientific literature: A natural process. Science **208**(4442), 369–371 (1980)
71. Reed Elsevier: Annual Reports and Financial Statements 2008—Information solutions for professionals. In. London, UK, (2008)
72. European Commission: Study on the economic and technical evolution of the scientific publication markets in Europe—Final Report. In: Science and society. pp. 1–112. Brussels, Belgium, (2006)
73. Elsevier: Products. http://elsevier.com/wps/find/subject_area_browse.cws_home?collapseAll=y&sh1State=0&allParents=y (2010). Accessed 25 Feb 2010
74. Braun, T., Zsindely, S.: Growth of scientific literature and the Barnaby Rich effect. Scientometrics **7**(3–6), 529–530 (1985). doi:10.1007/BF02017165
75. Björk, B.-C., Roos, A., Lauri, M.: Scientific journal publishing: yearly volume and open access availability. Information Research (2009)
76. Rowland, J.F.B.: Why are new journals founded? J. Documentation **37**(1), 36–40 (1981)
77. Tijssen, R.J.W.: Cartography of science: Scientometric mapping with multidimensional scaling methods. Science studies. DSWO Press, Leiden (1992)
78. Harnad, S.: Free at Last: The Future of Peer-Reviewed Journals. D-Lib Magazine (1999)
79. Schrader, U., Hennig-Thurau, T.: VHB-JOURQUAL2: Method, results, and implications of the German academic association for business research's journal. BuR—Bus. Res. **2**(2), 180–204 (2009)
80. ZBW—German National Library of Economics: ECONIS Select Business Administration—Bibliometrics in Business Economics: 1. Ranking of Journals. http://www.zbw-kiel.de/e_catalogues/e_econis_select/e_econis_select_docs_2008/e_mg_bwl_zeitschriften_ranking.htm (2008). Accessed 02 Mar 2010
81. Bräuninger, M., Haucap, J.: What Economists Think of Their Journals and How They Use Them: Reputation and Relevance of Economics Journals. Discussion paper of the Helmut Schmidt University—University of the Federal Armed Forces Hamburg (2002)
82. Hennig-Thurau, T., Walsh, G., Schrader, U.: VHB-JOURQUAL: Ein Ranking von betriebswirtschaftlich-relevanten Zeitschriften auf der Grundlage von Expertenurteilen. Schmalenbachs Zeitschrift für betriebswirtschaftliche Forschung **56**(6), 520–545 (2004)
83. Azar, O.H., Brock, D.M.: A citation-based ranking of strategic management journals. J. Econ. Manage. Strategy **17**(3), 781–802 (2008)
84. Ritzberger, K.: A ranking of journals in economics and related fields. Ger. Econ. Rev. **9**(4), 402–430 (2008)
85. Vieira, P.C.C.: An economics journals' ranking that takes into account the number of pages and coauthors. Appl. Econ. **40**(7), 853–861 (2008)
86. Serenko, A., Bontis, N.: A citation-based ranking of the business ethics scholarly journals. Int. J. Bus. Gov. Ethics **4**(4), 390–399 (2009)
87. Franke, N., Schreier, M.: A meta-ranking of technology and innovation management/entrepreneurship journals. Die Betriebswirtschaft **68**(2), 185–216 (2008)

88. Franke, N., Schreier, M.: Die Qualität der Zeitschriften im Bereich Technologie-/ Innovationsmanagement und Entrepreneurship: ein integrierendes und kalibriertes "Meta-Ranking" (2010)
89. Garfield, E.: Citation indexes for science—A new dimension in documentation through association of ideas. Science **122**(3159), 108–111 (1955)
90. Garfield, E.: Journal impact factor: a brief review. Can. Med. Assoc. J. **161**(8), 979–980 (1999)
91. Thomson Reuters: Analytical Resources. In, vol. AG0901008. pp. 1–2. Philadelphia, (2009)
92. Seglen, P.O.: Why the impact factor of journals should not be used for evaluating research. BMJ **314**(7079), 497 (1997)
93. Schutte, H.K., Svec, J.G.: Reaction of Folia Phoniatrica et Logopaedica on the current trend of impact factor measures. Folia Phoniatrica et Logopaedica **59**(6), 281–285 (2007)
94. Schutte, H.K., Svec, J.G.: 60 Years of Folia—Review and comparison of two general journals in our field. Folia Phoniatrica et Logopaedica **60**(6), 273–278 (2008)
95. Thomson Reuters: Journal Citation Reports® Notices. http://admin-apps.isiknowledge.com/JCR/static_html/notices/notices.html (2009). Accessed 02 Mar 2010
96. McVeigh, M.E., Mann, S.J.: The journal impact factor denominator—Defining citable (Counted) items. JAMA **302**(10), 1107–1109 (2009)
97. Williams, W.W.: Institutional propensities to publish in academic journals of business administration: 1979–1984. Q. Rev. Econ. Bus. **27**(1), 77–94 (1987)
98. Lepak, D.: Editor's comments: What is good reviewing? Acad. Manag. Rev. **34**(3), 375–381 (2009)
99. Schminke, M.: From the editors: Tensions. Acad. Manag. J. **45**(3), 487–490 (2002)
100. Pondy, L.R.: The reviewer as defense attorney. In: Cummings, L.L., Frost, P.J. (eds.) Publishing in the Organizational Sciences, pp. 183–194. SAGE Publications, Thousand Oaks (1995)
101. The Nobel Foundation: The Nobel Prize in Physiology or Medicine 1977. http://nobelprize.org/nobel_prizes/medicine/laureates/1977/ (2010). Accessed 14 Mar 2010
102. Yalow, R.S.: Radioimmunoassay: a probe for the fine structure of biologic systems. Science **200**(4347), 1236–1245 (1978). doi:10.1126/science.208142
103. Martin, B.: The use of multiple indicators in the assessment of basic research. Scientometrics **36**(3), 343–362 (1996)
104. van Raan, A.: In matters of quantitative studies of science the fault of theorists is offering too little and asking too much. Scientometrics **43**(1), 129–139 (1998)
105. Braun, T.: Editorial. Scientometrics **43**(1), 3 (1998)
106. Leydesdorff, L.: Theories of citation? Scientometrics **43**(1), 5–25 (1998)
107. Kostoff, R.: The use and misuse of citation analysis in research evaluation. Scientometrics **43**(1), 27–43 (1998)
108. Cronin, B.: Metatheorizing citation. Scientometrics **43**(1), 45–55 (1998)
109. Egghe, L.: Mathematical theories of citation. Scientometrics **43**(1), 57–62 (1998)
110. Scharnhorst, A.: Citation-networks, science landscapes and evolutionary strategies. Scientometrics **43**(1), 95–106 (1998)
111. Small, H.: Citations and consilience in science. Scientometrics **43**(1), 143–148 (1998)
112. Vinkler, P.: Comparative investigation of frequency and strength of motives toward referencing. The reference threshold model. Scientometrics **43**(1), 107–127 (1998)
113. Arunachalam, S.: Citation analysis: Do we need a theory? Scientometrics **43**(1), 141–142 (1998)
114. Rousseau, R.: Citation analysis as a theory of friction of polluted air? Scientometrics **43**(1), 63–67 (1998)
115. Garfield, E.: Random thoughts on citationology its theory and practice. Scientometrics **43**(1), 69–76 (1998)
116. Fujigaki, Y.: The citation system: Citation networks as repeatedly focusing on difference, continuous re-evaluation, and as persistent knowledge accumulation. Scientometrics **43**(1), 77–85 (1998)

117. Makino, J.: Productivity of research groups-relation between citation analysis and reputation within research communities. Scientometrics **43**(1), 87–93 (1998)
118. Smith, L.C.: Citation analysis. Libr. Trends **30**(1), 83–106 (1981)
119. International Committee of Medical Journal Editors: Uniform Requirements for Manuscripts Submitted to Biomedical Journals: Ethical Considerations in the Conduct and Reporting of Research: Authorship and Contributorship. http://www.icmje.org/ethical_1author.html (2009). Accessed 14 Mar 2010
120. Council of Science Editors: Authorship Task Force. http://www.councilscienceeditors.org/services/authorship.cfm (2010). Accessed 15 Mar 2010
121. Oxford Dictionary of English: In. Oxford University Press, Oxford, (2006)
122. Osareh, F.: Bibliometrics, citation analysis and co-citation analysis: a review of literature I. Libri **46**(3), 149–158 (1996)
123. Pritchard, A.: Statistical bibliography or bibliometrics? J. Documentation **25**(4), 348–349 (1969)
124. Norton, M.J.: Introductory Concepts in Information Science, 4th edn. Information Today, Medford (2008)
125. Broadus, R.N.: Toward a definition of "bibliometrics". Scientometrics **12**(5–6), 373–379 (1987)
126. Diodato, V.P.: Dictionary of Bibliometrics. The Haworth Press, Binghamton (1994)
127. Hirsch, J.E.: An index to quantify an individual's scientific research output. PNAS **102**(46), 16569–16572 (2005)
128. van Raan, A.F.J.: Comparison of the Hirsch-index with standard bibliometric indicators and with peer judgement for 147 chemistry research groups. Scientometrics **67**(3), 491–502 (2006)
129. Moed, H.F., Burger, W.J.M., Frankfort, J.G., van Raan, A.F.J.: The use of bibliometric data for the measurement of university research performance. Res. Policy **14**(3), 131–149 (1985)
130. Moed, H.F.: Citation analysis in research evaluation. Information science and knowledge management vol. 9. Springer, Dordrecht (2005)
131. Moed, H.F.: Bibliometric rankings of world Universities. In., vol. CWTS Report 2006-01, pp. 1–36. Centre for Science and Technology Studies (CWTS), Leiden University, Leiden, (2006)
132. Gilbert, G.N.: Measuring the growth of science—A review of indicators of scientific growth. Scientometrics **1**(1), 9–34 (1978)
133. European Commission: Towards a European Research Area: Science, Technology and Innovation—Key Figures 2007. In, vol. EUR 22572. vol. ISBN 92-79-03450-2, pp. 1–108. Office for Publications of the European Communities, Luxembourg, (2007)
134. Studt, T.: Measuring Innovation… Gauging Your Organizations's Success. R&D Magazine February 2005, 42–44 (2005)
135. Dalpé, R.: Bibliometric analysis of biotechnology. Scientometrics **55**(2), 189–213 (2002)
136. Gingras, Y.: Revisiting the "Quiet Debut" of the double Helix: A bibliometric and methodological note on the "Impact" of scientific publications. J. Hist. Biol. **43**(1), 159–181 (2010)
137. Hicks, D.M., Katz, J.S.: Where is science going? Sci. Technol. Human Values **21**(4), 379–406 (1996)
138. Hornbostel, S., Böhmer, S., Klingsporn, B., Neufeld, J., von Ins, M.: Funding of young scientist and scientific excellence. Scientometrics **79**(1), 171–190 (2009)
139. Gittelman, M., Kogut, B.: Does good science lead to valuable knowledge? Biotechnology firms and the evolutionary logic of citation patterns. Manage. Sci. **49**(4), 366–382 (2003)
140. Zucker, L.G., Darby, M.R., Armstrong, J.S.: Commercializing knowledge: University science, knowledge capture, and firm performance in biotechnology. Manage. Sci. **48**(1), 138–153 (2002)
141. Hornbostel, S., von Ins, M.: Leistungsmessung in der Forschung—Erfordernis oder Vermessenheit? Laborwelt **10**(5), 34–35 (2009)
142. van Raan, A.F.J.: Scientometrics: State-of-the-art. Scientometrics **38**(1), 205–218 (1997)

143. van Eck, N., Waltman, L., Noyons, E., Buter, R.: Automatic term identification for bibliometric mapping. Scientometrics **82**(3), 581–596 (2010)
144. Sigmund, P., Wallin, J.: Evaluation by citation: An imperfect system. Physics Today **62**(12), 10–11 (2009). doi:http://dx.doi.org/10.1063/1.3272998
145. Adams, J.D.: Fundamental stocks of knowledge and productivity growth. J. Polit. Econ. **98**(4), 673–702 (1990)
146. Tansey, M., Stembridge, B.: The challenge of sustaining the research and innovation process. World Patent Inf. **27**(3), 212–226 (2005)
147. Griliches, Z.: Patent statistics as economic indicators: A survey. J. Econ. Lit. **28**, 1661–1707 (1990)
148. EPC: Convention on the Grant of European Patents (European Patent Convention). In: The Contracting States of the European Patent Convention (ed.). European Patent Office, Munich, (2007)
149. Crespi, R.: An analysis of moral issues affecting patenting inventions in the life sciences: A European perspective. Sci. Eng. Ethics **6**(2), 157–180 (2000)
150. Beyleveld, D.: Regulating morality through patent law. Critique of the EC Directive. Law Human Genome Rev. **12**, 141–171 (2000)
151. van Overwalle, G.: Bio-patents, law and ethics. Critical analysis of the EU biotechnology directive. Law Human Genome Rev. **19**, 187–203 (2003)
152. Macer, D.R.J.: The Pharmacogenomics Journal. Patent or perish? An ethical approach to patenting human genes and proteins **2**(6), 361–366 (2002). doi:10.1038/sj.tpj.6500140
153. Council of the European Union: Addendum to the note: Proposal for a Council Regulation on the Community patent—General approach. In., vol. 16113/09 ADD 1. Brussels, Belgium, (2009)
154. Teece, D.J.: Profiting from technological innovation: Implications for integration, collaboration, licensing and public policy. Res. Policy **15**(6), 285–305 (1986)
155. Brouwer, E., Kleinknecht, A.: Innovative output, and a firm's propensity to patent: An exploration of CIS micro data. Res. Policy **28**(6), 615–624 (1999)
156. Schumpeter, J.: Theorie der wirtschaftlichen Entwicklung—Eine Untersuchung über Unternehmergewinn, Kapital, Kredit, Zins und den Konjunkturzyklus, 5th edn. Duncker and Humblot, Berlin (1952)
157. Gassmann, O., Bader, M.A.: Patentmanagement: Innovationen erfolgreich nutzen und schützen. Springer, Berlin (2006)
158. Independence Hall Association: National Constitution Center. http://www.ushistory.org/tour/tour_ncc.htm (2010). Accessed 22 Mar 2010
159. Constitutional Convention: The United States Constitution. In. John Carter, Providence (RI), USA, (1787)
160. Diallo, B.: Historical perspectives on IP protection for software in selected countries worldwide. World Patent Inf. **25**, 19–25 (2003)
161. WIPO: International Patent Classification. (2009)
162. Debackere, K., Verbeek, A., Luwel, M., Zimmermann, E.: Measuring progress and evolution in science and technology—II: The multiple uses of technometric indicators. Int. J. Manage. Rev. **4**(3), 213–231 (2002)
163. WIPO: Adoption of the simplification of the International Patent Classification System http://www.wipo.int/classifications/ipc/en/news/2010/news_0007.html (2010). Accessed 23 Mar 2010
164. Hanel, P.: Intellectual property rights business management practices: A survey of the literature. Technovation **26**(8), 895–931 (2006). doi:10.1016/j.technovation.2005.12.001
165. Criscuolo, P.: The 'home advantage' effect and patent families. A comparison of OECD triadic patents, the USPTO and the EPO. Scientometrics **66**(1), 23–41 (2006)
166. McCalman, P.: Reaping what you sow: an empirical analysis of international patent harmonization. J. Int. Econ. **55**(1), 161–186 (2001). doi:http://dx.doi.org/10.1016/S0022-1996(01)00091-5

167. Watanabe, C., Tsuji, Y.S., Griffy-Brown, C.: Patent statistics: deciphering a 'real' versus a 'pseudo' proxy of innovation. Technovation **21**(12), 783–790 (2001)
168. Varsakelis, N.C.: The impact of patent protection, economy openness and national culture on R&D investment: a cross-country empirical investigation. Res. Policy **30**(7), 1059–1068 (2001)
169. Wu, Y.-C.J., Lee, P.-J.: The use of patent analysis in assessing ITS innovations: US, Europe and Japan. Transp. Res. Part A: Policy Pract. **41**(6), 568–586 (2007)
170. Harhoff, D., Hoisl, K., Reichl, B., van Pottelsberghe de la Potterie, B.: Patent validation at the country level–The role of fees and translation costs. Research Policy **38**(9), 1423–1437 (2009)
171. van Pottelsberghe de la Potterie, B., Francois, D.: The Cost Factor in Patent Systems. J. Ind. Competition Trade **9**, 329–355 (2009)
172. Preschitschek, N., Bresser, D.: Nanotechnology patenting in China and Germany—a comparison of patent landscapes by bibliographic analyses. J. Bus. Chem. **7**(1), 3–13 (2010)
173. EPO: Patents around the world. http://www.epo.org/topics/patent-system/patents-around-the-world.html (2009). Accessed 22 Mar 2010
174. Bentwich, M.: Changing the rules of the game: addressing the conflict between free access to scientific discovery and intellectual property rights. Nat. Biotechnol. **28**(2), 137–140 (2010)
175. Tessenohn, J.A., Yamamoto, S.: Accelerated patent examination procedures spur Japanese university innovation. Nat. Biotechnol. **27**(9), 815–818 (2009)
176. Hogan, J.P., Banks, R.L.: Solid polymers of olefins and production of such polymers. United States Patent
177. Münch, V.: Patente, Marken, Design von A bis Z. Wiley-VCH Verlag, Weinheim (2010)
178. Soete, L., Wyatt, S.: The use of foreign patenting as an internationally comparable science and technology output indicator. Scientometrics **5**(1), 31–54 (1983)
179. OECD: Compendium of Patent Statistics. In: Dernis, H., Guellec, D., Zuniga-Lara, M.-P. (eds.). (2008)
180. Arundel, A., Kabla, I.: What percentage of innovations are patented? empirical estimates for European firms. Res. Policy **27**(2), 127–141 (1998)
181. Ernst, H.: Patenting strategies in the German mechanical engineering industry and their relationship to company performance. Technovation **15**(4), 225–240 (1995)
182. Bröring, S.: The front end of innovation in converging industries: The case of nutraceuticals and functional foods. DUV, Wiesbaden (2005)
183. Hussinger, K.: Is silence golden? Patents versus Secrecy at the firm level. In: Discussion Paper. vol. No. 04–78, pp. 1–24. ZEW—Centre for European Economic Research, Mannheim, Germany, (2004)
184. WIPO: World Patent Report—A Statistical Review. In., vol. WIPO Publication No. 931(E). Geneva, Switzerland, (2008)
185. Mansfield, E.: Patents and innovation: an empirical study. Manage. Sci. **32**(2), 173–181 (1986)
186. de Rassenfosse, G., van Pottelsberghe de la Potterie, B.: A policy insight into the R&D-patent relationship. Res. Policy **38**(5), 779–792 (2009)
187. Wurzer, A.: Unternehmensfinanzierung mit Patenten—Biotech-Patente schaffen Werte und verbessern das Unternehmensrating. CHEManager 1/2008, 3 (2008)
188. WIPO: WIPO Intellectual property handbook: policy, law and use. In., vol. WIPO Publication No. 489 (E). Geneva, Switzerland, (2004)
189. Scherer, F.M.: Patents: What Do We Know, What Must We Learn? Paper presented at the Mimeo, AEA Meeting, Luxembourg, November 1996
190. Marco, A.C.: The dynamics of patent citations. Econ. Lett. **94**(2), 290–296 (2007)
191. Sapsalis, E., van Pottelsberghe de la Potterie, B., Navon, R.: Academic versus industry patenting: An in-depth analysis of what determines patent value. Res. Policy **35**(10), 1631–1645 (2006)

192. Lanjouw, J.O., Schankerman, M.: Patent quality and research productivity: measuring innovation with multiple indicators. Econ. J. **114**(495), 441–465 (2004)
193. Bass, S., Kurgan, L.: Discovery of factors influencing patent value based on machine learning in patents in the field of nanotechnology. Scientometrics **82**(2), 217–241
194. Ashton, W.B., Sen, R.K.: Using patent information in technology business planning. Res. Technol. Manage. **31**(6), 42–46 (1988)
195. Grupp, H.: Foundations of the economics of innovation: theory, measurement, and practice. Edward Elgar, Cheltenham, UK; Northampton (MA), USA (1998)
196. Sass, S.: A patently false patent Myth. Skeptical Inquirer **13**, 310–313 (1989)
197. WIPO: World Patent Report—A Statistical Review. In., pp. 7–69. Geneva, Switzerland, (2008)
198. EPO: Facts and figures 2007. In: Schröder, O. (ed.). European Patent Office, Munich, Germany, (2007)
199. Mowery, D.C., Ziedonis, A.A.: Academic patent quality and quantity before and after the Bayh-Dole act in the United States. Res. Policy **31**(3), 399–418 (2002)
200. Fai, F., von Tunzelmann, N.: Industry-specific competencies and converging technological systems: evidence from patents. Struct. Change Econ. Dyn. **12**(2), 141–170 (2001)
201. Agrawal, A., Henderson, R.: Putting patents in context: Exploring knowledge transfer from MIT. Manage. Sci. **48**(1), 44–60 (2002)
202. White, E., Jakielski, K.: The Hydrogen Revolution—An evaluation of patent trends in the fuel cell industry. In: White Paper. vol. C4250, pp. 1–12. Thomson Scientific, London (2004)
203. Ernst, H., Fabry, B., Soll, J.H.: Enhancing market-oriented R&D planning by integrated market and patent portfolios. J. Bus. Chem. **1**(1), 2–13 (2004)
204. Fagerberg, J., Srholec, M.: National innovation systems, capabilities and economic development. Res. Policy **37**(9), 1417–1435 (2008)
205. Fu, X., Yang, Q.G.: Exploring the cross-country gap in patenting: A Stochastic Frontier Approach. Res. Policy **38**(7), 1203–1213 (2009)
206. Guan, J.C., Liu, S.Z.: Comparing regional innovative capacities of PR China-based on data analysis of the national patents. Int. J. Technol. Manage. **32**(3–4), 225–245 (2005)
207. Teichert, T., Ernst, H.: Assessment of R&D Collaboration by Patent Data. In: Kocaoglu, D., Anderson, T. (eds.) Technology and Innovation Management, PICMET '99, Portland (OR), USA 1999, pp. 78–86
208. Chang, P.-L., Wu, C–.C., Leu, H.-J.: Using patent analyses to monitor the technological trends in an emerging field of technology: a case of carbon nanotube field emission display. Scientometrics **82**(1), 5–19 (2010)
209. lo Storto, C.: A method based on patent analysis for the investigation of technological innovation strategies: European medical prostheses industry. Technovation **26**(8), 932–942 (2006)
210. Choi, C., Kim, S., Park, Y.: A patent-based cross impact analysis for quantitative estimation of technological impact: The case of information and communication technology. Technol. Forecast. Soc. Chang. **74**(8), 1296–1314 (2007)
211. von Wartburg, I., Teichert, T., Rost, K.: Inventive progress measured by multi-stage patent citation analysis. Res. Policy **34**(10), 1591–1607 (2005)
212. Trajtenberg, M.: A penny for your quotes: patent citations and the value of innovations. RAND J. Econ. **21**(1), 172–187 (1990)
213. Henderson, R., Jaffe, A.B., Trajtenberg, M.: Universities as a source of commercial technology: A detailed analysis of university patenting, 1965–1988. Rev. Econ. Stat. **80**(1), 119–127 (1998)
214. Mowery, D.C., Oxley, J.E., Silverman, B.S.: Technological overlap and interfirm cooperation: Implications for the resource-based view of the firm. Res. Policy **27**, 507–523 (1998)
215. Nerkar, A., Paruchuri, S.: Evolution of R&D capabilities: The role of knowledge networks within a firm. Manage. Sci. **51**(5), 771–785 (2005). doi:10.1287/mnsc.1040.0354

216. Meyer, M.: Patent citations in a novel field of technology—What can they tell about interactions between emerging communities of science and technology? Scientometrics **48**(2), 151–178 (2000)
217. Lin, B.-W., Chen, C.-J., Wu, H.-L.: Predicting citations to biotechnology patents based on the information from the patent documents. Int. J. Technol. Manage. **40**(1/2/3), 87–100 (2007)
218. Pilkington, A., Lee, L.L., Chan, C.K., Ramakrishna, S.: Defining key inventors: A comparison of fuel cell and nanotechnology industries. Technol. Forecast. Soc. Chang. **76**(1), 118–127 (2009)
219. Meyer, M.: Does science push technology? Patents citing scientific literature. Res. Policy **29**(3), 409–434 (2000)
220. Lee, P.-C., Su, H.-N., Wu, F.-S.: Quantitative mapping of patented technology—The case of electrical conducting polymer nanocomposite. Technol. Forecast. Soc. Chang. **77**(3), 466–478 (2010)
221. Huang, M.-H., Chiang, L.-Y., Chen, D.-Z.: Constructing a patent citation map using bibliographic coupling: A study of Taiwan's high-tech companies. Scientometrics **58**(3), 489–506 (2003)
222. Backhaus, K., Weiber, R.: Forschungsmethoden der Datenauswertung. In: Backhaus, K. (ed.) Handwörterbuch der Betriebswirtschaftslehre. Schäffer-Poeschel Verlag, Stuttgart (2007)
223. Frawley, W.J., Piatetskyshapiro, G., Matheus, C.J.: Knowledge discovery in databases—An overview. Ai Magazine **13**(3), 57–70 (1992)
224. Keim, D.A.: Information visualization and visual data mining. IEEE Trans. Visual Comput. Graph. **8**(1), 1–8 (2002)
225. Ruan, D., Chen, G., Kerre, E.E., Wets, G. (eds.): Intelligent data mining: techniques and applications. Springer, Berlin, (2005)
226. Williams, G.J., Simoff, S.J.: Data mining: theory, methodology, techniques, and applications. Springer, Berlin, (2006)
227. Bergmann, I., Butzke, D., Walter, L., Fuerste, J.P., Moehrle, M.G., Erdmann, V.A.: Evaluating the risk of patent infringement by means of semantic patent analysis: the case of DNA chips. R&D Manage. **38**(5), 550–562 (2008)
228. Lee, S., Lee, S., Seol, H., Park, Y.: Using patent information for designing new product and technology: keyword based technology roadmapping. R&D Manage. **38**(2), 169–188 (2008)
229. Moehrle, M.G., Walter, L., Geritz, A., Müller, S.: Patent-based inventor profiles as a basis for human resource decisions in research and development. R&D Manage. **35**(5), 513–524 (2005)
230. Desrochers, P.: On the abuse of patents as economic indicators. Q. J. Austrian Econ. **1**(4), 51–74 (1998)
231. Valentin, F., Jensen, R.L.: Reaping the fruits of science: Comparing exploitations of a scientific breakthrough in European innovation systems. Econ. Syst. Res. **14**(4), 363–388 (2002)
232. Daizadeh, I.: Using intellectual property to map the organisational evolution of firms: Tracing a biotechnology company from startup to bureaucracy to a multidivisional firm. J. Commercial Biotechnol. **13**(1), 28–36 (2006)
233. Ernst, H.: Patentinformationen für die strategische Planung von Forschung und Entwicklung. Betriebswirtschaftslehre für Technologie und Innovation. Deutscher Universitäts Verlag, Wiesbaden (1996)
234. Canongia, C., Antunes, A., Pereira, M.D.F.: Technological foresight—the use of biotechnology in the development of new drugs against breast cancer. Technovation **24**(4), 299–309 (2004)
235. Schmookler, J.: Invention and Economic Growth. Harvard University Press, Cambridge (1966)
236. Fendt, H.: Strategische Patentanalyse—Blick in die Zukunft. Wirtschaftswoche **37**(29), 40–45 (1983)

237. Ernst, H.: The use of patent data for technological forecasting: The diffusion of CNC-technology in the machine tool industry. Small Bus. Econ. **9**(4), 361–381 (1997)
238. Gambardella, A., Torrisi, S.: Does technological convergence imply convergence in markets? Evidence from the electronics industry. Res. Policy **27**, 445–463 (1998)
239. Duysters, G., Hagedoorn, J.: Technological convergence in the IT industry: The role of strategic technology alliances and technological competencies. Int. J. Econ. Bus. **5**(3), 355–368 (1998)
240. Pennings, J.M., Puranam, P.: Market convergence and firm strategy: new directions for theory and research. In: ECIS Conference, The Future of Innovation Studies, Eindhoven, vol. 20. 23 Sept 2001
241. Ernst, H.: Patent portfolios for strategic R&D planning. J. Eng. Tech. Manage. **15**(4), 279–308 (1998)
242. Brockhoff, K.K.: Instruments for patent data analyses in business firms. Technovation **12**(1), 41–58 (1992)
243. Intenational Monetary Fund: IMF-FSB Early Warning Exercise. http://www.imf.org/external/np/exr/facts/ewe.htm (2009). Accessed 03 Apr 2010
244. Bussiere, M., Fratzscher, M.: Towards a new early warning system of financial crises. In: Working Paper Series. vol. 145. European Central Bank, Frankfurt, (2002)
245. Ansoff, H.I.: Managing strategic surprise by response to weak signals. Calif. Manage. Rev. **18**(2), 21–33 (1975)
246. Armstrong, J.S.: The forecasting canon: nine generalizations to improve forecast accuracy. Foresight **1**(1), 29–35 (2005)
247. Hollanders, H., van Cruysen, A.: Rethinking the European innovation scoreboard: A new methodology for 2008–2010. In. pp. 1–44. PRO INNO Europe, (2008)
248. Armstrong, J.S.: Standards and Practices for Forecasting. (2001)
249. Armstrong, J.S.: Research on forecasting: a quarter-century review, 1960–1984. Interfaces **16**(1), 89–103 (1986)
250. Anonymous: Revision of principle on statistical significance. http://www.forecastingprinciples.com/index.php?option=com_content&task=view&id=337&Itemid=1 (2010). Accessed 03 Apr 2010

Chapter 4
Conceptual Framework and Research Design

After the practical relevance of an anticipation of areas of convergence has been underlined in the last chapter, the following paragraphs will discuss the theoretical reasoning for it. More precisely, it will focus on the questions: *Should a firm really be interested in an anticipation of converging industries, if its decisions were grounded on a sound theory? And, can a firm's reactions to convergence be explained from the perspective of its resources?* This will lead over to the formulation of research questions for the analysis, results, and discussion sections of this study. In conclusion, the setting of (potentially) converging industries in the area of NFF, in which these analyses have been carried out will be introduced. As it is not the aim or the focus of the study at hand to comprehensively cover NFF, their most important parameters will be discussed on the basis of the illustrative example of phytosterols, part of the second largest sector[1] of the Functional Foods market.[2]

4.1 Conceptual Framework: The Resource-Based View in Converging Industries

Several organizational theories exist, trying to explain differences in firms' actions and performance. In line with the reasoning in prior convergence-related works cf., e.g., [4, pp. 144ff; 2, pp. 93ff.],[3] the present study uses the **resource-based view (RBV)**[4] as the theoretical foundation for its conceptual framework. It is a theory of

[1] It is estimated that digestive/gut health is the largest sector of NFF, followed by the heart health market with a wide selection of products aiming especially at lowering the risk of cholesterol-related cardiovascular diseases [cf. 1].

[2] For a more detailed discussion of NFF as a setting of convergence see especially [2, pp. 131ff.; 3]. For further literature on NFF see Sect. 4.3.

[3] For a detailed discussion of different theories' possible contribution in explaining strategic actions in the light of convergence (and especially market convergence), see [4, pp. 35ff.].

[4] It is still being discussed whether the resource-based view should actually be considered (and called) the RBT. As RBV is the more widely used expression, it will be used in the study at hand. For a detailed discussion regarding RBV/RBT see, e.g., [5, pp. 11ff.]. See also Acedo, Barroso and Galan, who delimit the general framework of the RBT from the main trends: RBV, KBV (knowledge-based view), and the relational view [6, p. 621].

C.-S. Curran, *The Anticipation of Converging Industries*,
DOI: 10.1007/978-1-4471-5170-8_4, © Springer-Verlag London 2013

strategic management, which is based on economic theories as well as organizational theories. While economic theories are outlined to discuss the behavior of firms in external markets, organizational theories analyze firms as a complex construct of individuals and groups. It, thus, focuses on their internal structures and the relationships between each of the subdivisions. [cf. 7, p. 953] The RBV has been constantly developed and refined further since the first works of Selznick, Penrose, and Wernerfelt and has by now become one of the "most influential and cited theories in the history of management theorizing" [8, p. 350].[5] Selznick coined the term of the "specific competence" of an organization [11, p. 42],[6] laying the foundation for Penrose's perception of the firm as a "collection of productive resources" [12, pp. 24ff.]. Later Wernerfelt investigated the appropriateness of analyzing firms from a resource perspective instead of the output perspective [cf. 13]. In summary,[7] the RBV views a firm as an individual *collection of resources*, which will be different between individual firms and may be heterogeneous within industries[8] but rather persistent over time, as these resources will often not be easily exchangeable or tradable between firms [cf. 19, p. 101]. Thus, the RBV constitutes a firm-specific perspective stressing the importance of tangible and intangible assets as a central source of a firm's *sustained competitive advantage* [cf. 19, pp. 101ff.; 20, p. 637]. According to the RBV, if this sustained competitive advantage is to be achieved, these resources must be rare, inimitable, nonsubstitutable and valuable [cf. 8, p. 350].[9] The existence of a direct and positive relationship between strategically important resources and performance has been widely contended since the early 1990s. To resolve this dispute, Crook et al. *meta*-analyze 125 studies on RBV representing 127 samples with more than 29,000 organizations and find a clear and positive effect of the strategic resources–performance relationship. Consequently, they call on scholars,

[5] In this article, the authors provide a detailed overview of the RBV's possible weaknesses and critiques. In summary, they draw the conclusion that the RBV can withstand five of the eight discussed critiques quite well, while the explanations on resource, value, and competitive advantage should be widened in future developments of this theory. In any case, the 'ultimate challenge' to the RBV lies in its conclusions being proven to be right or wrong for larger 'populations of firms' and not so much in the verification of single constructs [cf. 9, p. 530; 10, p. 1142].

[6] He actually sees the building of special values and a distinctive competence as a prime function of leadership [cf. 11, p. 27].

[7] It is not intended in the present study to provide a comprehensive introduction to the RBV. In fact, this section is meant to briefly introduce the RBV as a theory to explain firms' actions in a setting of convergence and as a foundation for the present study's research questions laid out in Sect. 4.2. For more in-depth discussions of the RBV see, e.g., [5, 6, 8, 14–17].

[8] Although firms are expected to exhibit greater similarity regarding their resource profile in intra-industry than in inter-industry settings cf., e.g., [2, p. 94; 18, p. 83].

[9] A resource's value will most likely lie in reducing costs or increasing the value of a product or process to customers; competitors will experience difficulties in trying to substitute it with another resource or to imitate it; and it will be too rare to be equally sourced by competitors [cf. 10, p. 1142].

managers, and shareholders to strongly consider the importance of strategic resources for their respective organization [cf. 10, pp. 1141ff.].

In contrast to the historical perception of *resources* being a wildcard for any input into organizational processes, the RBV focuses on their strategic importance and thus on so-called 'strategic resources'.[10] They are generally divided into *tangible* and *intangible assets*.[11] While tangible assets are physical capital resources such as production facilities, access to raw materials, or equipment, intangible assets comprised human and organizational capital resources, such as know-how, experience, controlling and coordination systems, or innovation culture (Fig. 4.1) [cf. 19, p. 101; 18, pp. 83f.]. [12] In contrast to the *resource-dependence theory*, resources within the RBV are generally viewed to not be (easily) sourced from outside the organization [cf. 17, pp. 466f.].[13] Thus, they especially encompass intangible resources like unique knowledge, patents, and reputation [cf. 7, p. 954; 10, p. 1142; 17, pp. 463ff.]. [14] Amit and Schoemaker discern between resources and competencies. In their view, resources are factors owned or controlled by the firm and converted into products or services. These include intangible assets, such as know-how, patents, licenses and human capital, as well as financial or physical assets such as property, plants, and equipment. Contrastingly, **capabilities** are based on a firm's ability to deploy these resources to the desired end [cf. 23, p. 35]. These capabilities have been divided into two broad classes, *component competence* and *architectural competence* by Henderson and Cockburn [cf. 24, pp. 65ff.].[15] While *component capabilities* refer to firms' abilities which are "fundamental for day-to-day problem solving" [24, p. 65], *architectural* or *integrative capabilities* mirror firms' use of "resources and component capabilities to support organizational renewal" [20, p. 640].

Research intensive industries, such as the pharmaceutical industry, are deemed particularly suitable objects of investigation, as their success is highly dependent on their stock of knowledge and technological capabilities. Due to the pharmaceutical industry's strong focus on intellectual property protection through

---

[10] For a review of different definitions of resources and authors' use of them see, e.g., [17, pp. 463ff.].

[11] While this study mainly uses the term 'resources', the term 'assets' will be used interchangeably, where appropriate. This has also been done by, e.g., [2, 18, 21, 22].

[12] Naturally, not all of these assets are of strategic importance.

[13] zu Knyphausen-Aufseß also delineates both theories from each other, stressing the general outside perspective of the organization.

[14] While knowledge may often refer to hardly explicable knowledge or trade secrets, DeCarolis and Deeds view patents as "representative of stocks of organizational knowledge" [cf. 7, p. 958].

[15] In contrast to this study and other cited works, they actually view 'component competence' as including resources and not as a capability to deploy resources. Thus, in the remainder of the text 'competence' will be replaced by 'capabilities' to account for this difference.

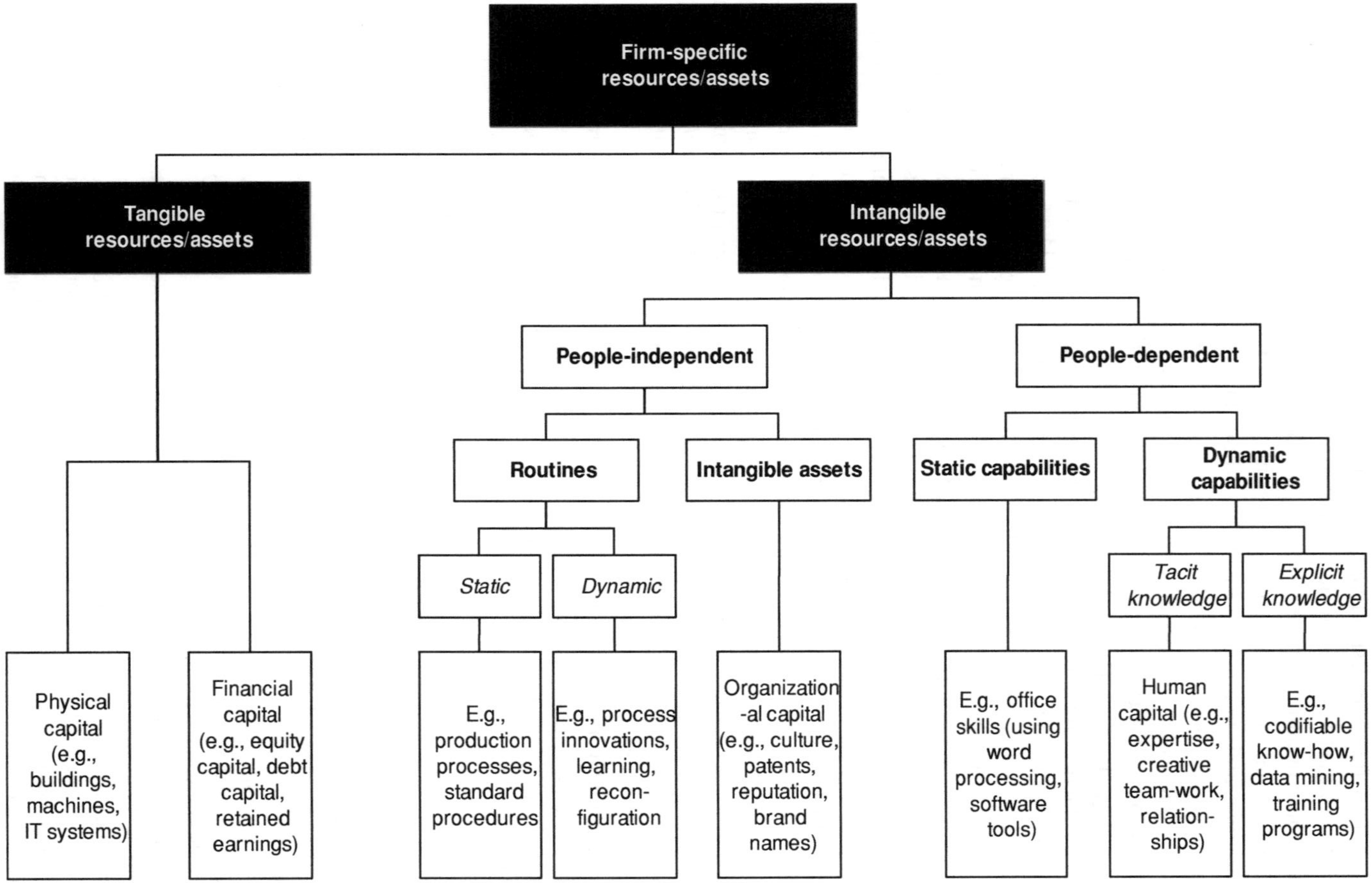

Fig. 4.1 Resource categorization according to Nothnagel. (*Source* Adapted from [25, p. 28] with kind permission from Springer Science+Business Media B.V.)

patents and also the well-defined stages of the development process,[16] the RBV perspective can be readily operationalized. Consequently, different pharmaceutical companies can be validly compared [cf. 20, pp. 637f]. Following the differing propensity to patent discussed in Sect. 3.3.1, the industries in the central focus of this study, namely the food and the pharmaceutical industries, can be expected to be characterized by different strategic resources. For instance, in the pharmaceutical industry patents will be a core resource, whereas brands[17] and reputation will traditionally be much more important in the highly marketing-dependent[18] food industry [cf., e.g., 2, pp. 96ff]. Notwithstanding these differences, the RBV is equally valid for both industries, as the strategic resource–performance relationship does not significantly differ between industry sectors [cf. 10, p. 1149].[19]

In Yeoh and Roth's view, the **pharmaceutical industry** possesses two main component capabilities: internal R&D efforts and therapeutic market focus [cf. 20, pp. 640f].[20] They also identify two main integrative capabilities in a pharmaceutical company's drug approval success and its emphasis on radical innovation [cf. 20, p. 641].[21] Against this background, Peteraf provides a strong argument for the importance of an early identification of convergence: "Such [nontradeable]

[16] Generally, due to tight regulations in respect to pharmaceutical drugs, the development process is highly structured into different stages of discovery and pre-clinical trials as well as four "phases" during clinical trials. As most of the actions within drug development need to be registered with the authorities, this process can be more easily monitored than in most other industries [cf., e.g., 26, pp. 205f].

[17] [cf. 19, p. 115], for a discussion of reputation as a source for competitive advantage. According to his line of reasoning, reputation (which is a large part of a brand name's value) will only serve as a resource, if not all firms have good reputations and they are thus rare and inimitable.

[18] See, e.g., [2, pp. 137f], for an overview of competency differences between the food and the pharmaceutical industries. Of course, the pharmaceutical industry also puts a lot of effort into marketing its products, especially when selling (or competing with) so-called "generic drugs" (generics). Generics are pharmaceutical drugs that are largely 'imitations' of "original drugs", after expiry of their patent protection. As generics manufacturers are facing substantially lower development costs, these generics sell at considerably lower prices. This leads to an increased competition between all the producers of drugs with the respective indication and mostly to increased marketing efforts.

[19] See also [7, p. 954].

[20] According to the reasoning of Yeoh and Roth, internal R&D efforts are more efficient and in the long run also more productive, as economies of experience can be acquired. This argument appears reasonable for well established therapeutic areas, but less convincing against a background of drying product pipelines and a tendency of large pharmaceutical companies to gain access to new knowledge and new drug candidates by acquisition of especially smaller biotech companies. In respect to therapeutic market focus, they highlight the importance of such a focus on single therapeutic markets, as understanding of complex diseases and treatments as well as the respective markets makes shifting from one to the other (or adding a new one) a rather costly adventure.

[21] Drug approval success is seen as mirroring a company's accumulated R&D competence, which lead to an approved drug in more cases than in a less successful company. Generally, the probability of any substance reaching approval as a pharmaceutical drug is about 0.01 %.

assets tend to defy imitation because they have a strong tacit dimension and are socially complex […]. Their development is 'path dependent' in the sense that it is contingent upon preceding levels of learning, investment, asset stocks and development activity" [14, p. 183]. Therefore, a company aiming at competing with incumbents in a different industry in an area of future convergence will need to start developing these assets at the earliest possible moment. This is all the more important bearing in mind that it needs to be not too path dependent in order to not block a transfer of its experience to new (therapeutic) areas [cf. 20, p. 642].[22] In this context, it is highly relevant, whether firms' efforts to adjust their stock of knowledge responding to substantial changes in their environment (such as convergence incidents) are depending on their existing stock of knowledge.[23] Helfat finds that indeed the accumulation of new knowledge depends on several resources of the firm, including their stock of knowledge in complementary domains [cf. 27, p. 339]. In addition, some scholars identify general differences between, for example, pharmaceutical companies. As such, chemical-based pharmaceutical companies are viewed to be relying more on 'mature' theoretical and practical knowledge of process technology. Conversely, biotechnology-based pharmaceutical companies are dealing with even more complex knowledge, sometimes even described as "more of an 'art' than a science" [28, p. 85].[24]

Grounded on the theoretical framework of the RBV, this theory can be applied in a setting of anticipating converging industries. A firm seen as a bundle of resources may be performing well[25] in its traditional path. In light of convergence, it will be facing decisively changing parameters. Based on its respective bundle of resources it may or may not be prepared sufficiently for the altered convergence situation. Barney emphasizes the effects "unanticipated changes in the economic structure of an industry" may have [19, p. 103]. These changes may lead to a nullification of some of the hitherto strategically important resources of a firm—or even turn them into weaknesses.[26] At the same time, this importance for a sustained competitive advantage may stay untouched or be increased [cf. 19, p. 103]. As it is not the purpose of this study to scrutinize the performance and resource characterization of individual firms, the derived research questions will be

(Footnote 21 continued)
(Relative) emphasis on radical innovation refers to more effort put into developing new compound entities rather than modifying or combining existing products.

[22] Who conclude their study with the assessment that the "pharmaceutical industry requires firm strategies that capitalize on resources and capabilities" [20, p. 649].

[23] A firm's knowledge stock is deemed to be critical for a sustained competitive advantage, especially in dynamic environments [cf. 7, p. 965].

[24] See also [7, p. 964].

[25] Acknowledging the fact, that terms like 'good performance' or 'success' are rather fuzzy and have been defined and especially operationalized very differently in literature, they are only used for illustrative purposes. No specific definition and operationalization is therefore provided here.

[26] When core competencies become weaknesses they are sometimes called core rigidities cf., e.g., [29, p. 28].

focusing more broadly on the consequences for the anticipation concept. The RBV can suit particularly well as a theoretical foundation for this study,[27] as it accounts for the uncertainty regarding the future value of a resource (and the asymmetrically distributed information about it). Thus, a firm will have an *ex ante* source of a sustained competitive advantage, if its managers assess the future value of a resource better than those of competing firms [cf. 8, pp. 350f.].[28]

4.2 Research Questions: The Anticipation of Converging Industries

However enticing this potential competitive advantage may be, how are managers supposed to assess the future value of their resources grounded on the RBV? Ansoff introduced an approach for strategic management based on *weak signals*. As a reaction to experiences during the oil crisis of 1973/1974,[29] he develops an approach for 'before the fact problem treatment' harnessing firms against *strategic surprises*. In contrast to after the fact reactions of, e.g., a fire brigade sentenced to await a potential emergency call first, firms should invest in procedures to spot weak signals as early as possible. In his seminal contributions, he requires information to satisfy two prerequisites. On the one hand, it is supposed to be available at an early point in time to allow for well-timed preparations of counter measures. On the other hand, they are supposed to be detailed enough to allow for a proper assessment of impact on the firm and the impact of potential responses [cf. 32, p. 22; 33, p. 136]. Strikingly, he explicitly highlights "alien technology", "previously unknown competitor[s]" and "new political coalition[s]" [32, p. 23] as potential sources of strategic surprises where traditional forecasting is likely to fail. All of these are of great relevance in settings of convergence. In analogy to his idea of a *graduated response*, this study follows a step-wise approach to the

[27] In contrast to the RBV with its proven strong empirical support, managers should not "invest heavily in guidance grounded in a theoretical perspective that has only modest support" [10, p. 1151].

[28] This better assessment does, of course, not have to be grounded on an in-depth rational analysis. While it may as well be the result of a 'lucky guess', firms should strive to employ methods ridding them of coincidental factors.

[29] The 1973/1974 oil crisis was marked by exploding costs for oil and oil-derived fuels as a reaction to rationing by the affected governments. This was necessary as crude oil supplies were cut back by an embargo of the Arabian OPEC (Organization of the Petroleum Exporting Countries) members to the USA, Japan, and several European countries. Caused by opposition to the (anticipated) US and European support of Israel in the Yom Kippur War, it is interpreted as a first massive manifestation of 'the oil weapon'. Most scholars attribute this embargo not only to a political reasoning, but also the target to increase earnings created by oil exports and to shift bargaining power from buyers to sellers of oil. Besides the broad political effects, the oil crisis also led to a major global economic crisis in the following. For an in-depth analysis of the triggers, drivers, and effects of the oil crisis see, e.g., [30, 31].

anticipation of convergence based on potentially rather weak signals, which should be conducted continuously instead of adhering to fixed intervals [cf. 32, p. 31; 33, p. 136]. Accordingly, continuous monitoring is especially important where the expected impact of strategic surprises is thought to be major but urgency tends to be rather low (as is the case when looking at incidents in the future) [cf. 33, p. 140]. Building on the necessity to forearm against future 'surprises' and in respect to Ansoff's information criteria, the first research question is derived:

Q.1: Can convergence be anticipated on the basis of weak signals in publicly available data?

As discussed in Sect. 2.2, convergence is triggered and driven by different parameters and is happening on different levels. First, the drivers can be clustered into technological, managerial, industrial, political, and societal drivers, each of which can be varyingly connected to firms' resources. Technological challenges can be mastered if a firm is able to deploy the necessary technological capabilities and resources. This is based partly on explicit technological know-how (which, e.g., constitutes patents) and partly on less codifiable (technological) knowledge of the firm. Similarly, managerial drivers largely lie within the firms. Firms with the necessary resources are hence able to strengthen such a convergence trend. The same holds true for industrial and political drivers, which are to a certain extent results of developments outside the firm—but at the same time every firm is part of its industry and therefore (depending on its resources) can influence these industry sector developments.[30] Especially in cases like NFF, where newly introduced regulations make product approval more intricate, a firm's resources in the form of approval procedures know-how, experienced and dedicated units as well as well-established relationships with the respective authorities will sometimes decide on the success or failure of this product. The degree of fruitful reaction to societal drivers is reliant on detailed knowledge of the firm's customers and overall societal trends. Also, the higher valued its brands are, the more likely will customers follow through with product changes or new releases.[31]

Second, convergence may happen in sciences, technologies, markets, or whole industries. Again, whether a firm will be able to use convergence on each of these levels as a source for an increased competitive advantage is largely based on its

[30] This is particularly the case, where companies will collectively engage in industrial organizations to agree on industry standards, but also to influence public opinion as well as individual politicians and governments.

[31] See, e.g., customers' reactions to Apple's various iPad and iPhone models. Many consumers have pre-ordered them without ever having seen it (other than on pictures and videos). Of course, this was partly due to fascination with the product and its functionalities. But unarguably, peoples' trust in Apple and the functionalities of its products also played a considerable role. Similar reactions to brands can be observed prior to the introduction of new models of some exclusive luxury car makers.

resources related to them. Regarding science convergence, internal scientific capabilities as well as established relationships with scientists at universities and other research institutions are decisive for an effective seizing of opportunities. The resources of a firm are accordingly often crucial in respect to the other three levels. Consequently, a firm needs to know which kinds of resources it will need and whether they can be determined through different sources of a convergence assessment. This leads to the second research question:

Q.2: Can the assessment of different levels of convergence be based on analyzing samples derived from different data sources?[32]

The last theoretical research question concerns the search strategy. Bearing in mind Ansoff's call for a continuous monitoring of weak signals for early detection, this will often be difficult to achieve on the basis of (traditional) search terms. The earlier the moment at which detection shall be conducted, the more unlikely it will be that an investigator will know a proper search terminology. Accordingly, broader search areas are favorable, like patent classification or journal topics. Even more so, when there is no indication for any specific area of convergence, but where firms rather try to get a 360° picture of their fields of science, technology, market, and industry. Consequently, the third research question asks:

Q.3: Is a keyword-based anticipation of convergence possible and can this also be achieved employing broader classes of articles or patents?

Next to these three central and more theoretical research questions, the next paragraph will derive more research questions, focusing on the specific research setting of the present study. This can only be done after an introduction to this setting. Therefore, the next part will provide an overview of the situation in the area of *Nutraceuticals and Functional Foods* as a development at the outskirts of the research-intensive pharmaceutical industry and the (traditionally) more brand-driven food industry. As some of the respective functional ingredients are also used in 'functional cosmetics', the remainder of this chapter will also briefly cover *Cosmeceuticals.*[33]

[32] This refers mainly to scientific publications and patent documents as proxies for different steps of the convergence process. It is decided by two factors. First, only if science convergence can at all be tracked in scientific publications, they can also be used to assess the degree of convergence on different levels. Second, if scientific publications are to serve as a precursor of developments in technologies and eventually industries, science convergence must be traceable in them earlier than in patent documents.

[33] It will be part of the results and discussion sections to answer the question whether one converged industry of cosmetics, food, and pharmaceuticals is likely to be formed at the intersection of these three distinct industries.

4.3 Research Setting: The Area of Nutraceuticals and Functional Foods

4.3.1 Defining and Delineating Functional Foods and Nutraceuticals

"Food is no longer viewed merely as a means to satisfy hunger, prevent diet-deficiency diseases or to provide the essential building blocks of nutrition [...] for the maintenance and/or repair of body tissue; it has become the primary vehicle to transport us along the road to optimal health and wellness" [34, p. 499S]. This apt quotation already contains a lot of the facets of functional foods. They are meant to provide more than 'just' nutritious value. At the same time, they are an expression of changing values and expectations in respect to foods and eating, at least for people who can afford it. Besides their intensively studied physiological effects, it also underlines the often exaggerated expectations of consumers looking for effortless ways to achieve 'optimal health and wellness'. Against a background of increasing concern regarding the health conditions of people in 'developed countries',[34] many of the major diseases could be prevented or minimized by comparably easy measures including a proper diet cf., e.g., [36, p. 31; 37, p. 1]. While many of these diseases could be fought based on a healthy diet of conventional foods, there is a rising demand for NFF.

At least as many definitions exist for foods with an additional physiological effect (other than its primary nutritional effects) as there are alternative expressions for it cf., e.g., [38, p. 358]. These include 'Functional Foods' and 'Nutraceuticals', but also, e.g., 'designer foods', 'health foods', 'pharmafoods', 'medifoods', 'vitafoods', 'dietary supplements', 'fortified foods', and even 'beauty foods' [cf. 39, 40, p. 12; 41, p. 10]. On the one hand, most of them will have an intended different field of application. On the other hand, they are often used synonymously. The most widely accepted terms are the ones used in this study—*Nutraceuticals* (a term derived by combination of *nutrition* and *pharmaceuticals*) and *functional foods* (FF).

[34] What exactly constitutes a 'developed country' is a considerably contentious issue in a longstanding debate. Within this study, it is used to describe countries that are generally regarded as having a highly industrialized economy and a very high standard of living. Ample definitions and indicators are employed by many organizations. One largely accepted is the human development index (HDI) used by the United Nations Development Programme and comprised proxies for measuring long and healthy lives, access to knowledge, and a decent standard of living. In the 2009 version, 38 countries are listed as having a very high human development status (HDI ranks 1–38), including, e.g., most of Europe; Australia and New Zealand; Hong Kong, Japan, Republic of Korea and Singapore; Israel, Kuwait and the United Arab Emirates; Canada and the USA [cf. 35]. In the context of food and eating habits, other countries would have to be included in so far as parts of their population were leading a lifestyle typical for developed countries. See the remainder of this paragraph for a further short explanation.

Chadwick et al. have slightly altered the already rather comprehensive definition of **Functional Foods** introduced by *The European Commission Concerted Action on Functional Food Science in Europe* (FUFOSE). According to this definition:

A food can be regarded as 'functional' if it is satisfactorily demonstrated to affect beneficially one or more target functions in the body, beyond adequate nutritional effects, in a way that is relevant to either an improved state of health and wellbeing and/or reduction of risk of disease. FF must remain foods, mostly composed of bulk ingredients and they must demonstrate their effects in amounts that can be expected for normal consumption patterns: they are not pills or capsules, but part of a normal diet.

A functional food can be a natural food, a food to which a component has been added, or a food from which a component has been removed by technological or biotechnological means. It can also be a food where the nature of one or more components has been modified, or a food in which the bioavailability of one or more components has been modified, or any combination of these possibilities. The process of how the functional component was added, removed or modified is not relevant for the definition of FF. A functional food might be functional for all members of a population or for particular groups of the population, which might be defined, for example, by age or by genetic constitution" [36, p. 32].[35]

One deviation from this exhaustive definition will be made in the present study: natural foods with no added, removed or modified components are not deemed to be FF, even if they exhibit influence on target functions in the human body (e.g., coffee or chocolate).[36] In line with this delineation of FF from Nutraceuticals, the Bureau of Nutritional Sciences, of the Food Directorate of Health Canada proposes the following shorter definitions of NFF:

> A *nutraceutical* is a product isolated or purified from foods that is generally sold in medicinal forms not usually associated with food. A nutraceutical is demonstrated to have a physiological benefit or provide protection against chronic disease.
>
> A *functional food* is similar in appearance to, or may be, a conventional food, is consumed as part of a usual diet and is demonstrated to have physiological benefits and/or reduce the risk of chronic disease beyond basic nutritional functions [43, p. 3].[37]

Therefore, Nutraceuticals are closer to pharmaceutical drugs than FF, as they do not provide a nutritional function anymore. As in the context of analyses in the

---

[35] This definition is a modification of the FUFOSE definition to be found in an article attributed to Diplock et al. [42, p. S6].

[36] Notwithstanding good reasons for including such food products into a definition of NFF, their existence is not a result of innovative activities caused by drivers of convergence. Even if their marketing would be altered (which is often being done in respect to health beneficial natural ingredients) to highlight the physiological effects, these activities could not be observed on the basis of an anticipation approach. Such shifts in marketing foci would hardly be reported in any of the suggested data sources, with the likely exception of companies' websites or similar marketing channels.

[37] Accentuation according to original document.

area of NFF, this distinction between Nutraceuticals and FF is of minor importance, NFF will be used in the remainder of this study as a synonym for either one of these terms.[38]

4.3.2 Scientific and Technological Background of NFF

The key to the commercial success of individual NFF products is said to lie in the existence of a sound research documenting effects beneficial to consumers' health [cf. 34, p. 504S; 44]. Hippocrates already demanded about 2,500 years ago: "Let food be your medicine and medicine be your food" Cited after [36, p. 39].[39] This quotation is a substantiation of the fact that humans have always used foods and parts thereof for nonnutritional purposes. For a long time, this has been done on the basis of trial and error and expertise passed on from generation to generation. Today, broad acceptance of *health claims*[40] should be (and has been) based on proof of efficacy[41] in randomized, placebo-controlled human clinical intervention trials [cf. 34, p. 504S] and on later research on effectiveness under real-life conditions.[42] Then NFF can be used preventively to reduce the risks of very many diseases and sometimes even alleviate their negative effects.[43]

The main **areas of application** are depicted in Table 4.1 and can be clustered into five groups according to their body area: bone, brain, gut, heart, and the whole body. Earlier works have highlighted the particular prevalence of NFF targeting body functions in bone, gut and heart [cf. 2, p. 145].[44] At the same time, while

---

[38] This is done mainly because the lines between conventional foods and Functional Foods as well as between Nutraceuticals and pharmaceutical drugs are similarly fine. Accordingly, an exclusion of Nutraceuticals from the further analyses could be justifiably challenged as being basically arbitrary. Furthermore, there is no reason to believe that fading boundaries between the food and the pharmaceutical industries would per se favor the appearance of either of the two product groups.

[39] See also [45, p. 62].

[40] See Sect. 4.3.3.3 for the concept and use of 'health claims' in NFF.

[41] Efficacy is a term mainly used in the pharmaceutical sciences. Efficacy is tested in respect to the question whether a treatment does more good than harm when delivered under optimum conditions. In contrast to that, effectiveness is tested under real-world conditions. Consequently, efficacy is necessary but not sufficient to achieve effectiveness [cf. 46, p. 451; see also 47, pp. 1261ff.].

[42] The lack of effectiveness based on typical diets is one major criticism in respect to NFF, even where authors accept the validity of a claimed proof of efficacy.

[43] The largely preventive nature of NFF is a widely accepted difference to pharmaceutical drugs. While they may as well have the target to provide future benefits, many drugs are used for an immediate effect. While this immediate effect may be achieved by consumption of NFF as well (see, e.g., the description of phytosterols in Sect. 4.3.3), it is almost always targeted at the prevention of future adverse health effects [cf. 2, p. 144; 48, p. 408].

[44] According to Leatherhead Food International, consumer sales in the USA are comprised about 40/35/25 % heart/gut/bone-related NFF.

Table 4.1 Overview of NFF by area of application and targeted disease or body function

Body area	Example of diseases/influenced body functions	Literature
Body (whole)	Antioxidant activity	[50, p. 342] [37, p. 7]
	Cancer (especially colon cancer)	[50, p. 342] [51, p. 18] [36, pp. 40ff.] [37, p. 7] [52, p. 113]
	Diabetes	[51, p. 18]
	Inflammation/Infection (especially arthritis, urinary tract infections, vaginitis)	[50, pp. 342ff.] [37, p. 7] [52, p. 113]
	Obesity	[51, p. 18] [36, pp. 40 ff.]
Bone	Osteoporosis	[50, p. 342] [51, p. 18] [36, pp. 40ff.] [39] [37, p. 7] [52, p. 113]
Brain	Psychological functions/behavior	[51, p. 18]
Gut	Gastro-intestinal functions/diarrhea	[50, p. 345] [51, p. 18] [52, p. 113]
Heart	Cardiovascular diseases/atherosclerosis/ hypertension/blood lipid profile	[50, p. 342] [36, pp. 40ff.] [39] [37, p. 7]

Source Own figure. Various articles exist for each of the possible applications, the cited references are more general articles/books on NFF providing overviews of certain application areas and serving well as starting points for further reading

classifications are based on different rationales, the targeted diseases/physiological functions are largely the same.[45] In all cases, ingredients (and thus their finished products as well) are claimed to have long-term positive effects on the health or performance level of certain body parts. While it is beyond the scope of this study

[45] A particularly detailed list of functions is provided in Ref. [49, p. S410]. And 24 different functions are listed to be considered under Chinese food regulation: immune regulation, postponement of senility, memory improvement, promotion of growth and development, anti-fatigue, body weight reduction, oxygen deficit tolerance, radiation protection, anti-mutation, anti-tumor, blood lipid regulation, improvement of sexual potency, blood glucose regulation, gastrointestinal function improvement, sleep improvement, improvement of nutritional anemia, protection of liver from chemical damage, lactation improvement, improvement for beauty, vision improvement, promotion of lead removal, removal of 'intense heat' from the throat and moistening of the throat, blood pressure regulation, and enhancement of bone calcification.

to cover all areas/types of NFF, it is at the same time striking to see the number of acclaimed functional ingredients and their versatility in respect to their targets.

The, thus far, introduced or discussed NFF can also be grouped by components class (see Table 4.2). This table, taken from a publication by the International Food Information Council Foundation, provides a comprehensive overview of the different components and classes of components to be used in NFF. Furthermore, potential sources and benefits are provided. Against this background of a myriad of different ingredients, it becomes again obvious of how diverse potential sources as well as targets are. However, in contrast to most pharmaceutical drugs, a selection of different (biological) sources does also mean a greater variance in chemical properties—resulting in potentially greater differences in aspired physiological effects.

Accordingly, the question arises whether NFF can be sensibly used to sustainably influence (mild) medical conditions. Even if they can be employed to that end, necessary R&D activities, testing, and production processes will be considerably more challenging and costly. This constitutes the specific setting of the NFF area within the pharmaceutical and the food industries. Corroborating the discussion on aspects and consequences of industry convergence, this underlines the intermediate position of the NFF sector between both 'extremes' (i.e. industries). Bröring highlights these differences very concisely in respect to differing technological, market, and regulatory competencies [cf. 2, p. 138][46].

In the remainder of this chapter, one substance class will serve as the anchor for a discussion of NFF in an industrial setting. Due to their broad applications in foods and cosmetics, phytosterols can work well as an example of two recent developments of industry convergence—the emerging sectors of NFF and of cosmeceuticals. This is done for two reasons. First, as several stakeholders have discussed potential benefits and drawbacks of each of the aforementioned components and even target areas, it would go beyond the scope of the study at hand to comprehensively review the respective extent bodies of the literature for all of the components. Second, to fully understand the degree of connectedness between both industries as well as underlying scientific disciplines, it appears to be necessary to cover at least one example in greater detail.[47]

---

[46] These differences will also be briefly discussed in Sect. 4.3.4.

[47] Many of the aspects considering individual substance classes and their use in NFF are similar when considering different classes; hence a discussion of one class may serve well as an example for most of the classes. Furthermore, phytosterols were chosen because of their use as an example in the remainder of the study. See particularly Sect. 5.1 for the results on convergence anticipation in the area of phytosterols and Sects. 6.1.1, 6.1.2, 6.1.3 for a discussion of these results. For a more comprehensive coverage of different substances in NFF, their commercial viability as well as regulatory activities and public perception cf., e.g., [36, 51, 53]. For a comprehensive analysis of phytosterols in respect of chemical, biological, medical, regulatory, and marketing aspects see also [54].

Table 4.2 Overview of NFF by component class

Component class/component[a]	Possible sources	Potential benefits
Carotenoids		
β-carotene	Carrots, pumpkin, sweet potato, and cantaloupe	Neutralizes free radicals, which may damage cells; bolsters cellular antioxidant defenses; can be made into vitamin A in the body
Lutein, zeaxanthin	Kale, collards, spinach, corn, eggs, and citrus	May contribute to maintenance of healthy vision
Lycopene	Tomatoes and processed tomato products, watermelon, and red/pink grapefruit	May contribute to maintenance of prostate health
Dietary (functional and total) fiber		
Insoluble fiber	Wheat bran, corn bran, and fruit skins	May contribute to maintenance of a healthy digestive tract; may reduce the risk of some types of cancer
Beta glucan[b]	Oat bran, oatmeal, oat flour, barley, and rye	May reduce risk of coronary heart disease (CHD)
Soluble fiber[b]	Psyllium seed husk, peas, beans, apples, and citrus fruit	May reduce risk of CHD and some types of cancer
Whole grains[b]	Cereal grains, whole wheat bread, oatmeal, and brown rice	May reduce risk of CHD and some types of cancer; may contribute to maintenance of healthy blood glucose levels
Fatty acids		
Monounsaturated fatty acids (MUFAs)[b]	Tree nuts, olive oil, and canola oil	May reduce risk of CHD
Polyunsaturated fatty acids (PUFAs)—omega-3 fatty acids—ALA	Walnuts, flax	May contribute to maintenance of heart health; may contribute to maintenance of mental and visual function
PUFAs—omega-3 fatty acids—DHA/EPA[b]	Salmon, tuna, marine, and other fish oils	May reduce risk of CHD; may contribute to maintenance of mental and visual function
Conjugated linoleic acid (CLA)	Beef and lamb; some cheese	May contribute to maintenance of desirable body composition and healthy immune function
Flavonoids		
Anthocyanins—cyanidin, delphinidin, and malvidin	Berries, cherries, and red grapes	Bolsters cellular antioxidant defenses; may contribute to maintenance of brain function
Flavanols—catechins, epicatechins, epigallocatechin, and procyanidins	Tea, cocoa, chocolate, apples, and grapes	May contribute to maintenance of heart health
Flavanones—hesperetin, naringenin	Citrus foods	Neutralize free radicals, which may damage cells; bolster cellular antioxidant defenses
Flavonols—quercetin, kaempferol, isorhamnetin, and myricetin	Onions, apples, tea, and broccoli	Neutralize free radicals, which may damage cells; bolster cellular antioxidant defenses

(continued)

Table 4.2 (continued)

Component class/component[a]	Possible sources	Potential benefits
Proanthocyanidins	Cranberries, cocoa, apples, strawberries, grapes, wine, peanuts, and cinnamon	May contribute to maintenance of urinary tract health and heart health
Isothiocyanates		
Sulforaphane	Cauliflower, broccoli, broccoli sprouts, cabbage, kale, and horseradish	May enhance detoxification of undesirable compounds; bolsters cellular antioxidant defenses
Minerals		
Calcium[b]	Sardines, spinach, yogurt, low-fat dairy products, fortified foods, and beverages	May reduce the risk of osteoporosis
Magnesium	Spinach, pumpkin seeds, whole grain breads and cereals, halibut, and brazil nuts	May contribute to maintenance of normal muscle and nerve function, healthy immune function, and bone health
Potassium[b]	Potatoes, low-fat dairy products, whole grain breads and cereals, citrus juices, beans, and bananas	May reduce the risk of high blood pressure and stroke, in combination with a low-sodium diet
Selenium	Fish, red meat, grains, garlic, liver, and eggs	Neutralizes free radicals, which may damage cells; may contribute to healthy immune function
Phenolic acids		
Caffeic acid, ferulic acid	Apples, pears, citrus fruits, some vegetables, and coffee	May bolster cellular antioxidant defenses; may contribute to maintenance of healthy vision and heart health
Plant stanols/sterols		
Free stanols/sterols[b]	Corn, soy, wheat, wood oils, fortified foods, and beverages	May reduce risk of CHD
Stanol/sterol esters[b]	Fortified table spreads, stanol ester dietary supplements	May reduce risk of CHD
Polyols		
Sugar alcohols[b]—Xylitol, sorbitol, mannitol, and lactitol	Some chewing gums and other food applications	May reduce risk of dental caries
Prebiotics		
Inulin, fructo-oligosaccharides (FOS), polydextrose	Whole grains, onions, some fruits, garlic, honey, leeks, fortified foods, and beverages	May improve gastrointestinal health; may improve calcium absorption
Probiotics		

(continued)

Table 4.2 (continued)

Component class/component[a]	Possible sources	Potential benefits
Yeast, *Lactobacilli, Bifidobacteria*, and other specific strains of beneficial bacteria	Certain yogurts and other cultured dairy and nondairy applications	May improve gastrointestinal health and systemic immunity; benefits are strain-specific
Phytoestrogens		
Isoflavones—daidzein, genistein	Soybeans and soy-based foods	May contribute to maintenance of bone health, healthy brain and immune function; for women, may contribute to maintenance of menopausal health
Lignans	Flax, rye, and some vegetables	May contribute to maintenance of heart health and healthy immune function
Soy protein		
Soy protein[b]	Soybeans and soy-based foods	May reduce risk of CHD
Sulfides/thiols		
Diallyl sulfide, allyl methyl trisulfide	Garlic, onions, leeks, and scallions	May enhance detoxification of undesirable compounds; may contribute to maintenance of heart health and healthy immune function
Dithiolthiones	Cruciferous vegetables	May enhance detoxification of undesirable compounds; may contribute to maintenance of healthy immune function
Vitamins		
A[c]	Organ meats, milk, eggs, carrots, sweet potato, and spinach	May contribute to maintenance of healthy vision, immune function, and bone health; may contribute to cell integrity
B1 (thiamin)	Lentils, peas, long-grain brown rice, and brazil nuts	May contribute to maintenance of mental function; helps regulate metabolism
B2 (tiboflavin)	Lean meats, eggs, and green leafy vegetables	Helps support cell growth; helps regulate metabolism
B3 (niacin)	Dairy products, poultry, fish, nuts, and eggs	Helps support cell growth; helps regulate metabolism
B5 (pantothenic acid)	Organ meats, lobster, soybeans, and lentils	Helps regulate metabolism and hormone synthesis
B6 (pyridoxine)	Beans, nuts, legumes, fish, meat, and whole grains	May contribute to maintenance of healthy immune function; helps regulate metabolism

(continued)

Table 4.2 (continued)

Component class/component[a]	Possible sources	Potential benefits
B9 (folate)[b]	Beans, legumes, citrus foods, green leafy vegetables, fortified breads, and cereals	May reduce a woman's risk of having a child with a brain or spinal cord defect
B12 (cobalamin)	Eggs, meat, poultry, and milk	May contribute to maintenance of mental function; helps regulate metabolism and supports blood cell formation
Biotin	Liver, salmon, dairy, eggs, and oysters	Helps regulate metabolism and hormone synthesis
C	Guava, sweet red/green pepper, kiwi, citrus fruit, and strawberries	Neutralizes free radicals, which may damage cells; may contribute to maintenance of bone health and immune function
D	Sunlight, fish, fortified foods and beverages such as milk, juices, and cereals	Helps regulate calcium and phosphorus; helps contribute to bone health; may contribute to healthy immune function; and helps support cell growth
E	Sunflower seeds, almonds, hazelnuts, and turnip greens	Neutralizes free radicals, which may damage cells; may contribute to healthy immune function and maintenance of heart health

Notes [a] Examples are not an all-inclusive list
[b] FDA approved health claim established for component
[c] Preformed vitamin A is found in foods that come from animals. Provitamin A carotenoids are found in many darkly colored fruits and vegetables and are a major source of vitamin A for vegetarians
Source Adapted from [55, p. 2f.]

4.3.3 *Phytosterols*

4.3.3.1 Chemical Structure and Biological Function

Phytosterols (or plant sterols) are triterpenes, constituting important structural components of plant membranes. As such, they consist of a cyclopentanophenantrine ring structure, carrying a hydroxyl functional group at the C-3 position. Their structure is very similar to the well-known sterol *cholesterol*, but with differing side-chain configurations. In analogy to the function of cholesterol in animal cell membranes, phytosterols stabilize phospholipid bilayers in plant cell membranes. These plant cell membranes contain little or no cholesterol, but several types of phytosterols. In deviation from the cholesterol structure they carry a methyl or ethyl group at C-24 (Fig. 4.2) [cf. 56, pp. 457ff.]. While the main function of most phytosterols is controlling membrane fluidity and permeability, some as well have specific functions in signal transduction [cf. 57, p. 939][48]. Of the more than 250 different phytosterols most contain 28 or 29 carbon atoms and one or two double bonds [cf. 56, p. 457; 57, p. 939; 62, p. 217].

In addition to *plant sterols* also *plant stanols* exist, a subgroup of fully saturated phytosterols. For instance, the difference between the depicted *campesterol/campestanol* is the existence/nonexistence of a double bond between carbons 5 and 6 (C-5, 6) in the B-ring of the sterol nucleus (i.e., the second ring), as exemplified in Fig. 4.2. Different approaches exist regarding the *nomenclature* of phytosterols.[49] While phytostanols are often referred to as merely a subclass of phytosterols cf., e.g., [56], some authors also view them as an additional class cf., e.g., [63]. The European Food Safety Authority (EFSA) uses plant sterols as the more general term and phytosterols/-stanols as the respective subgroups [cf. 64, p. 7]. For increased legibility, the present study follows Ostlund in using phytosterols as an all-inclusive term for all different plant sterols and stanols, if not stated otherwise [cf. 65, p. 536].

Phytosterols can be divided further into three groups based on the number of functional groups at C-4. 4-dimethylsterols carry two methyl groups, 4-monomethyl one and the 4-desmethylsterols none. As the di- and monomethylsterols are metabolic intermediates in the pathway leading to desmethylsterols, they are mostly found in low concentrations in most plant tissues and are also of lesser

[48] For a more detailed description of biological functions of phytosterols see, e.g., [cf. 56, p. 939, 57–61].

[49] In addition to different classification hierarchies of phytosterols and phytostanols, nomenclature of phytosterols is particularly confusing as researchers and companies follow international attempts at standardization only in part. Two main streams are following the IUPAC-IUB [IUPAC = International Union of Pure and Applied Chemistry; IUB = International Union of Biochemistry, now the International Union of Biochemistry and Molecular Biology (IUBMB)] recommendations of 1976 and 1989, which are numbering the carbon atoms of the C-24 functional group as C-28, 29, and $C24^1$, 24^2, respectively. This study will instead use the common (trivial) names, as is mainly done by regulators and companies as well [cf. 56, p. 460].

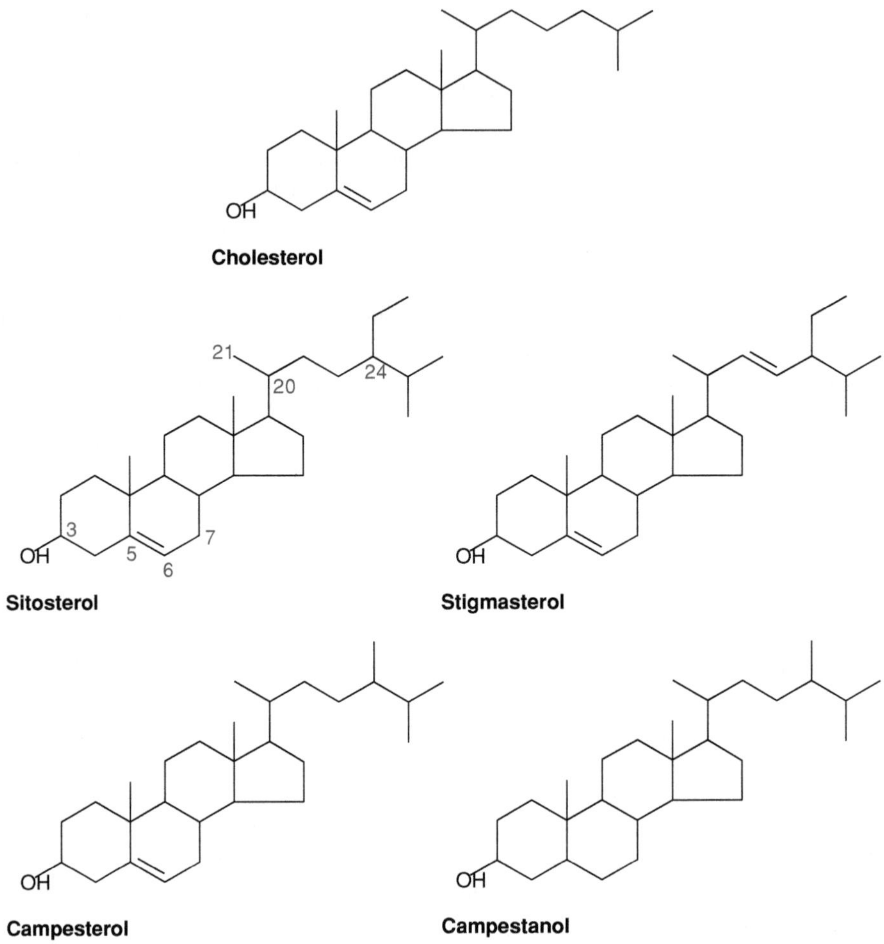

Fig. 4.2 Chemical structure of cholesterol and typical phytosterols/-stanols. *Source* Own figure. For further information on, e.g., stereochemistry of phytosterols cf. [56]

importance [cf. 56, p. 460].[50] Highest concentrations of phytosterols are found in corn, sunflower, and safflower oil with 9.52, 7.25 and 4.44 g/kg edible portion, respectively [cf. 65, pp. 537ff.].[51] Phytosterol *content* is much lower in cereal

---

[50] In plant tissues, phytosterols occur in these four forms: as the free alcohol, fatty acids, steryl glycosides, and acylated steryl glycosides. While the hydroxyl group at C-3 is underivatized in the free alcohol, it is covalently bound to other constituents in the three latter phytosterol conjugates [cf. 56, pp. 465ff.]. They also report on differences in phytosterol forms in individual species and tissues.

[51] In contrast, bananas, apples and tomatoes, for example, only contain 160, 120, and 70 mg/kg edible portion, respectively. Furthermore, see [57, p. 947], who also compare crude oil contents with up to 15.57 and 32.25 g/kg phytosterol content in crude corn oil and rice bran oil,

grains, vegetables, and fruits (in most cases below 1 g/kg fresh weight). The highest concentration in these products can be found in corn (1.78 g/kg), rye (0.91–1.10 g/kg) and barley (0.59–0.83 g/kg) [cf. 57, p. 949].[52]

The most dominant phytosterols are presented in Fig. 4.2: *sitosterol*, *campesterol,* and *stigmasterol* [cf. 57, p. 939].[53] Overall, the distribution of individual phytosterols in plants is very much dependent on plant families and even species.[54] Phytosterols may be found in different isomeric forms (especially regarding the functional groups at C-24, which may be a 24α or a 24β epimer), but due to the dominance of the three main phytosterols stereochemistry has been rarely considered in animal and human studies [cf. 65, p. 536].[55]

4.3.3.2 Physiological and Health-Related Effects

Normal human diets contain phytosterols as **natural components** in amounts of approximately 250 mg/d, mainly from vegetable oils, cereals, fruits, and vegetables. Consequently, daily intake for vegetarians may reach almost twice the level [cf. 56, p. 483; 57, p. 940].[56] As phytostanols occur in considerably lower concentrations in plants, their intake reaches only about a tenth (about 25 mg/d); mainly from corn, wheat, rye, and rice [cf. 56, p. 483]. In most western societies, this amount is about equivalent to the consumption of cholesterol (approximately 300 mg/d) [cf. 56, p. 483]. While the nutritional information on food products often insinuates absence of cholesterol in plants, it may account for up to 5 % of total sterol content in certain plants or tissues. However, in comparison with the amount of cholesterol taken up through consumption of meat and dairy products, this cholesterol source is negligible.

(Footnote 51 continued)
respectively. Unsurprisingly, phytosterol concentrations vary greatly with different production processes and steps. Phytosterols are also commonly derived from tall oil, a by-product in the production of wood pulp from coniferous trees. For a list of commercial suppliers of phytosterols and their sourcing from tall oil or vegetable oil, see [66, pp. 3ff].

[52] Piironen et al. provide a comprehensive overview of phytosterol contents in various foods.

[53] According to Calpe-Berdiel et al., these three phytosterols represent more than 95 % of total phytosterol dietary intake [cf. 63, p. 19].

[54] [cf. 56, pp. 465ff] for a more detailed description of phytosterol composition in different species and organs.

[55] For a comprehensive overview of phytosterol stereochemistry see, e.g., [56].

[56] Ostlund reports on a greater variance in the estimated daily intake, ranging from 167 to 437 mg [cf. 65, p. 537]. Even higher values are provided by Poli et al. for people in Mediterranean countries (500–600 mg/d) [cf. 67, p. S10].

Because of the high structure analogy to **cholesterol**, phytosterols compete with the absorption of dietary cholesterol in the intestines, and (depending on their concentration) thus lead to lower blood concentrations of LDL[57]-cholesterol.[58] Bearing in mind that high concentrations of LDL-cholesterol are a risk factor for cardiovascular diseases[59] such as near term CHD, several studies have undertaken research on possible uses in primary and secondary prevention of CHD [cf. 2; 65; 70, p. 132]. The potential negative effect of LDL-cholesterol is caused by the fact that high levels of cholesterol ('stored' in the form of LDL-cholesterol) in the serum support the formation of atherosclerotic plaques in the whole body. These waxy plaques can then build up in the artery walls, and thus narrow the artery and make it less flexible—potentially leading to heart attacks or strokes from clogged arteries. The ratio of so-called 'bad' LDL-cholesterol over supposedly 'good' HDL-cholesterol serves as a predictor for the susceptibility to cardiovascular diseases, with it being about 3.5 in healthy humans [cf. 68, p. 801]. Phytosterols could also help patients suffering from chronically high concentrations of LDL-cholesterol due to one of the most frequent monogenic hereditary disorders, the familial hypercholesterolemia [cf. 70, pp. 42ff.]. Traditionally, elevated levels of cholesterol are combated by dietary restrictions (limit the cholesterol intake) or by the use of drugs such as several statins.[60]

Phytosterols as a means to lower serum cholesterol levels were already discussed in the 1950s, either as a supplement or a dedicated drug. The first to discuss phytosterol feeding accordingly was Peterson in 1951 [cf. 72]. Shortly afterward, considerable amounts of sitosterol were found to be active in rabbits [cf. 73, pp. 696ff] and in humans [cf. 74, pp. 702ff]. The pharmaceutical firm Eli Lilly marketed it as a pharmaceutical drug under the brand name *Cytellin*® between 1954 and 1982, with dosages normally 6–18 g/d and sometimes even up to 45 g/d [cf. 56, p. 484; 75, p. 2485]. These high doses were necessary for satisfactory

---

[57] LDL = low density lipoprotein. LDLs are the parts of IDLs (intermediate density lipoproteins) not taken up by the liver and serve as the most important carriers of cholesterol in the blood. These IDLs are formed from VLDLs (very low density lipoproteins), which serve as a transporter of unnecessary cholesterol and triacylglycerine from the liver into the blood. In contrast to LDLs, which are carrying cholesterol to peripheral tissues and regulate the *de-novo* synthesis of cholesterol there, HDLs (high density lipoproteins) take up cholesterol from dead cells and transport it back to the liver or transfer it to VLDLs or LDLs [cf. 68, pp. 800f].

[58] For a more precise description of the molecular actions of phytosterols in cholesterol metabolism and a review of recent studies, see [63].

[59] For Europe, the British Heart Foundation has estimated cardiovascular diseases to cause costs of approximately €110 billion in 2006, nearly 10 % of total healthcare expenditures, underlining the medical and economic importance of potential improvements to heart health [cf. 69, p. 21].

[60] Statins block the enzyme HMG-CoA reductase (3-hydroxy-3-methylglutaryl-coenzyme A reductase or HMGR), the rate-controlling enzyme in the metabolic pathway producing cholesterol in the liver. [cf. 68, pp. 804f] These statins include different active pharmaceutical ingredients such as rosuvastatin, lovastatin, and atorvastatin. Atorvastatin is marketed by Pfizer under the brand name LIPITOR®, constituting the best-selling prescription pharmaceutical product in the world (2009 sales at US$ 11.4 billion), despite lower sales than in previous years (US$ 12.4 billion in 2008 and 12.7 billion in 2007) [cf. 71, p. 21].

efficacy due to the poor solubility and bioavailability of free phytosterols, leading to a replacement of these drugs and supplements by statin-based drugs. This perception of very limited use of phytosterols only changed with their utilization in NFF [cf. 56, p. 484].[61] In their 1995 landmark study, Miettinen et al. found sitostanol ester enriched margarine to be effectively reducing LDL-cholesterol levels in mildly hypercholesterolemic patients [cf. 76].[62]

Very many studies have been conducted in respect to the **efficacy of different forms** and types of phytosterols.[63] The question of which of the different substances would be most successful in achieving efficacy has been extensively debated. In this discussion, four main forms of phytosterols were tested and compared: free sterols and stanols as well as steryl and stanyl esters. It has been argued that the physiologically active forms are the unesterified, free sterols and stanols. Accordingly, esterified forms would be active due to them being enzymatically hydrolyzed in the small intestine [cf. 56, p. 485]. Moreau et al. summarize the findings of the aforementioned studies in two findings: First, "[...] stanyl and steryl esters appear to be roughly equivalent in their cholesterol-lowering efficacy" [56, p. 485]. Second, "while more research remains to be done to completely resolve this question, a current analysis of recent research appears to indicate that free sterols and stanols and steryl and stanyl esters, formulated in a manner that makes them "equally bioavailable," all may give similar effects on LDL-C [LDL-cholesterol] levels" [56, p. 485].[64] Consequently, the effect does appear not to be primarily dependent on the type of phytosterol chosen, but rather on securing bioavailability of the respective substance.[65]

With a **dose** of 2 g/d (3 g/d) of phytosterols the risk of heart disease is said to be reduced by 25 % (15–40 %), depending on the individual parameters of the patient (e.g., diet, age, and weight), [cf. 56, p. 488][66] leading the FDA to conclude that "there is significant scientific agreement that plant sterol/stanol esters from certain sources will help reduce serum cholesterol and that such reductions may reduce the

---

[61] See Sect. 4.3.3.3 for a discussion of phytosterol use in NFF.

[62] In their study reduction amounted to 14.1 % within 12 months of 2.6 g/d sitostanol ester consumed.

[63] For a comprehensive discussion of the different studies on 'health-promoting effects of phytosterols' and their subtypes see, e.g., [56, 57, 77].

[64] For a further discussion of this finding see also [78].

[65] It would go beyond the scope of this study to discuss in length the different positive and negative aspects of each individual substance. In addition to efficacy, bioavailability, and ease of preparation other factors such as raw material availability and prices would have to be taken into account as well. See, e.g., [56] For example, absorption and resulting availability of phytosterols is one central point in efficacy discussions already since 1959 [cf. 79].

[66] Poli et al. also stress the fact that dosage in excess of about 2.5 g/d does not provide additional benefits [cf. 67, p. S11]. This significant reduction is underlined by other studies investigating the effects of lowered cholesterol levels on heart disease prevention. For instance, Law et al. found that a long-term reduction in serum concentration of 10 % led to a lowering of the risk of ischemic heart disease of between 20 and 50 % [cf. 80, p. 367].

risk of CHD" [77, p. 54701].[67] Chen et al. underline the additional effects of a diet as recommended by health professionals for blood cholesterol management and phytosterol consumption. For the reduction in LDL-cholesterol, they can attribute 72.7 % to phytosterol intake and 27.3 % to a change in diet [cf. 81, pp. 278ff.].[68] Interestingly, Ostlund stresses one major shortcoming in probably all the hitherto completed studies. The effects of phytosterols are only considered in relation to an already phytosterol-containing baseline diet. Accordingly, all findings are only in respect to the effect of *additional* phytosterol intake and cannot be interpreted in comparison to "the unknown basal effect of dietary phytosterols" [cf. 65, pp. 542f.]. This cognition is particularly important in the light of a great variety in individual diets as well as food products with lowered phytosterol content in the end product due to removal of phytosterols to produce supplements or enriched food products.

In addition to the discussed effects on CHD, phytosterols are also investigated for potential anticancer and antioxidant activity. Also, since the early 1990s further biological functions have been investigated, including anti-inflammatory, antipyretic, and antidiabetic properties [cf. 62, p. 222]. However, research is still limited in all five areas, making future research necessary to judge reasonably on the significance of these early results [cf. 56, p. 489].[69]

The most highly contended issue still is the question of **potential adverse effects** of phytosterols on human health, especially when consumed regularly in higher doses. Due to their structural similarity to cholesterol, concern arose in respect to potentially high serum levels of phytosterols leading to atherogenic effects. As studies found such a side effect only for patients suffering from the autosomal recessive disorder *phytosterolaemia* (also called *sitosterolaemia*),[70] the FDA has recognized steryl and stanyl esters to be *generally recognized as safe* (GRAS) [cf. 77, pp. 54688ff.]. In spite of these positive findings, it is relatively undisputed that a phytosterol rich diet reduces the absorption of fat-soluble *β-carotene* (provitamin A), *lycopene,* and *α-tocopherol* (vitamin E[71]). All of these

---

[67] The original quotations are two identical sentences with either sterol or stanol. Based on the data available at the time of the ruling, the FDA required these health claims to recommend a daily intake of 3.4 g stanyl esters and only 1.3 g of steryl esters (or the respective amounts in free stanols/sterols) [cf. 77, p. 54712]. According to Moreau et al. this was caused by the on average higher dosages administered in stanol studies compared to such with sterols [cf. 56, p. 488].

[68] However, they also stress the limitations of their study in respect to generalizability due to patient sample composition.

[69] For potential anticancerogenic effects see also, e.g., [82–84].

[70] This inherited sterol storage disease with a strong predisposition to premature coronary atherosclerosis is very rare. In fact, research showed a reduction of serum sterol levels under administration of the unabsorbable sitostanol, possibly by competitive inhibition of sterol absorption [cf. 85, pp. 181ff.]. For other potentially affected patients see also [86].

[71] Natural vitamin E includes actually eight distinct molecules: α-, β-, γ-, and δ-tocopherol as well as α-, β-, γ-, and δ-tocotrienol [cf. 87, p. 692]. However, only α-tocopherol meets human vitamin E requirements and is thus often referred to as 'the' vitamin E [cf. 88, p. 5].

are important antioxidants.[72] Hence reported reductions of up to 25 % may be potentially negative for patients' health. As some researchers argue that a key role for these vitamins may be the protection of LDL-cholesterol from oxidation and as carotenoids and tocopherols are often associated with LDL particles, an adjustment for the reduction through reduced (by the sterols or stanols) LDL-cholesterol levels may be appropriate. This would lead to only insignificant reductions of tocopherol levels [cf. 56, pp. 489f.].[73] Contrastingly, the level of β-carotene would still be reduced by between 7 and 19 % [cf. 56, p. 490; 67, p. S11; 90, p. 333]. On the one hand, a potential undersupply with β-carotene could be corrected relatively easily by an increased intake of carotenoid-rich foods like apricots, beet greens, carrots, kale, pumpkin, or spinach [cf. 56, p. 490; 84, p. 150; 91, pp. 175ff; 92, p. 2336].[74] On the other hand, researchers like Thurnham call for "much more intensive experiments to evaluate the postprandial levels of β-carotene and retinol palmitate" [93, p. 254]. In his view, such studies are necessary to judge on phytosterols' potential interference with the absorption of carotenoids [cf. 93, p. 254]. Furthermore, he underscores the importance of a continued healthy diet when using FF products [cf. 94, p. 256].

4.3.3.3 Use in NFF

Due to their aforementioned properties (especially their potential to lead to lower LDL-cholesterol levels) phytosterols receive considerable interest for employment in Nutraceuticals and especially FF. In *Nutraceuticals,* they are largely marketed over the Internet to cure prostate illnesses and lower cholesterol levels like, for instance, fairvital's *Beta-Sitosterin* (120 mg/d), or to improve the 'sense of well-being in the bladder of men above 50 years of age' like, e.g., Natura Vitalis®' *Phytosterol* (9.6 mg/d of phytosterols) [cf. 95, 96].[75] Strikingly, the field of use in drugs and Nutraceuticals is different to that in FF, resulting in drastically lower daily dosages as well. However, in contrast to pharmaceutical drugs targeting other medical conditions, some Nutraceuticals are still also targeting high LDL-cholesterol levels [cf. 96], despite dosages about 20 times lower than the recommended daily intake. Unlike FF and pharmaceutical drugs, Nutraceuticals are often

[72] Antioxidants may protect cells in the human bodies against free radicals that stem from, e.g., the breakdown of consumed food, smoke, or radiation and are believed to play a role in heart disease, cancer, and other diseases [cf. 89].

[73] Slightly different findings are reported by Chen et al. who find qualitatively comparable but larger deviations in tocopherol and carotene levels [cf. 81, pp. 277ff.].

[74] Moreau et al. also mention the possibility that free sterols and stanols could have a lesser adverse effect on the absorption of antioxidants.

[75] Quotation translated by the author. They are also used in low concentrations (e.g., thrice daily 20 mg) pharmaceutical drugs like, e.g., in Germany Schwarz Pharma's Harzol®, Triastonal® and Sitosterin Prostata-Kapseln or Sandoz' Azuprostat Sandoz® (twice daily 65 mg) for the treatment of benign prostatic hyperplasia [cf. 97].

produced by small and hardly known companies, selling their products directly or indirectly through internet-based distribution channels.

As FF phytosterols are produced in a wide selection of products, from the more typical yellow fat spreads and (fermented) milk type products to smoothies [cf. 98] or cookies [cf. 99].[76] Within the European Union foods with phytosterols are classified as **novel foods**, being marketed under the 'novel foods regulation' (EC) 258/97.[77] By May 2007, nine products had been authorized and a further three were still awaiting decision [cf. 100, p. 10]. Including Raisio's *Benecol®* brand,[78] which was on the market prior to (EC) 258/97 and thus did not need approval, and those only needing notification, 21 firms were on the approval list with their product ranges by February 2008 [cf. 64, p. 9]. Next to these, about 70 further manufacturers had applied for authorization for their "substantially equivalent products" [cf. 100, pp. 9ff.]. Amongst them are companies from all industry segments, including Unilever (*pro.activ*), Coca Cola, Teriaka (*Diminicol™*), Archer Daniels Midland (ADM, *CardioAid™*), Enzymotech (*CardiaBeat™*), Forbes Medi-Tech (*Reducol™* and *Phyto-S Sterol™*), Degussa Food Ingredients (*Cholestatin™*), and Cognis (*Vegapure®*) [cf. 66, p. 5]. The approval for some of the 21 companies is specifically targeted at only one application, while other companies aim at a whole array of products. As already mentioned, Raisio does not need approval for its product range. Danone's *Danacol®* ingredient is approved for fermented milk type products only and Oy Karl Fazer may use phytosterols just in rye bread. The remaining 18 companies may offer their products in between three and nine of the listed ten applications [cf. 64, p. 9]:

- Yellow fat spreads—15 companies with approval.
- Milk type products—17 companies with approval.
- Fermented milk type products (e.g., yoghurt type products)—19 companies with approval.
- Milk-based fruit drinks—eight companies with approval.
- Soya drinks—12 companies with approval.
- Cheese type products—15 companies with approval.
- Salad dressings—11 companies with approval.
- Spice sauces—11 companies with approval.
- Rye bread—eight companies with approval.
- Rice drinks—one company with approval.

Cargill (*Corowise™*), Cognis, and Forbes Medi-Tech command the broadest approved application range—the nine applications other than rice drinks. The only company with an approval in the area of rice drinks is Teriaka. In line with earlier

---

[76] Starling covers the introduction of Right Direction Foods' *Right Direction Cookies*, which did interestingly not carry the FDA approved health claim due to trans fat and sugar content.

[77] See Sect. 4.3.4.2 for a brief overview of regulation of NFF in Japan, the USA and the EU.

[78] In other countries, like the USA, the Benecol brand is marketed by McNeil Nutritionals (part of Johnson & Johnson) [cf. 101].

ruling in the USA, the European Commission prohibits the use of **health claims** on food, unless they are authorized by the Commission and are included in a list of permitted claims (Regulation (EC) No. 1924/2006). Based on this regulation, McNeil Nutritionals and Unilever were authorized in late 2009 to use the (revised) health claim of "Plant stanol esters [plant sterols in the case of Unilever] have been shown to lower/reduce blood cholesterol. High cholesterol is a risk factor in the development of coronary heart disease", as long as they provide information to consumers that at least 2 g/d of plant sterols/stanols need to be ingested [cf. 102, p. L 277/9]. These companies' *Benecol* and *pro.activ* (respectively) product ranges were the first to obtain approval for their disease risk reduction claim under regulation EC/1924/2006 [cf. 103].[79] While Raisio's/McNeil Nutritionals' Benecol spread based on phytostanol esters was the first FF with phytosterols at all (product launch in Finland was in 1995), Unilever's pro.activ/Promise activ/Take Control spread was the first sterol based (1999 and 2000 in the USA and Europe, respectively) [cf. 66, p. 9].

According to Poli et al. "food matrices do not appear to affect significantly the cholesterol-lowering efficacy of phytosterols at equivalent doses" [67, p. S11]. While this can be interpreted in favor or against the use of phytosterol-enriched NFF,[80] they do stress that consumption has been demonstrated to be more effective when linked to principal meals. AbuMweis et al. report on partially different findings. In their meta analysis of 59 randomized clinical trials, they *do* find an influence of the *food carrier*. Phytosterols incorporated into fat spreads, mayonnaise, salad dressing, milk, or yoghurt-reduced LDL-cholesterol levels much more decisively than those entailed in other food products, like chocolate, orange juice, cheese, meat, or cereal bars [cf. 104, p. 7]. However, splitting up a daily dose into several smaller doses during meals does not appear to have any effect [cf. 67, p. S11]. Furthermore, more drastically than Poli et al. AbuMweis et al. present their conclusion that the intake does only result in considerable cholesterol level reductions if taken with lunch or the principal meal. In contrast, no effect could be proven for consumption once per day in the morning. [cf. 104, p. 7]. This finding should be considered when evaluating the sense of enriched fat spreads—mostly consumed with breakfasts (at least in many Western societies). Consequently, the question arises whether a FF should be preferred over a Nutraceutical preparation (or a drug), as relatively exact dosing will most likely be more difficult with FF products. An alternative solution in FF would be the consumption of exact daily doses through, e.g., yoghurt drinks.[81] A study by Ruiu et al. proves efficacy of

---

[79] Although EFSA did also voice a positive opinion in regard to Danone's Danacol products, these are still awaiting a decision by the European Commission. EFSA is in the process of reviewing in total more than 175 claims in relation to cholesterol and CHD.

[80] Apparently, it does in fact not make a difference whether phytosterols are consumed in the form of Nutraceuticals or FF, or as pharmaceutical drugs. Hence, it would be up to the consumer to decide on the most convenient (and economical) way of securing a daily dose of phytosterols.

[81] Yoghurts (drinks) have experienced considerably higher growth rates due to the elevated interest by consumers [cf. 105].

phytosterol enriched yoghurt drinks, albeit the reduction in LDL-cholesterol levels is decisively lower than caused by other products. Reasoning for the only 4.9 % reduction may be based on the consumption of only 1.6 g/d (compared to largely used 2.5 g/d) and outside mealtimes in the morning (compared to the suggested connection to principal meals) [cf. 106, pp. 158f.].

Besides their possible applications in pharmaceuticals and foods, phytosterols are also commonly topically applied in cosmetics, e.g., as hair care agents, moisturizers, skin barrier strengtheners, or in treating cellulite as well as in deodorizing preparations.[82]

4.3.3.4 Company Activities in NFF

Naturally, not all the firms offer their complete product range in every of the European Union's member states. Especially, broad ranges can be found on the market shelves in Finland, Belgium, Germany, and the United Kingdom. Additionally, the FDA has given GRAS status to further products like coffee, snack bars, pasta, and egg products [cf. 112, p. 15; 113]. Furthermore, phytostanol esters are used in a variety of additional products, such as turkey liver sausages, chicken balls, instant oat meal, or candy chews [cf. 66, pp. 10f.]. One main *driver* behind the growing interest in commercializing phytosterol-enriched products is the considerably higher margin compared to conventional food products. Yellow fat spreads with phytosterols may sell at prices of more than five times the price of the cheapest conventional spread available in the same supermarket. With estimates for 2007 of a 72,000 t and €700 million market of phytosterol enriched yellow fat spreads in Europe alone, food companies can certainly hope to increase their overall margins by adding such products to their portfolio [cf. 64, p. 14].[83] This is especially promising in the light of an estimated 50–60 % of Europeans (and other countries with comparable lifestyles) not achieving their consensus cholesterol goals [cf. 117, p. 2]. Contrastingly, market shares for each of the FF product classes still remain in the lower one digit percentages in most cases and countries [cf. 64, p. 13]. In addition to the traditional food matrix of spreads, especially (drinkable and spoonable) yoghurts enriched with phytosterols have gained

[82] cf., e.g., [107, 108–111]. See also Sect. 4.3.5 for a short excursus on Cosmeceuticals.

[83] However, market size and growth rates are reasonably disputed. A different estimate by Frost and Sullivan sees the market in 2007 at €420 million, growing at 20 % per year [cf. 114, 115]. According to [1], Leatherhead Food Research even forecasted a 10 % decline in the UK cholesterol-lowering spread market in 2009 after an allegedly similar development in 2008, despite overall growth in the FF market. His line of reasoning is built on the existence of contradictory messages from governments, companies, medical authorities, and the media as well as several competing products, like spreads with polyunsaturated or omega-3 fatty acids. Two further estimates see the global market for end-products containing sterol-based esters at €500 million in 2005 and the global market for phytosterol-based end products at US$ 805 million (US$ 600 million for Europe, US$ 130 million for Japan and US$ 75 million for the USA) [cf. 99, 116].

momentum in the last years, experiencing above average growth rates in the cholesterol-lowering foods market (worth €142 million in the UK alone) [cf. 98]. According to Starling, Benecol commands nearly 50 % of the 'one-shot drinks' market, growing between 16.8 and 31.1 % in contrast to 4.5 % growth in Benecol spreads [cf. 105]. At the same time, Unilever's proactive is reported to hold more than 50 % of the overall phytosterol FF market in the UK, despite suffering from a decline in sales in the recent past. Globally, the market should be well in excess of €1 billion, with proactive at €546 million and Benecol at €197 million [cf. 118].

Activities of **pharmaceutical companies** in FF have thus far been very limited. Besides activities of consumer goods and pharmaceutical company Johnson & Johnson (and its McNeil Nutraceuticals subsidiary), one exception was Novartis' (Consumer Health division) decision to obtain an exclusive license for the worldwide marketing of Forbes Medi-Tech's *Phytrol* (consumer brand Reducol) ingredients. In 2000, Novartis entered into a 50/50 joint venture with Quaker Oats forming Altus Foods, the first joint venture of a global pharmaceutical player and an established food company [cf. 56, p. 491; 119, p. 359] In 1998, Novartis had combined the three branches of *Medical Nutrition*, *OTC* ("over the counter" = prescription free pharmaceutical drugs), and *Health and Functional Food* into the new division of Consumer Health. This was driven by anticipated synergies in R&D activities as well as marketing efforts of these largely consumer-driven businesses. Additionally, Novartis wanted to gain access to the far larger part of the 'healthcare continuum'—the food market. While this was estimated to be comprised about US$ 2 trillion in 1997, worldwide pharmaceuticals' sales only reached approximately US$ 300 billion [cf. 119, pp. 359f.].[84] Strikingly, despite Novartis' estimate of the US FF market at US$ 10–20 billion and with growth rates of about 10 % per year, Novartis and PepsiCo (having acquired Quaker Oats) agreed to close the joint venture in 2002 [cf. 121]. This decision was unsurprising, as Novartis had already stopped activities with its own *Aviva* products about half a year earlier [cf. 122]. Consequently, in February 2002 Novartis announced that it would completely sell the Health & Functional Food part of its Consumer Health division [cf. 123].

4.3.3.5 Consumers of Phytosterol-Enriched NFF

In general, most buyers of these products are over 55 years of age, with companies like McNeil Nutritionals now increasingly targeting 'younger' audience (over 45 years) with different, dual benefit products (e.g., phytosterol enriched smoothies rich in fruits) [cf. 98]. **Marketing** of NFF is a complex task. On the one hand, the target group is a subgroup of the general population with a mild medical condition. On the other hand, "consumers don't want to be told directly that they

[84] Schaffnit-Chatterjee estimates the global market size for processed foods (constituting about 75 % of total world food sales) in 2009 to have reached US$ 3 trillion [120, p. 13].

are managing a condition".[85] Consequently, Cargill markets its phytosterols under the *CoroWise* label to be used also by their customers, insinuating that consumers picking CoroWise branded products act 'wisely' [cf. 124]. It took companies years of substantial marketing efforts to get the message of phytosterols' potential benefits across to consumers. For instance, companies in the USA were initially not allowed to advertize their products clearly as FF, hampering market success. Only after Lipton/Unilever's and McNeil Nutraceuticals' petitioning, the FDA decided in late 2000 to allow companies to advertise a proven connection between the functional ingredients and lowered LDL-cholesterol levels [cf. 125]. Even that may not be sufficient, as in the case of Mars' CocoaVia® (chocolate snacks with phytosterols), which was discontinued shortly after initial introduction [cf. 126, 127] Despite a global functional chocolate market of nearly US$ 400 million and 34 new products, both in 2009, many companies struggle to get comparably complex health-related messages across to consumers [cf. 105].

In the light of decisive premiums to be paid for FF in comparison to conventional foods, the use of approved health claims seems to be vital for the success of FF. However, many consumers appear to be still overwhelmed by the plentitude of choices (e.g., competing products for heart health, such as polyunsaturated and omega-3 fatty acids) and information sources (governments, medical experts, companies, media) [cf. 1]. This combination of elevated price levels, contradictory and complex messages, and the targeting of consumers with medical conditions (or perceived conditions) have led to peculiar **consumer demographics**. According to Lea and Hepburn, who present the results of Unilever's post-launch monitoring (PLM)[86] in five major EU markets, about three-quarters of purchasers of phytosterol-enriched yellow fat spreads are living in one- or two-person households. About 90 % do not have children living with them and about 80 % are over 45 years of age. These numbers increase even further when looking at regular purchasers. Even where purchasers live in larger households, it is predominantly one person consuming the FF [cf. 128, p. 1216]. Interestingly, the household panel data also show that consumers tend to use other phytosterol-enriched product types (e.g., yoghurts) as well, where they are available [cf. 128, p. 1217].[87]

This does again constitute an important *caveat* when discussing optimal dosage of such functional ingredients. In contrast to prescription drugs and even OTC pharmaceuticals—how can possible agglomeration effects of a combination of FF products be controlled? Furthermore, EFSA reports that up to 40 % of consumers may not belong to the target group, including a few children. On the level of understanding for the importance and complexity of product labeling, it concludes: "The special information required on product labels for these types of food is

[85] The quotation is by Pam Stauffer, global communication manager at Cargill Health, and Food Technologies [cf. 124].

[86] PLM is used to distinguish this monitoring for signals of unexpected health effects from post-marketing surveillance (PMS), a much more rigorous system used in the pharmaceutical industry [cf. 128, p. 1214].

[87] For a thorough assessment of consumer demographics and rationale see also [100, pp. 18ff.].

rarely understood. A high proportion of the consumers are not aware of the dietary guidelines in relation to phytosterol products and the importance to also consume sufficient fruit and vegetables to prevent a reduction in plasma carotenoids levels" [112, p. 2]. Consequently, in a large-scale German survey Niemann et al. find that 62 % of consumers purchased FF to combat 'elevated cholesterol levels'. The remaining nearly 40 % based their decision on a different reasoning, such as 'protecting their health', 'compensating for an unhealthy lifestyle' or 'because it tastes good'. In total, only 15 % acted on a recommendation by their doctor, pharmacist or nutritionist and only 21 % of the purchasers had discussed their consumption with their doctor—in spite of the clear recommendation on the products [cf. 100, pp. 23ff.]. The misunderstanding or ignoring of product labeling becomes even more obvious when analyzing answers regarding its central issues. Only 4 % of consumers (nearly irrespective of education level) could correctly state the maximum recommended daily intake, only 27 % did not believe phytosterol products to be suitable for all consumers and only 4 % were able to state correctly why they should regularly eat fruit and vegetables [cf. 100, pp. 29ff.]. Against this background, it remains questionable whether the general public can be expected to reliably follow more complex recommendations in respect to FF products.

4.3.4 Industry and Regulatory Background of NFF

4.3.4.1 Industry Background

As is the case for **market estimates** of phytosterol-enriched NFF, estimates for the complete (N)FF market vary greatly. Partly, this is caused by differences in delineating Nutraceuticals from FF, conventional foods and drugs. Estimates for the FF market (probably excluding Nutraceuticals) range from US$ 33 billion in 2000 to US$ 61 billion in 2003, or even US$ 100 billion in 2001 [cf. 2, p. 131; 129, p. 52; 130, p. 221; 131, pp. 457f.]. The three markets of the USA, Japan and the EU are believed to constitute up to 90 % of the global sales volume. In most of the markets, FF sales will represent a share of the complete food market in the lower single-digit numbers, but with above average growth rates [cf. 131, p. 457]. This growth is driven by aging and increasingly overweight societies as well as trends toward one-stop shopping. Not surprisingly, the number of products is increasing as well. In Japan, the number of FF products launched between 1988 and 1998 alone amounted to 1,700. In 2005, more than 500 products on the market were labeled food for specified health uses (FOSHU) [cf. 131, p. 457]. Since 1990 more than 5,500 FF have been introduced to this prime FF market. It all started with the introduction of Otsuka Pharmaceuticals' dietary fiber soft drink '*Fibe Mini*' in 1988 [cf. 51, p. 136]. When looking at single market segments, the number of products launched between January 2007 and August 2009 in respect to heart health alone is reported to be over 150. Palmer estimates that this segment should reach US$ 15 billion by 2014 (growing 40 % from the 2009 figures) [cf. 1]. Tentatively, the share of functional food products in countries' general food market is even smaller in countries with traditionally more healthy diets (as, for instance, most southern European countries) [cf. 1; 131, p. 457].

Theoretically, NFF should be highly interesting and rewarding for the food industry as well as for the pharmaceutical industry. The latter (having even introduced the first FF product) could employ its strong R&D background and its experience in conducting clinical trials and seeking approval on the basis of extensive scientific literature. **Pharmaceutical** companies are used to explain complex modes of action and resulting effects. In the light of a much larger food market and in combination with considerably lower development costs and especially reduced development cycle times, one should expect decisive interest and activities from this industry, despite potentially lower margins [cf. 2, pp. 135ff; 132, p. 717].

The **food** industry, on the other hand, is struggling from slim margins in intense price competitions with the few global competitors and the often highly consolidated retail industry. Accordingly, the food industry could use its involvement to increase their overall profit margins. Their most important strengths lie in its strong brands, its consumer market insights, access to distribution channels and its being used to short R&D cycles [cf. 2, pp. 134ff; 132, p. 717].

Consequently, one should expect strong tendencies of both industries to engage in NFF development and marketing and also to form alliances with incumbents from the respective other industry. With companies similar in size and often strong in similar markets, collaboration should not present an insuperable hurdle (as is more often the case when companies are very different in size and corporate culture). Companies from both industries could learn from each other and be better prepared for the convergence field. However, companies have thus far only rarely collaborated (e.g., Novartis and Quaker Oats) and overall the pharmaceutical industry appears to be rather reluctant to engage heavily in the development of NFF.

4.3.4.2 Regulation of NFF

Bearing in mind the potential to significantly alter medical conditions in consumers, regulation for NFF should be an important consideration.[88] While international incentives such as the *Codex Alimentarius*[89] have tried to harmonize food

---

[88] It is not the aim of this study to provide a comprehensive assessment of regulatory aspects of the NFF sector. Instead, this section is intended to provide a brief introduction to the importance of national and international regulation and its importance in the development of markets for NFF. For deeper insights into regulatory aspects see, e.g., [2, pp. 147ff.; 36, pp. 115ff.; 51, pp. 55f.; 133, pp. 1ff.; 134, pp. 9ff.].

[89] The Codex Alimentarius and its standards, guidelines, and related texts have been compiled by the Codex Alimentarius Commission, created jointly by Food and Agriculture Organization of

standards and to develop codes of practice, regulation is still very diverse around the globe. In line with being the first major market for NFF, **Japan** was also the country with the first dedicated regulation on NFF. In 1991, the Ministry of Health and Welfare introduced the FOSHU system, building on earlier regulations on FF presented in 1984. In 2001, FOSHU were integrated next to 'foods with nutrient function claims' (FNFC) into the new regulatory system of 'foods with health claims' (FHC). Finally, in 2005 the Japanese Government changed the FOSHU system to include three new subsystems (standardized FOSHU, qualified FOSHU, and disease risk reduction claims), yielding a highly complex assortment of different laws and regulations (in part by different ministries) applicable to foods and NFF. These laws include the important 'Pharmaceutical Affairs Law', controlling functional claims on food products. According to this law, the question whether a product is to be judged as a drug or nondrug is based on ingredients and potential efficacy claims [cf. 136, pp. 249ff.]. Consequently, foods can be separated into two classes: (1) 'foods for special dietary use' (FDSU) and (2) 'FHC'. While FDSU entail foods for patients, pregnant women and infants, they also partially cover FOSHU. These are also part of the FHC regulations, together with FNFC. Approval of new products under the FOSHU regulations is then contingent upon several requirements, including clear evidence of functionality, safety assessments, and enhancements of the state of health [cf. 136, pp. 263ff.]. While certain health claims are allowed under the Japanese system, one central peculiarity is the prohibition to use disease-reduction claims [cf. 36, p. 135].

In the **USA**, NFF are being monitored by the FDA, a sub-body of the US Department of Health and Human Services. Regulation is based on several laws, most prominently the Federal Food, Drug, and Cosmetic Act (FD&C Act) of 1938 (amended severally in subsequent years) dealing with all food products, cosmetics and pharmaceutical drugs. The most important amendments with relevance for NFF are the nutrition labeling and education act (NLEA) of 1990 and the dietary supplement health and education act (DSHEA) of 1994. The NLEA has introduced the possibility of health claims (e.g., "diets high in calcium may reduce risk of osteoporosis") into the American system, with the need to be approved by the FDA. Contrastingly, structure/function claims (e.g., "calcium builds strong bones") for '*food supplements*' under the DSHEA do not need authorization. The same holds true for nutrient content claims (e.g., "high content of calcium").

While it is easily understandable that health claims have to be authorized by the FDA, a rationale for leaving the other two types of claims in the sole responsibility of manufacturers is less obvious. It remains debatable whether consumers do really discern between the different types of claims. As the term '*functional food*' does not exist in the US system, FF are mostly viewed as '*food ingredients*' which in

(Footnote 89 continued)
the United Nations (FAO) and World Health Organization (WHO) in 1963. Its main targets are "protecting health of the consumers and ensuring fair trade practices in the food trade, and promoting coordination of all food standards work undertaken by international governmental and non-governmental organizations" [cf. 135].

contrast to supplements need to obtain GRAS status. In consequence, safety studies and clinical trials may be necessary, considerably increasing the necessary effort for the introduction of a food ingredient when compared to food supplements. This leads to a greater importance of supplements than food ingredients in the US market [cf. 2, pp. 156ff.; 36, pp. 135ff.; 45, p. 69].

Finally, in the **European Union**, common overarching principles in respect to food safety and further regulation and directives[90] are part of respective national legislation of the individual member state [cf. 137]. However, there has been considerable effort undertaken to begin to harmonize national laws. The EFSA is responsible for food safety and consumer protection and is in charge of the harmonized authorization of '*novel foods*' and food ingredients. Before they are placed on the market, EFSA testing has to demonstrate safety of the products. The most important EU regulation is "Regulation (EC) No 258/97 of the European Parliament and of the Council of 27 January 1997 concerning novel foods and novel food ingredients", affecting all "placing on the market within the Community of foods and food ingredients which have not hitherto been used for human consumption to a significant degree within the Community" and which belong to one of six categories, such as "consisting of or isolated from plants".[91] Food products already on the market prior to the entry into force of (EC) 258/97 (15 May 1997) may remain on the market under the "principle of mutual recognition" without approval [cf. 137].[92] In cases where a 'substantially equivalent' substance has already been approved, a notification of the authorities will be sufficient—in accordance with article 5 of (EC) 258/97.

Interestingly, supplements do not fall under the novel foods regulation by (EC) 258/97. Instead, they are dealt with in "Directive 2002/46/EC of the European Parliament and Council of 10 June 2002 on the approximation of the laws of Member States relating to food supplements" [cf. 138]. In contrast to novel foods, Article 6(2.) explicitly states "2. The labelling, presentation and advertising must not attribute to food supplements the property of preventing, treating or curing a human disease, or refer to such properties". Annexes I and II explicitly state which vitamins and minerals (substances) may be used in the manufacture of food supplements. All other substances had to be phased out by August 2006 or listed as well after submission of additional scientific material. Hence, new supplement substances will have to be approved before market entry.

The use of nutrition or **health claims** is controlled under "Regulation (EC) No 1924/2006 of the European Parliament and of the Council of 20 December 2006 on nutrition and health claims made on foods". It regulates that "Nutrition and health claims may be used in the labelling, presentation and advertising of foods placed

---

[90] While EU regulations are applied in all Member States without the necessity for any further implementation, directives are only setting objectives to be achieved by means deemed sensible by the respective Member State. Both are commonly adjusted by amendments or corrections.

[91] (EC) 258/97, Article 1(2.).

[92] Phytostanol esters as one example for a novel food already on the market have been discussed in the preceding chapter.

on the market in the Community only if they comply with the provisions of this Regulation".[93] It discerns between *nutritional claims* (particular beneficial nutritional properties), *health claims* (relationship between a food and health), and *reduction of disease risk claim* (consumption of food significantly reduces a risk factor in development of a human disease).[94] However detailed the European regulation may be, again it is difficult to distinguish between the different types of products (foods, supplements, drugs etc.), leaving considerable room for interpretation.

4.3.5 Excursus: Cosmeceuticals

A setting similar to that in NFF can be found in the sector of Cosmeceuticals, a term coined by US Society of Cosmetic Chemists co-founder Reed in 1961 and popularized in the following decades by American Klingman to describe cosmetics with drug-like functionalities on the skin [cf. 139, 140, p. 446]. In contrast to FF, apparently no 'Functional Cosmetics' terminology is in use. This may be caused by the way traditional cosmetics are (at least supposedly) influencing their micro- and macro-environment. In contrast to food products, conventional cosmetics are not only targeting one 'simple aspect' such as nutrition, but tend to be more '(multi-) functional' per se. Accordingly, a delineation of cosmeceuticals from topical drugs and conventional cosmetics is difficult. While cosmetics are mainly "meant to cover and beautify" [141, p. 155] pharmaceutical drugs are concentrating on the healing or changing of certain skin conditions (or other body parts relevant to the personal care industry) [cf. 141, p. 156]. Cosmeceuticals are positioned in between both product classes, but the borders are admittedly rather fuzzy. As a result, Krohn et al. *define* Cosmeceuticals as "topically applied cosmetic ingredients with claimed target-related bioactivity that provide functional and/or health benefits beyond the basic cosmetic function". [142, p. 1148]. There is a plethora of cosmeceutical ingredients, for instance, gingko (antioxidant activity for a smoother skin), panthenol (moisturizing and soothing irritations), boswellia (antiinflammatory and 'anti-aging'), and β-carotene [cf. 141, p. 157].[95]

It is reported to be the fastest growing segment of the 'natural personal care industry' with cosmetic-pharmaceutical hybrids aiming at enhancing both 'beauty' and health [cf. 141, p. 155]. While health benefits appear to be more disputed than for most products in the NFF sector,[96] some experts are calling for a harmonized

[93] (EC) 1924/2006, Article 3.

[94] cf. (EC) 1924/2006, Article 2(2.).

[95] Dureja et al. also provide a brief overview of common cosmeceutical contents. A further overview of types of cosmeceutical agents can be found in, e.g., [143, 144].

[96] See, e.g., Crompton 139 who also cites Nicholas Perricone, founder of NV Perricone MD Cosmeceuticals, opposing accusations of Cosmeceuticals' low credibility: "Cosmeceuticals are

regulation through bodies like the FDA [cf. 141, p. 158]. The broader the health claims marketed with the Cosmeceuticals, the more likely they would have to undergo scrutiny by drug administration bodies, leading to a potential convergence of regulations and standards in the future. Until now, they are not subject to mandatory approval by the FDA or comparable authorities in most countries, especially as they have not been acknowledged as a product class in their own right (in difference to NFF in many countries). While in Europe, for instance, cosmetic ingredients (and products) need to comply with Council Directive 76/768/EEC "on the approximation of the laws of the Member States relating to cosmetic products" and the subsequent amendments and corrections, this regulation mainly deals with forbidden substances, labeling, and animal testing. [cf. 145, pp. 1ff].[97] In consequence, reliability of 'health claims' regarding Cosmeceuticals are mostly much more questioned by consumers than in the case of NFF [cf. 141, p. 158]. While this may be partly rooted in the lack of regulation in the market, it will also be supported by the way most companies promote their products ("71 % of women confirmed visible results after regular use*. *Test with 184 women"[98]). While consumers are increasingly looking for new and clinically proven cosmetics, [cf. 147, p. 507] they commonly suffer from myths and misconceptions—particularly in respect to the safety and validity of claims [cf. 140, pp. 446ff.].

In many cases *claims* are based on small sample sizes, very short testing periods, and hardly quantifiable output measures (e.g., "users see/feel better/more beautiful"). Therefore, despite the nonapproval of cosmetic products, the FD&C Act is prohibiting the marketing of adulterated or misbranded cosmetics in the USA [cf. 148]. Promotion of a product with therapeutic claims may lead to the FDA considering this product as a drug and thus applying the appropriate regulatory demands, [cf. 149] especially since the FD&C Act does not recognize the term 'Cosmeceutical' (despite FDA's acknowledgment of industry's use of this term for "cosmetic products with medicinal or drug-like benefits") [cf. 150, 151]. Differentiation in the FD&C Act between a drug and a cosmetic is hence based on its intended use, whether it is marketed to 'alter consumer's appearance' (cosmetic) or to 'diagnose, cure, mitigate, treat, or prevent a disease' (drug).[99] As such, cosmetics are produced to change (mainly skin) properties for the duration of the application, while drugs aim at altering conditions more sustainably and for prolonged times. Hence, Draelos criticizes the fact that products in Cosmeceuticals are to date still classified on the basis of their claims. According to her line of reasoning a product 'eliminating wrinkles' would be classified as a *drug*, whilst the

(Footnote 96 continued)
science-based rather than marketing-driven, giving the customer results and value for their investment".

[97] This document constitutes the consolidated version of the original Council Directive 76/768/EEC and all its amendments and corrections up to mid2008 See also [142, p. 1148].

[98] Ref. [146]. While this advertisement is for one specific product, most other advertisements will sound similar, irrespective of company or country.

[99] [cf. 151]. For a more thorough description of the regulation system in the USA see also [140].

same product claiming to 'minimize the appearance of wrinkles' would be classified as a *cosmetic*. Naturally, this claim could not be picked completely arbitrarily, but Draelos argues that the current system of regulation does hamper the development and sound efficacy testing of more powerful Cosmeceuticals [cf. 152, pp. 433f.].

Notwithstanding the aforementioned frequent points of criticism, the global cosmetic and toiletry **market** (retail value of sold products) is estimated to be nearly €250 billion in 2008 by Euromonitor, with the US and Japanese sales slightly decreasing and emerging markets (e.g., Brazil and China) experiencing double-digit growth rates [cf. 153].[100] While Cosmeceuticals are still representing only a smaller share of the global market, it is decisively larger than in NFF. Depending on the definition of Cosmeceuticals, their market value in the USA alone has been valued at US$ 2.8 billion in 2001 [cf. 154, p. 1270][101]. On the basis of global cosmetic, toiletry and fragrance sales of US$ 124 billion in 2001 and under the assumption that the US market share was about one-third at that time, the Cosmeceuticals segment would represent about 7 % of all personal care sales [cf. 154, pp. 1264ff.]. Kumar identifies them as one of four major technology and innovation trends [cf. 154, p. 1270]. However, in the light of distinctively lower regulatory hurdles one could have expected an even larger share—especially when considering Morganti's appraisement: "We are living in the era of cosmeceuticals and nutraceuticals!" [156, p. 320].

4.3.6 Research Questions: The Areas of NFF and Cosmeceuticals

Based on the discussion of NFF and Cosmeceuticals as settings of industry convergence, four 'practical' research questions are derived in addition to the three 'theoretical' research questions. As can be seen especially from the more in-depth example of phytosterols, many indications for a blurring of boundaries between pharmaceutical drugs, Nutraceuticals and FF exist on the product or market side. Given the high research intensity necessary for a development of such products, one could expect an underlying convergence in science and technology. However, only very few signs of potential industry convergence are described—as would be identifiable in collaborations, alliances or M&A activities. Consequently, the fourth research question is derived:

Q.4: Are the food and the pharmaceutical industries converging in the area of NFF?

---

[100] For an overview of the personal care industry and the most important companies see, e.g., [154].

[101] A different estimate by Kline is US$ 57 billion globally in 2008, with Europe, the USA and Japan accounting for 35, 18 and 11 %, respectively [cf. 155].

A similar—although weaker—pattern can be found in the field of Cosmeceuticals. In some cases, like with phytosterols, both fields even employ similar or identical ingredients. Accordingly, it should be questioned whether there are analog developments between the personal care and the pharmaceutical industries:

Q.5: Are the personal care and the pharmaceutical industries converging in the area of Cosmeceuticals, perhaps even into one broad NFF/Cosmeceuticals area?

When discussing market activities as well as regulatory issues, especially in the three main regions of Japan, the USA and Europe, it becomes obvious that different patterns evolve in different countries and regions. However, in the light of mostly large and global players in the pharmaceutical and the food industries, developments in one region will mostly cause similar developments in the other regions as well. Hence, it should be expected that analyses in different countries would yield similar but most likely time-delayed (or accelerated) results. As this study is proposing an approach for an early anticipation of convergence also for practitioners' usage, it would be preferable if investigators could use data from countries they are most familiar with. As a result, the analyses section will use and compare patent data from Europe, the USA and Japan to answer the sixth research question:

Q.6: Does the analysis of activities in different countries lead to different results?

Against the backgrounds of NFF and Cosmeceuticals, this concept would ideally be generic enough to be employed not only in a setting based largely on the biological, chemical, and life sciences but also in other research-intensive fields, such as the most widely analyzed ICT industry. While some of the analyses will be based on keyword searches, others will be conducted keyword independently. Accordingly, the applicability of the concept in different areas would per se depend on database coverage and the existence of appropriate keywords and/or classification schemes. Even if the concept *can* be applied across industry and scientific discipline boundaries, it is not clear whether patterns of potential convergence would look the same or at least similar. If this was not the case, it would be considerably more difficult to use the developed approach as an early warning concept across whole industries and sectors. Thus, the final research question is as follows:

Q.7: Are patterns of convergence generalizable across industry boundaries?

References

1. Palmer, D.: Functional foods facing an uncertain future. Australian Food News (2009)
2. Bröring, S.: The Front End of Innovation in Converging Industries: The Case of Nutraceuticals and Functional Foods. DUV, Wiesbaden, Germany (2005)

3. Bröring, S., Cloutier, L.M., Leker, J.: The front end of innovation in an era of industry convergence: evidence from nutraceuticals and functional foods. R&D Manage. **36**(5), 487–498 (2006)
4. Stieglitz, N.: Strategie und Wettbewerb in konvergierenden Märkten. Deutscher Universitäts, Wiesbaden, Germany (2004)
5. Nothnagel, K.: Empirical Research within Resource-Based Theory: A Meta-Analysis of the Central Propositions. Strategisches Kompetenz-Management. Gabler GWV Fachverlage, Wiesbaden, Germany (2008)
6. Acedo, F.J., Barroso, C., Galan, J.L.: The resource-based theory: dissemination and main trends. Strateg. Manag. J. **27**(7), 621–636 (2006)
7. DeCarolis, D.M., Deeds, D.L.: The impact of stocks and flows of organizational knowledge on firm performance: an empirical investigation of the biotechnology industry. Strateg. Manag. J. **20**(10), 953–968 (1999)
8. Kraaijenbrink, J., Spender, J.C., Groen, A.J.: The resource-based view: a review and assessment of its critiques. J. Manage. **36**(1), 349–372 (2010). doi:10.1177/0149206309350775
9. Godfrey, P.C., Hill, C.W.L.: The problem of unobservable in strategic management research. Strateg. Manag. J. **16**(7), 519–533 (1995)
10. Crook, T.R., Ketchen Jr, D.J., Combs, J.G., Todd, S.Y.: Strategic resources and performance: a meta-analysis. Strateg. Manag. J. **29**(11), 1141–1154 (2008)
11. Selznick, P.: Leadership in Administration: A Sociological Interpretation. University of California Press, Berkeley and Los Angeles (CA), USA (1957)
12. Penrose, E.T.: The Theory of the Growth of the Firm. Oxford University Press, Oxford (1959)
13. Wernerfelt, B.: A resource-based view of the firm. Strateg. Manag. J. **5**(2), 171–180 (1984)
14. Peteraf, M.A.: The cornerstones of competitive advantage: a resource-based view. Strateg. Manag. J. **14**(3), 179–192 (1993)
15. Teece, D.J., Pisano, G., Shuen, A.: Dynamic capabilities and strategic management. Strateg. Manag. J. **18**(7), 509–533 (1997)
16. Silverman, B.S.: Technological resources and the direction of corporate diversification: toward an integration of the resource-based view and transaction cost economics. Manage. Sci. **45**(8), 1109–1124 (1999)
17. zu Knyphausen-Aufseß, D.: Auf dem Weg zu einem ressourcenorientierten Paradigma? Resource Dependence-Theorie der Organisation und Resource-based View des Strategischen Managements im Vergleich. In: Ortmann, G., Sydow, J., Türk, K. (eds.) Theorien der Organisation: Die Rückkehr der Gesellschaft. pp. 452–480. Westdeutscher, Wiesbaden, Germany (2000)
18. Herzog, P.: Open and Closed Innovation - Different Cultures for Different Strategies. Gabler, Wiesbaden (2008)
19. Barney, J.: Firm resources and sustained competitive advantage. J. Manage. **17**(1), 99–120 (1991). doi:10.1177/014920639101700108
20. Yeoh, P.-L., Roth, K.: An empirical analysis of sustained advantage in the U.S. pharmaceutical industry: impact of firm resources and capabilities. Strateg. Manag. J. **20**(7), 637–653 (1999)
21. Liebermann, M.B., Montgomery, D.B.: First mover (dis)advantages: retrospective and link with the resource-based view. Strateg. Manag. J. **19**(12), 1111–1125 (1998)
22. Afuah, A.: Innovation management: Strategies, implementation and profits. Oxford University Press, Oxford, UK (2003)
23. Amit, R., Schoemaker, P.J.H.: Strategic assets and organizational rent. Strateg. Manag. J. **14**(1), 33–46 (1993)
24. Henderson, R., Cockburn, I.: Measuring Competence? Exploring Firm Effects in Pharmaceutical Research. Strategic Management Journal **15**(S1), 63-84 (1994)

25. Nothnagel, K.: Empirical Research Within Resource-Based Theory: Methodological Challenges and a Meta-Analysis of the Central Propositions. Freie Universität, Berlin (2007)
26. Amir-Aslani, A., Mangematin, V.: The future of drug discovery and development: Shifting emphasis towards personalized medicine. Technol. Forecast. Soc. Chang. **77**(2), 203–217 (2010)
27. Helfat, C.E.: Know-how and asset complementarity and dynamic capability accumulation: the case of R&D. Strateg. Manag. J. **18**(5), 339–360 (1997)
28. Pisano, G.P.: knowledge, integration, and the locus of learning: an empirical analysis of process development. Strateg. Manag. J. **15**(S1), 85–100 (1994)
29. DeCarolis, D.M.: Competencies and imitability in the pharmaceutical industry: an analysis of their relationship with firm performance. J. Manag. **29**(1), 27–50 (2003). doi:10.1177/014920630302900103
30. Hohensee, J.: Der erste Ölpreisschock 1973/74: die politischen und gesellschaftlichen Auswirkungen der arabischen Erdölpolitik auf die Bundesrepublik Deutschland und Westeuropa. Historische Mitteilungen, Beihefte 17. Franz Steiner, Stuttgart, Germany (1996)
31. Merrill, K.R.: The Oil Crisis of 1973–1974: A Brief History with Documents, 1st edn. Bedford/St. Martin's, New York, USA (2007)
32. Ansoff, H.I.: Managing strategic surprise by response to weak signals. California Management Review **18**(2), 21–33 (1975)
33. Ansoff, H.I.: Strategic issue management. Strateg. Manag. J. **1**(2), 131–148 (1980)
34. Hasler, C.M.: The changing face of functional foods. J. Am. Coll. Nutr. **19**(5), 499S–506S (2000)
35. United Nations Development Programme: Human Development Report 2009—Overcoming barriers: Human mobility and development. New York, USA (2009)
36. Chadwick, R., Henson, S., Moseley, B., Koenen, G., Liakopoulos, M., Midden, C., Palou, A., Rechkemmer, G., Schröder, D., von Wright, A.: Functional Foods. Wissenschaftsethik und Technikfolgenbeurteilung, vol. 20. Springer, Berlin/Heidelberg, Germany (2003)
37. Wildman, R.E.C., Kelley, M.: Nutraceuticals and functional foods. In: Wildman, R.E.C. (ed.) Handbook of Nutraceuticals and Functional Foods, pp. 1–21. CRC Press, Boca Raton (FL), USA (2007)
38. Newsholme, E., Leech, T.: Functional Biochemistry in Health and Disease. Wiley, Chichester, UK (2010)
39. Anonymous: Is functional functioning? Dairy Ind. Int. **72**(11), 17 (2007)
40. Robertfroid, M.B.: Defining functional foods. In: Gibson, G.R., Williams, C.M. (eds.) Functional foods: Concept to product, pp. 9–27. Woodhead Publishing, Cambridge, UK (2000)
41. Kiefer, I., Burger, P., Blass, M., Berghofer, E., Hoppichler, F.: Functional Food—Lebensmittel mit Zusatznutzen? Journal für Ernährungsmedizin **4**(2), 10–15 (2002)
42. Anonymous: scientific concepts of functional foods in europe consensus document. Brit. J. Nutr. **81**(4), S1–S27 (1999). doi:10.1017/S0007114599000471
43. Health Canada: Policy Paper of the Therapeutic Products Programme and the Food Directorate from the Health Protection Branch: Nutraceuticals/Functional Foods and Health Claims on Foods. In., pp. 1-29. Ottawa (ON), Canada, (1998)
44. EUFIC.: Scientific Substantiation: A key ingredient for functional foods and health claims. http://www.eufic.org/article/en/artid/scientific-substantiation-functional-foods-health-claims/ (2003). Accessed 11 Nov 2010
45. Thompson, A.K., Moughan, P.J.: Innovation in the foods industry: functional foods. Innovation: Manag., Policy Pract. **10**(1), 61–73 (2008)
46. Flay, B.R.: Efficacy and effectiveness trials (and other phases of research) in the development of health promotion programs. Prev. Med. **15**(5), 451–474 (1986)

47. Glasgow, R.E., Lichtenstein, E., Marcus, A.C.: Why don't we see more translation of health promotion research to practice? Rethinking the efficacy-to-effectiveness transition. Am. J. Public Health **93**(8), 1261–1267 (2003). doi:10.2105/ajph.93.8.1261
48. Clydesdale, F.M.: Science, education, and technology: new frontiers for health. Crit. Rev. Food Sci. Nutr. **38**(5), 397–419 (1998)
49. Arai, S.: Global view on functional foods: Asian perspectives. Br. J. Nutr. **88**(Supplement 2), S139–S143 (2002)
50. Sanders, M.E.: Overview of functional foods: emphasis on probiotic bacteria. Int. Dairy J. **8**(5), 341–347 (1998)
51. Heasman, M., Mellentin, J.: The Functional Foods Revolution: Healthy people, healthy profits?. Earthscan Publications, London, UK (2001)
52. Prado, F.C., Parada, J.L., Pandey, A., Soccol, C.R.: Trends in non-dairy probiotic beverages. Food Res. Int. **41**(2), 111–123 (2008). doi:10.1016/j.foodres.2007.10.010
53. Wildman, R.E.C. (ed.): Handbook of Nutraceuticals and Functional Foods, 2nd edn. CRC Press, Boca Raton (FL), USA (2007)
54. Dutta, P.C. (ed.): Phytosterols as Functional Food Components and Nutraceuticals. Marcel Dekker, New York, USA (2004)
55. International Food Information Council Foundation: Functional Foods. In., pp. 1–4. Washington DC, USA. http://www.foodinsight.org/Content/6/functionalfoodsbackgrounder.pdf (2007). Accessed 10 Apr 2010
56. Moreau, R.A., Whitaker, B.D., Hicks, K.B.: Phytosterols, phytostanols, and their conjugates in foods: structural diversity, quantitative analysis, and health-promoting uses. Prog. Lipid Res. **41**(6), 457–500 (2002)
57. Piironen, V., Lindsay, D.G., Miettinen, T.A., Toivo, J., Lampi, A.-M.: Plant sterols: biosynthesis, biological function and their importance to human nutrition. J. Sci. Food Agric. **80**(7), 939–966 (2000)
58. Demel, R.A., De Kruyff, B.: The function of sterols in membranes. Biochimica et Biophysica Acta (BBA)—Reviews on. Biomembranes **457**(2), 109–132 (1976)
59. Hartmann, M.-A., Benveniste, P.: Plant membrane sterols: Isolation, identification, and biosynthesis. In: Methods in Enzymology, (vol. 148, pp. 632–650). Academic Press, New York (1987)
60. McKersie, B.D., Thompson, J.E.: Influence of plant sterols on the phase properties of phospholipid bilayers. Plant Physiol. **63**(5), 802–805 (1979). doi:10.1104/pp.63.5.802
61. Haughan, P.A., Lenton, J.R., Goad, L.J.: Sterol requirements and paclobutrazol inhibition of a celery cell culture. Phytochemistry **27**(8), 2491–2500 (1988)
62. Brufau, G., Canela, M.A., Rafecas, M.: Phytosterols: physiologic and metabolic aspects related to cholesterol-lowering properties. Nutr. Res. **28**(4), 217–225 (2008)
63. Calpe-Berdiel, L., Escolà-Gil, J.C., Blanco-Vaca, F.: New insights into the molecular actions of plant sterols and stanols in cholesterol metabolism. Atherosclerosis **203**(1), 18–31 (2009)
64. European Food Safety Authority (efsa).: Consumption of Food and Beverages with Added Plant Sterols in the European Union. Efsa J. (2008) **133**, 1–21. Parma, Italy (2008)
65. Ostlund, R.E.J.: Phytosterols in human nutrition. Annu. Rev. Nutr. **22**, 533–549 (2002)
66. Cantrill, R.: Phytosterols, phytostanols and their esters (Chemical and Technical Assessment). In., pp. 1–13. Joint FAO/WHO Expert Committee on Food Additives (JEFCA) (2008)
67. Poli, A., Marangoni, F., Paoletti, R., Mannarino, E., Lupattelli, G., Notarbartolo, A., Aureli, P., Bernini, F., Cicero, A., Gaddi, A., Catapano, A., Cricelli, C., Gattone, M., Marrocco, W., Porrini, M., Stella, R., Vanotti, A., Volpe, M., Volpe, R., Cannella, C., Pinto, A., Del Toma, E., La Vecchia, C., Tavani, A., Manzato, E., Riccardi, G., Sirtori, C., Zambon, A.: Non-pharmacological control of plasma cholesterol levels. Nutr., Metab. Cardiovas. Dis. **18**(2), S1–S16 (2008)

68. Berg, J.M., Tymoczko, J.L., Stryer, L.: Biochemie, 5th edn. Spektrum Akademischer, Berlin, Germany (2003)
69. Anonymous: Cholesterol buster. research*eu **57**(July 2008), 21 (2008)
70. Moruisi, K.G., Oosthuizen, W., Opperman, A.M.: Phytosterols/stanols lower cholesterol concentrations in familial hypercholesterolemic subjects: a systematic review with meta-analysis. J. Am. Coll. Nutr. **25**(1), 41–48 (2006)
71. Pfizer Inc.: 2009 Financial Report. New York, USA (2010)
72. Peterson, D.W.: Effect of soybean sterols in the diet on plasma and liver cholesterol in chicks. Proc. Soc. Exp. Biol. Med. **78**(1), 143–147 (1951)
73. Pollak, O.J.: Successful prevention of experimental hypercholesteremia and cholesterol atherosclerosis in the rabbit. Circulation **7**(5), 696–701 (1953)
74. Pollak, O.J.: Reduction of blood cholesterol in man. Circulation **7**(5), 702–706 (1953)
75. Jones, P.J.H.: Ingestion of phytosterols is not potentially hazardous. J. Nutr. **137**(11), 2485 (2007)
76. Miettinen, T.A., Puska, P., Gylling, H., Vanhanen, H., Vartiainen, E.: Reduction of serum cholesterol with sitostanol-ester margarine in a mildly hypercholesterolemic population. N. Engl. J. Med. **333**(20), 1308–1312 (1995). doi:10.1056/nejm199511163332002
77. Anonymous: Food labeling: health claims; plant sterol/stanol esters and coronary heart disease; Interim Final Rule. In: Department of Health and Human Services: Food and Drug Administration (ed.), vols. 65, 175, pp. 54685–54739. Office of the Federal Register, Washington (DC), USA, (2000)
78. Plat, J., Kerckhoffs, D.A.J.M., Mensink, R.P.: Therapeutic potential of plant sterols and stanols. Curr. Opin. Lipidol. **11**(6), 571–576 (2000)
79. Schön, H.: Sterol-balance experiments in humans. Nature **184**(4702), 1872–1873 (1959)
80. Law, M.R., Wald, N.J., Thompson, S.G.: By how much and how quickly does reduction in serum cholesterol concentration lower risk of ischaemic heart disease? Br. Med. J. **308**(6925), 367–372 (1994)
81. Chen, S., Judd, J., Kramer, M., Meijer, G., Clevidence, B., Baer, D.: Phytosterol intake and dietary fat reduction are independent and additive in their ability to reduce plasma LDL cholesterol. Lipids **44**(3), 273–281 (2009)
82. National Cancer Institute: Phytosterol.: http://www.cancer.gov/Templates/db_alpha.aspx?CdrID=44485 (2008). Accessed 18 April 2008
83. Woyengo, T.A., Ramprasath, V.R., Jones, P.J.H.: Anticancer effects of phytosterols. Eur. J. Clin. Nutr. **63**, 813–820 (2009)
84. Jones, P.J.H.a., AbuMweis, S.S.b.: Phytosterols as functional food ingredients: linkages to cardiovascular disease and cancer. Curr. Opin. Clin. Nutr. Metab. Care **12**(2), 147–151 (2009)
85. Lütjohann, D., von Bergmann, K.: Phytosterolaemia: diagnosis, characterization and therapeutical approaches. Ann. Med. **29**(3), 181–184 (1997)
86. Patel, M.D., Thompson, P.D.: Phytosterols and vascular disease. Atherosclerosis **186**(1), 12–19 (2006)
87. Sen, C.K., Khanna, S., Roy, S.: Tocotrienols in health and disease: the other half of the natural vitamin E family. Mol. Aspects Med. **28**(5–6), 692–728 (2007)
88. Traber, M.G., Atkinson, J.: Vitamin E, antioxidant and nothing more. Free Radical Biol. Med. **43**(1), 4–15 (2007)
89. US National Library of Medicine: Antioxidants.: http://www.nlm.nih.gov/medlineplus/antioxidants.html (2010). Accessed 26.05.2010
90. Neil, H.A.W., Meijer, G.W., Roe, L.S.: Randomised controlled trial of use by hypercholesterolaemic patients of a vegetable oil sterol-enriched fat spread. Atherosclerosis **156**(2), 329–337 (2001)
91. Holden, J.M., Eldridge, A.L., Beecher, G.R., Marilyn Buzzard, I., Bhagwat, S., Davis, C.S., Douglass, L.W., Gebhardt, S., Haytowitz, D., Schakel, S.: Carotenoid Content of U.S. foods: an update of the database. J. Food Compos. Anal. 12(3), 169–196 (1999)

92. Fernandes, P., Cabral, J.M.S.: Phytosterols: applications and recovery methods. Bioresour. Technol. **98**(12), 2335–2350 (2007)
93. Thurnham, D.D.I.: Functional foods: cholesterol-lowering benefits of plant sterols ? reply by Thurnham. Br. J. Nutr. **84**(02), 254 (2000). doi:10.1017/S0007114500001501
94. Thurnham, D.I.: Functional foods: cholesterol-lowering benefits of plant sterols. Br. J. Nutr. **82**(04), 255–256 (1999). doi:10.1017/S0007114599001440
95. Natura Vitalis.: Phytosterol—Kapseln u.a. mit Sägepalmextrakt. http://www.naturavitalis.de/index.php?page=shop.product_details&flypage=flypage.tpl&product_id=76&category_id=25&option=com_virtuemart&Itemid=2&gclid=CKOYs_W99KECFZaNzAodiCxoFA (2010). Accessed 31 May 2010
96. Fairvital.: Beta-Sitosterin 120 mg. http://www.fairvital.com/product_info.php?ref=6&products_id=56&gclid=CIPXpPS99KECFYWUzAodP1asEg (2009). Accessed 31 May 2010
97. Rote Liste Service: Rote Liste Online: Arzneimittelinformationen für Deutschland. http://www.rote-liste.de/Online/jumpsearch (2010). Accessed 31 May 2010
98. Starling, S.: Benecol smoothie courts younger cholesterol-lowering market. Food Drink Europe (2009)
99. Starling, S.: Sterol cookie launched in US. Functional Foods Nutraceuticals (2006)
100. Niemann, B., Sommerfeld, C., Hembeck, A., Bergmann, C.: Plant sterol enriched foods as perceived by consumers. In: BfR Wissenschaft. Federal Institute for Risk Assessment, Berlin, Germany, (2007)
101. McNeil Nutritionals.: Benecol—Proven to Reduce Cholesterol. http://www.benecolusa.com/index.jhtml (2010). Accessed 31 May 2010
102. Anonymous.: COMMISSION REGULATION (EC) No 983/2009 of 21 October 2009 on the authorisation and refusal of authorisation of certain health claims made on food and referring to the reduction of disease risk and to children's development and health. In: The Commission of the European Communities (ed.), Commission Regulation (EC) No. 983/2009. pp. L 277/273–L 277/212 (2009)
103. Stones, M., Starling, S.: Plant stanol and sterol claims now law in EU. NutraIngredients (2009)
104. AbuMweis, S.S., Barake, R., Jones, P.J.H.: Plant sterols/stanols as cholesterol lowering agents: a meta-analysis of randomized controlled trials. Food Nutr. Res. (2008). doi:10.3402/fnr.v52i0.1811
105. Starling, S.: Targeted and talkative: the Benecol success story. NutraIngredients (2009)
106. Ruiu, G., Pinach, S., Veglia, F., Gambino, R., Marena, S., Uberti, B., Alemanno, N., Burt, D., Pagano, G., Cassader, M.: Phytosterol-enriched yogurt increases LDL affinity and reduces CD36 expression in polygenic hypercholesterolemia. Lipids **44**(2), 153–160 (2009)
107. Ichiji, Y.: Cosmetics containing sulfated phytosterols. Japan Patent JP 2004250402, 20040909
108. Wachter, R., Tesmann, H., Behler, A., Maurer, K.-H.: Deodorizing Preparations. WO 98/17241
109. Meyer, H.: Cosmetic preparation for topical application, used for treating cellulite. Switzerland Patent WO 01/87257 A1
110. Eisfeld, W., Busch, P., Issberner, U., Biehl, P.: Haarpflegemittel mit natürlichen Ölen. Deutschland Patent DE 10126448 A1
111. Wachter, R., Salka, B., Magnet, A.: Phytosterole—pflanzliche Wirkstoffe in der Kosmetik. Parfümerie und Kosmetik **75**(11), 755–761 (1994)
112. European Food Safety Authority.: consumption of food and beverages with added plant sterols in the European Union. EFSA J. (2008) Nutraceuticals, pp. 1–21. Parma, Italy (2008)
113. U.S. Food and Drug Administration.: GRAS notice inventory: GRN No. 250—plant sterols and stanols from pine trees (2008)
114. Daniells, S.: Snack Size Science: Taking heart from phytosterol review. NutraIngredients (2009)
115. Daniells, S.: Phytosterols for cholesterol cuts supported by review. NutraIngredients (2009)

116. de Guzman, D.: Sterol investments on the rise. CMR ICIS publication 11–17 April 2005, **24** (2005)
117. Cognis: The future of healthy concepts. cognis connect—nutrition and health (2004)
118. Starling, S.: UK sterol foods market slows amid consumer confusion. NutraIngredients (2008)
119. Zeller, C.: Globalisierungsstrategien—Der Weg von Novartis. Springer, Berlin, Germany (2001)
120. Schaffnit-Chatterjee, C.: The global food equation: Food security in an environment of increasing scarcity. In: Schneider, S. (ed.) Current Issues, pp. 1–38. Deutsche Bank Research (2009)
121. Anonymous: Novartis, PepsiCo End Functional Foods Venture. http://www.allbusiness.com/retail-trade/food-stores/4245788-1.html (2002). Accessed 01 June 2010
122. Anonymous.: Novartis nimmt "Functional Food" vom Markt. Handelsblatt (2001)
123. Novartis: Novartis (1996–Gegenwart). http://www.novartis.ch/about-novartis/company-history/novartis-chronicle.shtml (2010). Accessed 01 June 2010
124. Runestad, T.: Cholesterol leads cardio category. Functional Ingredients (2009)
125. Ahlberg, P.: Phytosterol-based foods: Still seeking a foothold. http://www.just-food.com/analysis/still-seeking-a-foothold_id93676.aspx (2001). Accessed 30 June 2010
126. Starling, S.: Markets: functional chocolate finding its feet. Confectionery News (2009)
127. Mars: FAQs. https://cirkuhealth.com/FAQs.aspx (2010). Accessed 01 June 2010
128. Lea, L.J., Hepburn, P.A.: Safety evaluation of phytosterol-esters. Part 9: results of a European post-launch monitoring programme. Food Chem. Toxicol. **44**(8), 1213–1222 (2006)
129. Hilliam, M.: Functional food—how big is the market? World Food Ingredients **12**, 50–52 (2000)
130. Kotler, P., Armstrong, G., Wong, V., Saunders, J.: Principles of marketing, 5th European ed. Pearson Education, Harlow, UK (2008)
131. Siró, I., Kápolna, E., Kápolna, B., Lugasi, A.: Functional food. Product development, marketing and consumer acceptance—a review. Appetite **51**(3), 456–467 (2008)
132. Mark-Herbert, C.: Innovation of a new product category—functional foods. Technovation **24**(9), 713–719 (2004)
133. Bagchi, D. (ed.): Nutraceutical and functional food regulations in the United States and Around the World. Academic Press, San Diego (CA), USA (2008)
134. Boin, G., Conte-Salinas, N., Grube, M., Rusconi, G. (eds.): Functional Food: a synopsis of the legal framework in the European Union and selected other countries. Book on Demand, Norderstedt, Germany (2009)
135. Codex Alimentarius Commission.: Codex Alimentarius. http://www.codexalimentarius.net/web/index_en.jsp (2010). Accessed 03 June 2010
136. Ohama, H., Ikeda, H., Moriyama, H.: Health foods and foods with health claims in Japan. In: Bagchi, D. (ed.) Nutraceuticals and functional food regulations in the United States and Around the World, pp. 249–280. Academic Press, San Diego (CA), USA (2008)
137. European Commission.: General Food Law—Introduction. http://ec.europa.eu/food/food/foodlaw/index_en.htm (2010). Accessed 03 June 2010
138. European Commission: Food Supplements. http://ec.europa.eu/food/food/labellingnutrition/supplements/index_en.htm (2010). Accessed 03 June 2010
139. Crompton, S.: Hype or beautiful science? Times Online. http://women.timesonline.co.uk/tol/life_and_style/women/body_and_soul/article2645139.ece?print=yes&randnum=1208519464093 (2007). Accessed 18 Apr 2008
140. Newburger, A.E.: Cosmeceuticals: myths and misconceptions. Clin. Dermatol. **27**(5), 446–452 (2009)
141. Dureja, H., Kaushik, D., Gupta, M., Kumar, V., Lather, V.: Cosmeceuticals: an emerging concept. Indian J. Pharmacol. **37**(3), 155–159 (2005)

142. Krohn, M., Kleber, A., Schaffar, G., Dechert, U., Eck, J.: Now we are talking sense! Functional approaches to novel nutraceuticals and cosmeceuticals. Biotechnol. J. **3**(9–10), 1147–1156 (2008)
143. Bissett, D.L.: Common cosmeceuticals. Clin. Dermatol. **27**(5), 435–445 (2009)
144. Amer, M., Maged, M.: Cosmeceuticals versus pharmaceuticals. Clin. Dermatol. **27**(5), 428–430 (2009)
145. Council of the European Communities.: COUNCIL DIRECTIVE of 27 July 1976 on the approximation of the laws of the member states relating to cosmetic products, vol. 76/768/EEC (2008)
146. Beiersdorf: 2010 good-bye cellulite, hello bikini challenge. http://www.niveausa.com/highlights/local_highlight/local_gbc2010challenge (2010). Accessed 03 June 2010
147. Paul, L.C.: Cosmeceuticals and patents: the physician and dentist brands. Clin. Dermatol. **27**(5), 507–512 (2009)
148. U.S. Food and Drug Administration: FDA Authority Over Cosmetics. http://www.fda.gov/Cosmetics/GuidanceComplianceRegulatoryInformation/ucm074162.htm (2005). Accessed 03 June 2010
149. U.S. Food and Drug Administration: Cosmetic Labeling & Label Claims. http://www.fda.gov/Cosmetics/CosmeticLabelingLabelClaims/default.htm (2006). Accessed 03 June 2010
150. U.S. Food and Drug Administration: "Cosmeceutical". http://www.fda.gov/Cosmetics/ProductandIngredientSafety/ProductInformation/ucm127064.htm (2000). Accessed 03 June 2010
151. U.S. Food and Drug Administration: Is It a Cosmetic, a Drug, or Both? (Or Is It Soap?). http://www.fda.gov/Cosmetics/GuidanceComplianceRegulatoryInformation/ucm074201.htm (2002). Accessed 03 June 2010
152. Draelos, Z.D.: Cosmeceuticals: undefined, unclassified, and unregulated. Clin. Dermatol. **27**(5), 431–434 (2009)
153. de Guzman, D.: Global economy to limit cosmetic sales in 2009—Euromonitor. http://www.icis.com/Articles/2009/04/03/9206067/global-economy-to-limit-cosmetic-sales-in-2009-euromonitor.html (2009). Accessed 03 June 2010
154. Kumar, S.: Exploratory analysis of global cosmetic industry: major players, technology and market trends. Technovation **25**(11), 1263–1272 (2005)
155. Matthews, I.: The Growing Trend for Cosmeceuticals. http://www.in-cosmetics.com/page.cfm/Link=445/t=m (2010). Accessed 03 June 2010
156. Morganti, P.: Reflections on cosmetics, cosmeceuticals, and nutraceuticals. Clin. Dermatol. **26**(4), 318–320 (2008)

Chapter 5
Analysis and Results

This chapter will present the results of analyzes conducted on scientific publications and patent documents in two different databases and with three different software tools. In total six datasets are analyzed, of which two were compiled on the basis of a keyword-search (with the example of phytosterols) and the last four on the basis of a keyword-independent classification search (with examples from NFF and ICT). First, each sample will be briefly introduced with a short description of its originating database and used software. Second, it will be analyzed on the basis of indicators and procedures described in Chap. 3 to answer the questions derived from the theoretical foundation and the practical example of NFF in the preceding chapter. In the course of these analyzes, mapping techniques as well as "one-dimensional" and "two-dimensional" bibliometrics will be employed [cf. 1], cited after [2, p. 12].[1] Third, Chap. 6 will discuss all the results independently and then in context of the complete study.

5.1 Keyword-Based Analysis: Indications of Convergence Around Phytosterols

5.1.1 SciFinder Scholar

5.1.1.1 Searching CAplus and MEDLINE with SciFinder Scholar

SciFinder®[2] is a widely disseminated research discovery tool used by academic[3] and industrial scientists, primarily in the areas of chemistry and biology. It offers

[1] This distinction is attributed to van Raan, who discerns between counts of activity (one-dimensional) and co-occurrence measures (two-dimensional).

[2] Signs and indications of trademarks or registered names will be only mentioned during the first appearance of a new name. For easier reading, those signs will be omitted from the following part.

[3] By academics it is mostly used in the SciFinder Scholar version, which offers slightly reduced functionalities at a reduced rate for university students and faculty members. Within this study the SciFinder Scholar 2007 version was used.

C.-S. Curran, *The Anticipation of Converging Industries*,
DOI: 10.1007/978-1-4471-5170-8_5, © Springer-Verlag London 2013

access to scientific information contained in more than 10,000 currently published journals and to approximately 29 million patent documents from more than 60 patent authorities [cf. 3, 4]. SciFinder employs the two databases *CAplus*^{SM} and *MEDLINE*, with more than 32 and 18 million documents, respectively. MEDLINE, provided by the US National Library on Medicine, focuses on scientific articles from the biomedical and biotechnological field as of 1947. In comparison, CAplus, provided by CAS, is centered on patents [cf. 5][4] and scientific publications from chemistry and related fields, such as biotechnology, materials sciences and physics. Its full coverage starts 1907, but some documents date back to the middle of the nineteenth century. CAplus can also be used to retrieve information on chemical syntheses or substances, over 110 million entries on organic and inorganic substances as well as sequences are included in the database [cf. 6, 7].

CAplus can be searched using SciFinder or other products (such as STN Express® and STN® AnaVist™) and the information is supplemented by additional classification information added by CAS' own experts. Before turning to the generation of the sample, it should be borne in mind that SciFinder has been established in order to provide easy and quick access to information on possible interactions, reactions or processes. It is thus content-driven and not designed to be a tool of (advanced) statistical analysis [cf. 8].[5]

5.1.1.2 Sample

The first set of analyzes was conducted in early 2008 with scientific and patent documents on phytosterols. Accordingly, an initial SciFinder search was performed using *phytosterol*, *plant sterol* and its derivatives as search-terms.[6] This search yielded (after removal of duplicate entries) 7,455 references (N_{SF}),[7] ranging from 1897 to 2008.[8] These consisted of 6,212 scientific publications

[4] In respect to patents, CAS provides a list of IPCs with either guaranteed or selective coverage.

[5] For example, if a set of documents is selected and the number of citations to each of these documents is to be assessed, then this has to be done one by one. Using this selection and then looking for the citing documents will yield all documents citing any from the original sample. But it will not show whether, e.g., one document will cite more than one of the original documents. Thus, any statistical analysis will have to be carefully planned and arranged in order to derive unbiased results.

[6] In SciFinder the function "containing at least one of the concepts" was used, yielding all references that contain at least one of the search-terms.

[7] Technically, a search in SciFinder only yields references (and direct links) to scientific publications or patent documents, but not the original document. They contain broad information on the document like title, author/inventor/assignee, classifications, patent numbers, dates, journal and abstract. However, they do not contain the full document's content or even the document itself. Notwithstanding this decisive difference, the text will refer to these references as if they were the original documents for an increased legibility.

[8] A repetition of the same search at the beginning of 2010 yielded an additional 1,311 references for the time since 2008 alone, highlighting the still growing interest in phytosterols.

(83 %[9]; n_{SF_Pub}) and 1,243 patents (17 %; n_{SF_Pat}). An analysis of the filing organizations for the patents resulted in a list of 654 organizations, including countries, research institutions and companies.

To assess the degree of collaboration between companies from different industries, all non-companies were excluded from the 50 organizations with the highest numbers of patents. An inclusion of their respective subsidiaries led to a final sample of 28 companies with 451 patents ($n_{SF\_Pat\_C}$).[10] As phytosterols are functional ingredients not only used in NFF but also in Cosmeceuticals, the 28 companies were assigned to four industries as follows:

- **Personal care** nine companies selling or conducting R&D on phytosterols applied in Cosmeceuticals.
- **Food** nine companies either active in agricultural processing of food ingredients or selling food and food ingredients under their own brand names.
- **Pharmaceuticals** four companies specialized in developing and marketing pharmaceutical drugs.
- **Chemicals** six companies producing phytosterols and selling it in the B2B market.

After a grouping of the 451 patents according to these four industries, the subject areas of the whole sample were analyzed. Subject areas are assigned content-based by CAS' examiners. The used subject areas were then structured into four disjunctive clusters, with the first three clusters representing the core areas of *personal care*, *food* and *pharmaceuticals* and the last one including all the remaining subject areas.[11]

Accordingly, the scientific publications were also analyzed regarding their authoring organization and their subject cluster. Of the 6,212 scientific publications in n_{SF_Pub}, 4,364 (70 %; $n_{SF\_Pub\_i}$) could be assigned to identifiable subject areas, which represented at least 1 % of all publications.[12] This resulted in 111 ($n_{SF\_Pub\_C}$) publications submitted by one of the 28 sample firms. Figure 5.1 depicts the main steps in the preparation of the SciFinder sample and its subsamples.

---

9 For ease of reading, percentages will be mathematically rounded to whole numbers, wherever appropriate.

10 This sample represents a good 36 % of the initial set of patents.

11 There was intentionally no dedicated *chemicals* cluster arranged, as chemical topics cannot be sensibly separated from the three other industries. A lot of the chemical subject areas may as well be a core area of the chemical industry as of, e.g., the pharmaceutical industry. Therefore, all potentially *chemicals'* related subject areas were also assigned to the *Rest* cluster.

12 That is, each subject area that represented less than 1 % of all documents was omitted for handling reasons.

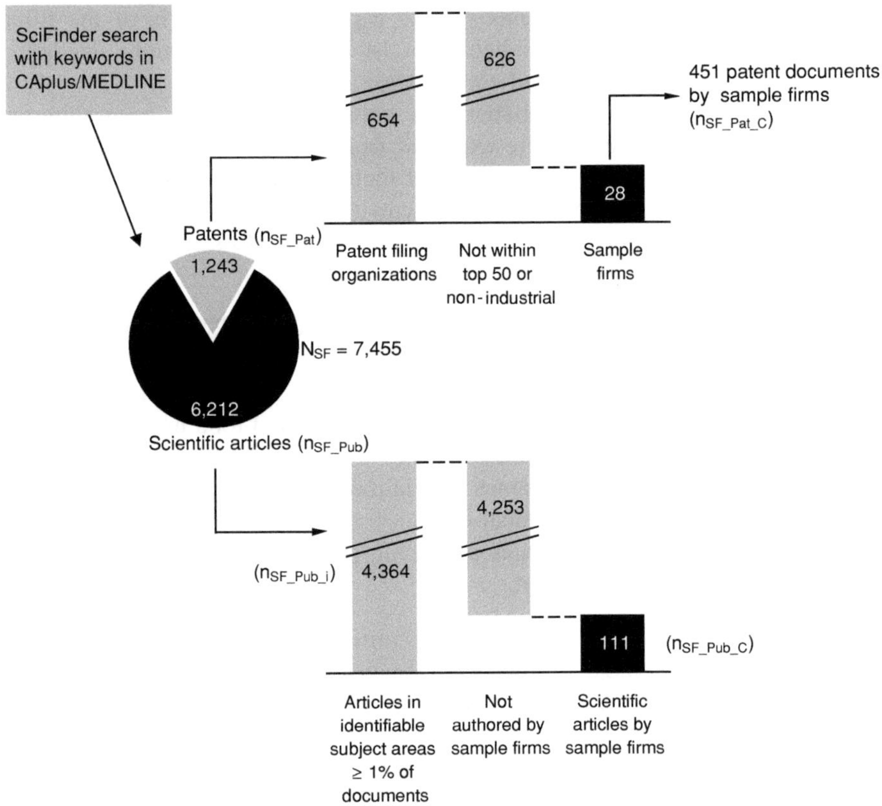

Fig. 5.1 Construction of SciFinder sample on phytosterols. (Graph prepared using think-cell® chart)

5.1.1.3 Scientific Articles

Looking at the development of subject areas since the first publication of articles on phytosterols (Fig. 5.2), a strong increase in articles since the mid 1990s can be seen. This holds especially true for articles in the food cluster. From only very few articles up to basically 1997, since then the number of articles has risen to about 150 in 2007. A similar pattern, although much weaker, can be seen in pharmaceuticals. The single-digit numbers have turned into nearly 50 articles in 2007. Contrastingly, personal care does not only show very few articles but also virtually no tendency to change this. The remainder of articles focuses mostly on biological and plant physiological topics.

Unsurprisingly, this minor importance of personal care in scientific articles is also reflected in the overall share of articles and the share of this subject area in the company sub-sample. As shown in Table 5.1, personal care represents only 1 % of all articles and only 6 % of articles in the sub-sample. The only subject cluster with even fewer articles in the sub-sample is pharmaceuticals (4, 12 % in all articles).

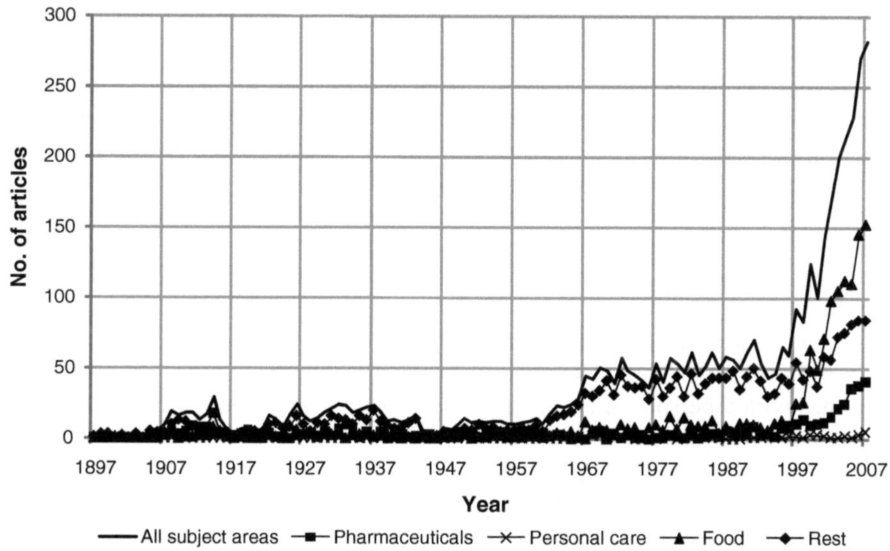

Fig. 5.2 Development of scientific articles on phytosterols ($n_{SF\_Pub\_i}$ = 4,364). *Source* Adapted from [9, p. 392]

Table 5.1 Scientific articles on phytosterols by industry and subject cluster

Subject cluster	Industries of sample:				Share of subject area in $n_{SF\_Pub\_C}$	Share of subject area in $n_{SF\_Pub\_i}$
	Personal care	Food	Pharmaceuticals	Chemicals		
Personal care (%)	40	0	0	11	6	1
Food (%)	27	77	25	67	66	31
Pharmaceuticals (%)	7	3	13	0	4	12
None of other three (%)	27	20	63	22	24	55
Number of articles	**15**	**79**	**8**	**9**	**111**	**4,364**

Due to rounding, shares do not necessarily add up to exactly 100 %. *Source* Adapted from [9, p. 391]

Overall, food is the most important subject (66 % in sub-sample and 31 % in all articles). Furthermore, the food industry is also the most active author of scientific articles, being responsible for 71 % of the articles in the sub-sample. While the personal care and food industries are most likely to author articles in their closely related field (40 and 77 %, respectively), this is different in the pharmaceutical and chemical industries. Their very few articles cover predominantly none of the three subject clusters and food, respectively.

5.1.1.4 Patents

In comparison to scientific articles, patents reflect a greater technological focus and orientation towards application. A closer look at the share of patents in all publications over time (Fig. 5.3) should thus shed some light on the degree of commercial interest in phytosterols. While the share does vary greatly from year to year, the trend line highlights the overall trend. With a local maximum around the Second World War, the patent share has been well below 10 % for more than 90 years, before increasing exponentially since the late 1980s. In 2007 it reached nearly 35 %.

When looking at the originating industry (Fig. 5.4) and the subject cluster (Fig. 5.5) of patents, a mixed pattern is visible. Within the group of 28 companies, the ones with a personal care or food background have turned out to be the most active actors. In particular since the early 2000s the substantial increase in patents is largely driven by these two industries. Looking at the subject clusters in the larger sub-sample, the general trend is similar but the three distinctively labeled clusters pharmaceuticals, personal care and food evolve rather parallel.

In analogy to the analysis of scientific articles in Sect. 5.1.1.3, analysis of patents by subject cluster and industry yields a comparable picture. Again, companies from the personal care, food and pharmaceutical (differing from scientific articles) industry file most of their patents in the subject clusters most closely

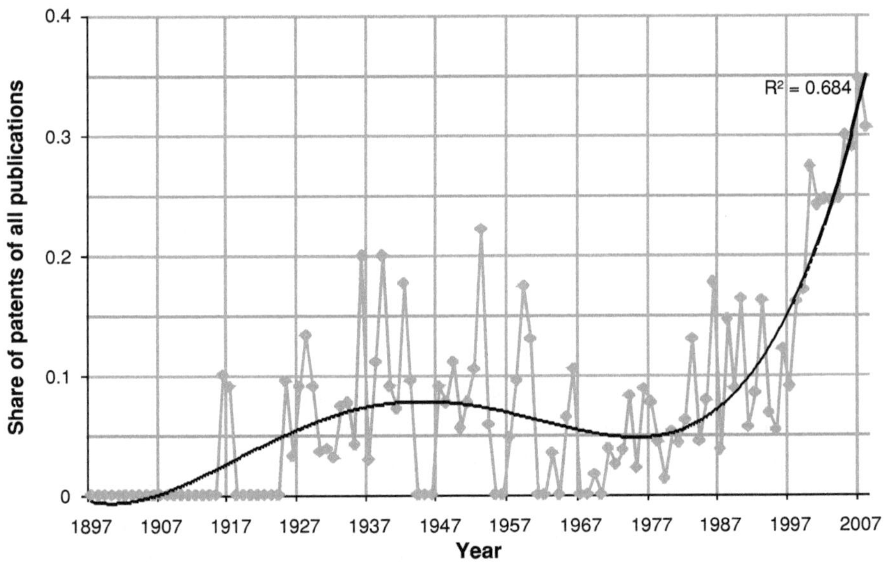

Fig. 5.3 Development of ratio patents/all publications on phytosterols ($n_{SF\_Pub\_i} = 4{,}364$). For easier visual inspection a fourth order polynomial trend line is included. *Sources* Adapted from [10, p. 7] and [11, p. 263]

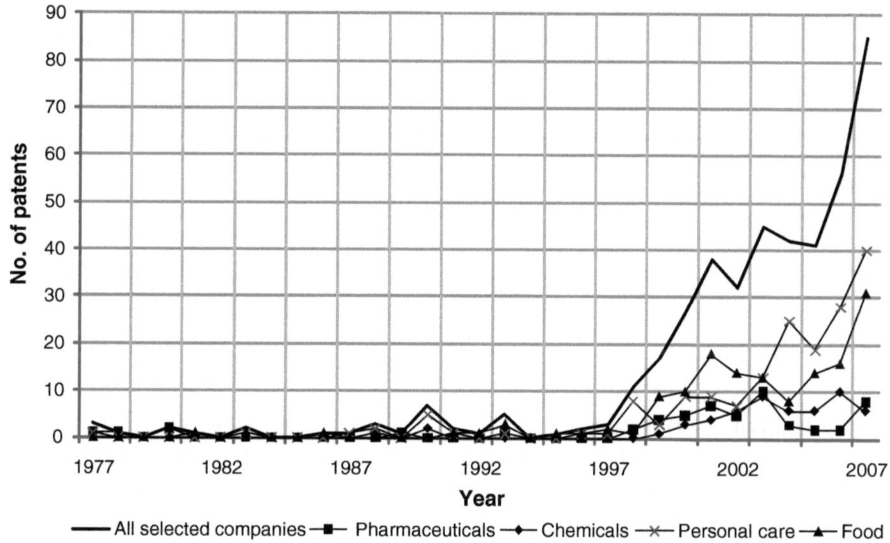

Fig. 5.4 Development of patenting behavior of industries regarding phytosterols 1977–2007 ($n_{SF\_Pat\_C}* = 430$) (Only 430 of the 451 patents are included as the remaining 21 patents stretch over the previous 80 years and add little to the development over time). *Source* Adapted from [10, p. 7]

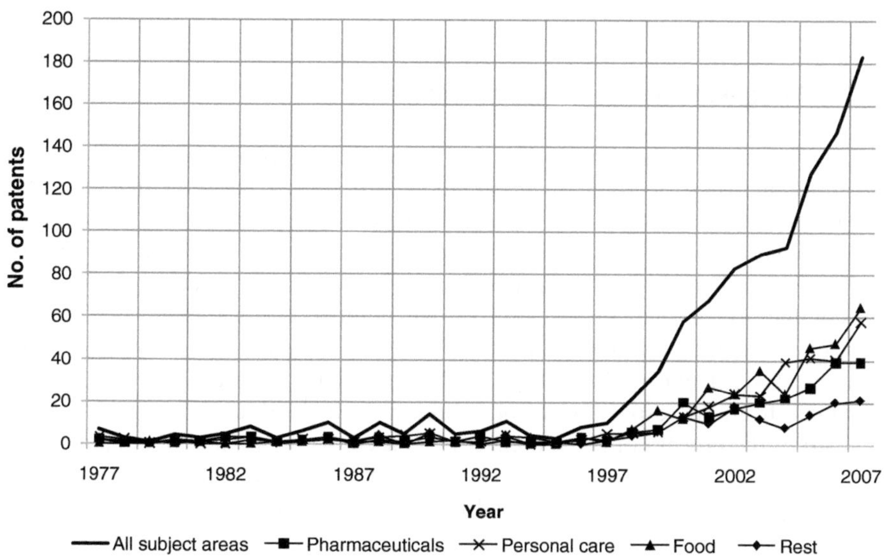

Fig. 5.5 Development of patenting behavior by subject area regarding phytosterols 1977–2007 ($n_{SF\_Pat}* = 1{,}036$) (1,036 of the 1,243 patents are included as 55 patents stretch over the previous 80 years and add little to the development over time, while the remaining 152 are assigned to subject areas below the 1 % threshold.) *Source* Adapted from [9, p. 392]

Table 5.2 Patents on phytosterols by industry and subject cluster

Subject cluster	Industries of sample:				Share of subject area in $n_{SF\_Pat\_C}$	Share of subject area in $n_{SF\_Pat\_i}$
	Personal care	Food	Pharmaceuticals	Chemicals		
Personal care (%)	87	8	19	33	45	29
Food (%)	1	71	16	33	30	31
Pharmaceuticals (%)	9	7	43	16	14	25
None of other three (%)	3	14	22	18	11	15
Number of publications	184	152	58	57	451	1,101

Due to rounding, shares do not necessarily add up to exactly 100 %. *Source* Adapted from [9, p. 391]

related to their core businesses (87, 71 and 43 %,[13] respectively). While companies from these three industries exhibit only little to medium activity in the more alien subjects, the chemical companies file one third of their patents regarding personal care and food each. The remaining third is nearly evenly split between pharmaceuticals and all the remaining subjects.

In contrast to scientific articles, not food but personal care is the most dominant subject cluster in the 451 selected patents. Nearly half of the patents across all four industries are filed in respect to personal care, nearly a third to food and 14 % to pharmaceuticals. This picture changes considerably when including patents not assigned to the 28 firms. Here personal care, food and pharmaceuticals are about evenly important (29, 31 and 25 %, respectively) (Table 5.2).

5.1.1.5 Weighted Average Year of Publication

The graphs on the development of scientific articles and patents over time grouped by either industry or subject cluster show differences between the individual groups. Notwithstanding these differences, a clear evaluation is difficult to derive, as nearly all of the curves share a common direction (i.e., the later the year, the higher the number of scientific articles/patents), with varying steepness. To reduce the complexity of such information to one number, the weighted average year (WAY) of scientific articles and patents in each of the industry sectors and subject clusters was calculated. With this WAY a measure for the age of an average document from the respective sector or area was obtained, enabling an assessment of a part of the process of convergence. It is based on the number of documents per year (thus weighted) multiplied by the respective year and divided by the total number of documents. The employed formula is as follows:

[13] Allowedly, 43 % is not the majority of patents from the pharmaceutical sector, but it demonstrates a similar tendency insofar as it is the single most important subject cluster.

Table 5.3 Weighted average year of scientific articles and patents, according to industry or subject cluster

	Industry						Subject cluster				
	All	Select.	Ph.	Ch.	PC	F.	All	Ph.	Rest	PC	F.
Art.	1982.3	2000.1	1996.3	2004.8	1991.1	2001.7	1988.2	1973.3	1978.1	1990.4	1993.4
Pat.	1998.2	2002.1	1999.3	2001.8	2002.7	2002.5	1999.8	1998.0	1995.0	2002.3	2001.4
Lag	15.9	2.0	3.0	−3.0	11.6	0.8	11.6	24.7	16.9	11.9	8.0

Art. scientific articles; *Pat.* patents; *All* all industry sectors or subject areas, respectively; *Select.* selected firms; *Ph.* pharmaceuticals; *Ch.* chemicals; *PC* personal care; *F.* food. *Source* Adapted from [9, p. 392]

Table 5.4 Weighted average year of patents, according to industry and subject cluster

	Subject cluster			
Industry sector	Rest	Pharmaceuticals	Personal care	Food
Chemicals	1995.0	2005.4	2003.8	2004.9
Pharmaceuticals	1992.7	2000.2	2003.5	2003.2
Personal care	1996.0	2002.7	2003.1	1989.0
Food	1999.7	2001.2	2000.2	2003.6

Source Adapted from [9, p. 393]

$$\mathrm{WAY}(p, s) = \left[\sum (n_i(p, s) * y_i(p, s))\right] / [N(p, s)]$$

with WAY*(p, s)* = weighted average year of patent or article *p* in subject cluster or industry sector *s*; n_i = number of *p, s* in each year *y*; *N(p, s)* = total number of *p* in *s* in all *i* years.

Hence the WAY will be the higher (a late average year of publication) the larger the share of documents is in later years when compared to earlier years. Table 5.3 presents the findings of this calculation. With one exception (the industry of chemicals), the WAY of scientific articles for each of the industries and each of the subject clusters lies earlier than the WAY of the respective patents.

Additionally, the time lag between both varies between 0.8 years (industry food) and 24.7 years (subject cluster pharmaceuticals). When looking deeper into the individual groups, it is found that generally patents from the "rest" cluster are the oldest ones (see Table 5.4). While in the pharmaceutical industry patents regarding the pharmaceuticals subject cluster are the second oldest ones, in the personal care and food sector it is those focusing on food (in the personal care sector even the, on average, oldest patents) and personal care, respectively.

5.1.2 STN AnaVist

5.1.2.1 Searching CAplus with STN AnaVist

In contrast to SciFinder, STN AnaVist is a frequency-based analysis and visualization software that is explicitly designed for interactive and statistical examinations [cf. 12, p. 314ff], [13, p. 282ff]. It presents a wide array of analysis possibilities, including the possibility to automatically organize selected documents in a three-dimensional topographical map (research landscape) on the basis of documents' contents. An underlying algorithm uses document similarity scores[14] to position documents relative to each other in a two-dimensional space, in which each document will be represented by one dot based on its x,y coordinates. Research clusters can then be evaluated on the basis of proximity (documents with greater overlap will be closer together than those with less content similarity) and the individual peak height (a peak will be higher the more documents it consists of). This clustering can be further refined by selecting subsets, adjusting the maximum appearance of clustering concepts[15] and grouping or exclusion of certain terms. Dots in such a landscape represent individual documents and can be colored differently based on the grouping of documents (e.g., by author name, IPC, country). Investigators can turn and tilt the landscape, zoom in and out and transform it into a two-dimensional landscape. Furthermore, tables and matrices can be aggregated with information from 13 different fields within documents, such as companies, inventors, years and IPCs [cf. 14].[16] Interactivity is especially supported in respect to a selection or highlighting of a subgroup of any content in landscapes or tables. This highlighting will be reflected in all of the other landscapes and tables of the same project as well. The answer set to be analyzed is created with STN Express, a software package for online search and post processing of scientific and technical databases [cf. 15].[17]

---

[14] The source of significant keywords and concepts for this similarity calculation can be chosen when initially importing the dataset into STN AnaVist. Based on the database, different clustering fields for data and text mining can be chosen. While the patent databases *PCTFULL*, *USPATFULL* and *WPINDEX* offer the possibility to use the IPC codes or (parts of) the claims, *CAplus* can be visualized on the basis of title and abstract content or CAplus' own technology indicators. IPC codes can be employed for patent documents in CAplus, but naturally not for scientific articles. Whatever choice is made, the clustering fields are permanent for a certain project and all visualizations created from it. Therefore, visualization of the current sample was based on the contents in title/abstract and the claims (in patents) of the documents.

[15] Terms may be included in a maximum of between 10 and 70 % of all documents. This concept frequency cannot be adjusted when using IPC code visualizations, which always use a frequency of 100 %.

[16] STN AnaVist, Version 2.0 was used in the course of these analyzes.

[17] STN Express, Version 8.4 for Windows was used in the course of these analyzes.

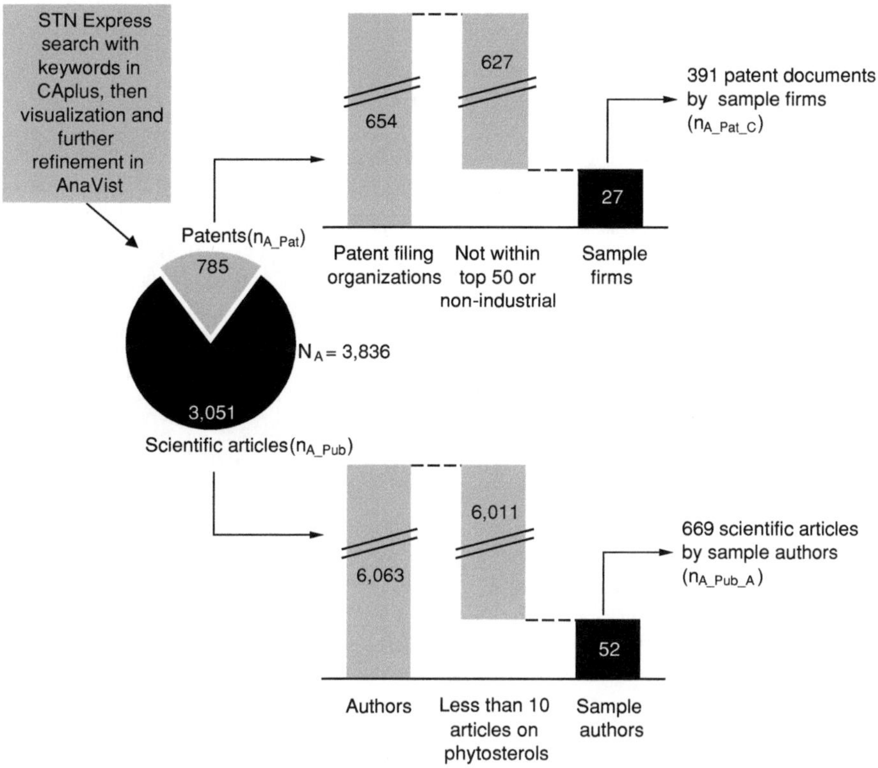

Fig. 5.6 Construction of STN AnaVist sample on phytosterols. (Graph prepared using think-cell® chart)

5.1.2.2 Sample

Based on the analysis described in Sect. 5.1.1, a second sample was compiled from the CAplus database through STN Express in late 2008. The same search-terms were used[18] for the compilation of this dataset. This resulted in a sample with 3,836 references (N_A), containing 3,051 scientific articles (80 %; n_{A_Pub}) and 785 patents (20 %; n_{A_Pat}). Albeit being substantially smaller than the SciFinder sample (due to differences in the employed databases), this sample did not structurally deviate from it. Subsequently, identification of the same 28 companies led to 391 patents ($n_{A\_Pat\_C}$) (Fig. 5.6).

To gain a first overview of the whole sample, it was visualized in a research landscape as described in the preceding paragraph. The resulting picture is

[18] The employed search-terms were ‘phytosterol’ and ‘plant(W)sterol’ and their derivatives. The *(W)* is a proximity operator requiring that answers will have the connected terms adjacent to each other in the specified order.

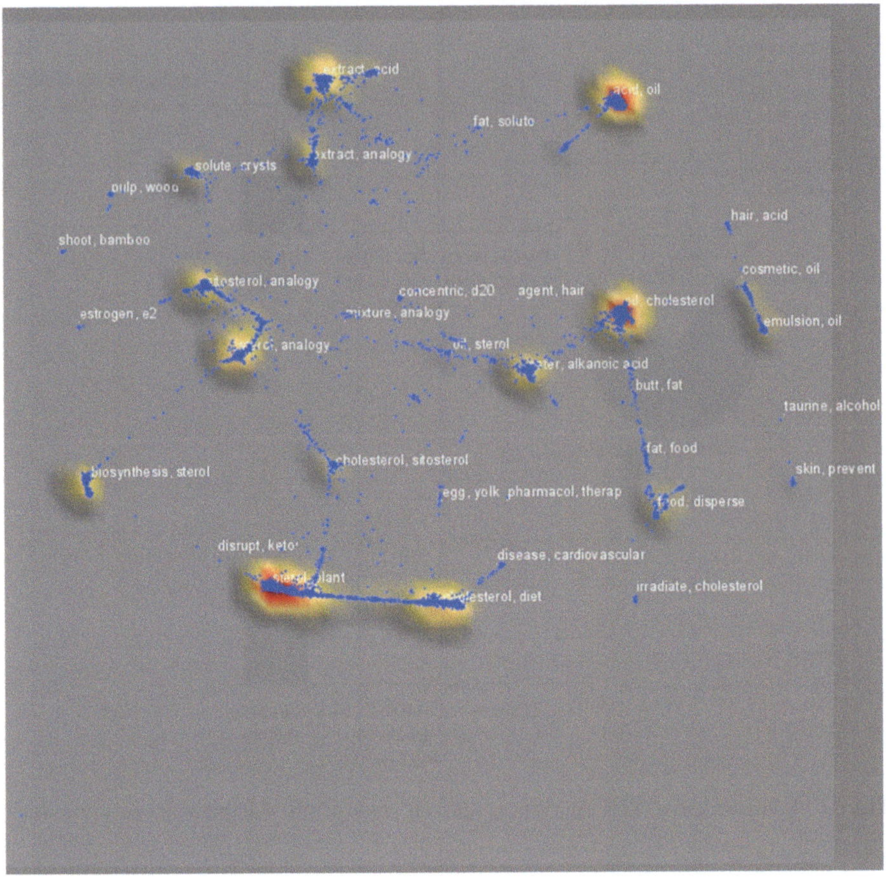

Fig. 5.7 Research landscape of the complete phytosterol sample ($N_A = 3{,}836$; concept frequency 50 %)

presented in Fig. 5.7 with a concept frequency of 50 %. While there are various distinct clusters to be identified, the dots highlight the fact that the positioning of documents in this x,y coordinate system is not limited to these research 'hotspots'. The darker centers of the clusters resemble higher document densities. This can be seen even better when looking at the three-dimensional map not from the top. This tilted view (without the individual dots) is shown in Fig. 5.8.

The importance of concept frequency for a distinction of research hotspots can be demonstrated when comparing Fig. 5.7 to the two research landscapes depicted in Fig. 5.9. While Fig. 5.9a (70 % concept frequency) exhibits no difference to the 50 % concept frequency landscape, a frequency reduction to 10 % yields a much more scattered picture (Fig. 5.9b). Based on the desired outcome, the concept frequency needs to be adjusted to the optimum level. To ensure comparability, an identical maximum occurrence of 50 % was employed for all of the research landscapes within the present study.

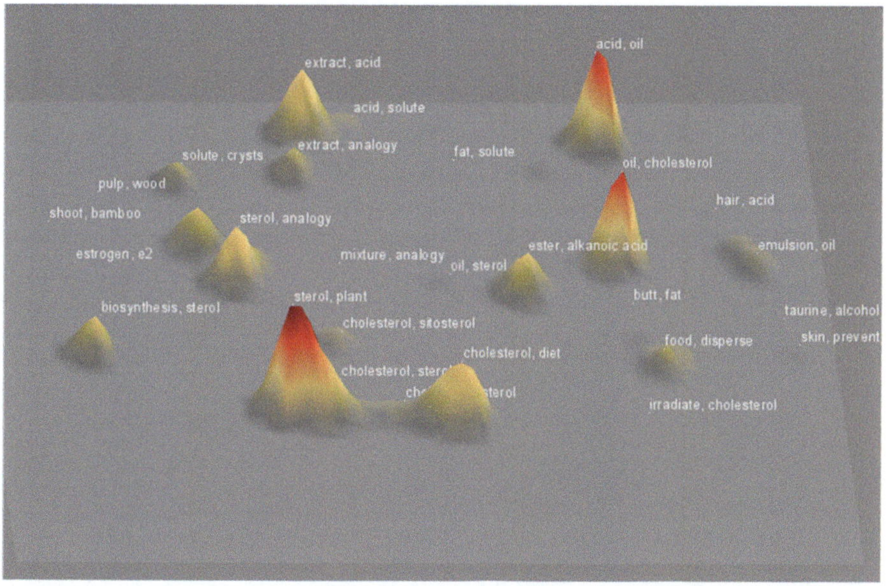

Fig. 5.8 Tilted research landscape of the complete phytosterol sample ($N_A = 3{,}836$; concept frequency 50 %)

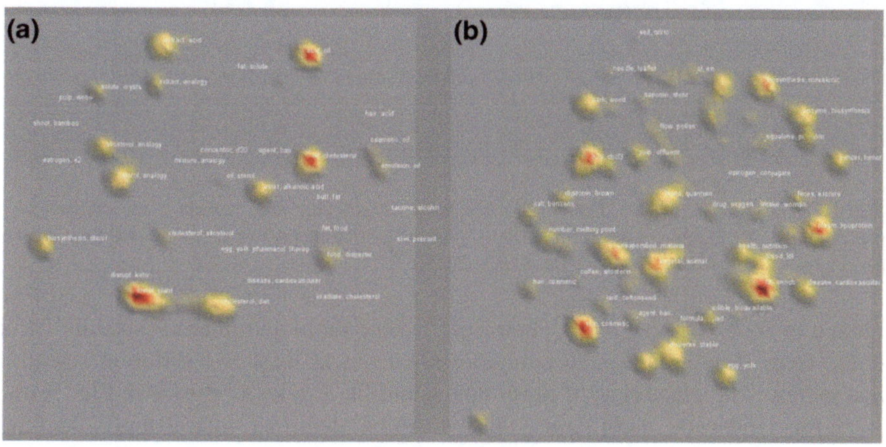

Fig. 5.9 Research landscapes of the complete phytosterol sample with a concept frequency of (**a**) 70 % (*left*) and (**b**) 10 % (*right*) ($N_A = 3{,}836$)

5.1.2.3 Scientific Articles

To analyze indications of science convergence in the field of phytosterols, the 52 authors of the author sample were manually assigned to the three scientific areas

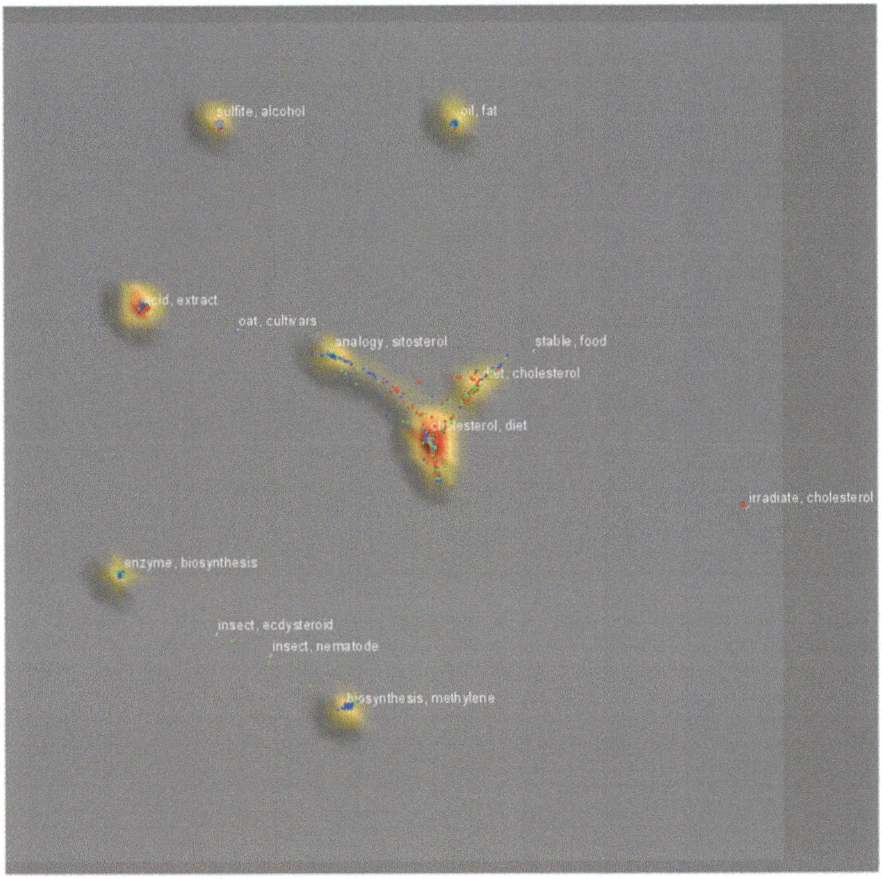

Fig. 5.10 Scientific articles from 1904 to 2008 of selected 52 authors with *colored dots* representing chemicals (*blue*), food (*green*), pharmaceuticals (*red*) or cross-disciplinary articles (*grey*) ($n_{A\_Pubs\_A} = 669$; concept frequency 50 %)

related to food[19] (green dots), chemicals[20] (blue dots) and pharmaceuticals[21] (red dots). Grey dots indicate the co-authorship from at least two different scientific disciplines. The clustering of these 669 scientific articles is shown in Fig. 5.10.

In this landscape, eight larger clusters can be identified. Interestingly, the three in the middle exhibit the highest document density and contain articles from all three disciplines. The two in the lower part consist mainly of chemicals and food

[19] The group of authors from the food area contains researchers from companies or universities/ research institutions in areas such as nutrition, animal feed, biology and agriculture.

[20] Besides chemistry in a narrower sense, researchers from biochemical institutions and alike were also included.

[21] Pharmaceuticals include affiliations within the pharmaceutical sciences as well as medical institutions.

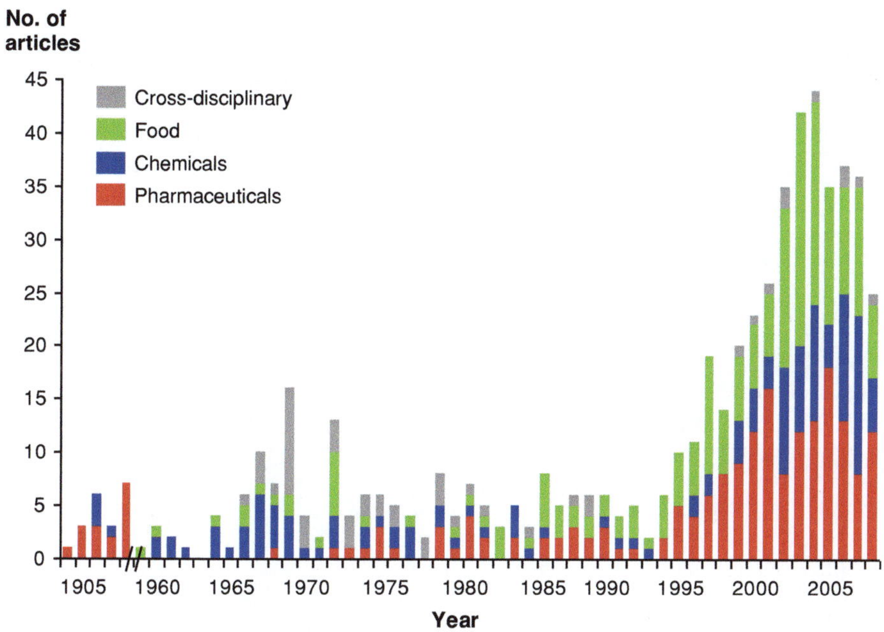

Fig. 5.11 Scientific articles from 1904 to 2008 of selected 52 authors with *colored dots* representing chemicals (*blue*), food (*green*), pharmaceuticals (*red*) or cross-disciplinary articles (*grey*) ($n_{A\_Pubs\_A} = 669$; concept frequency 50 %). (The depicted documents add up to only 577 as the years 1909–1958 have been omitted from this figure due to too small numbers of documents in the respective decades and in order to remain comparable to the results shown in Fig. 5.12. Graph prepared using think-cell® chart)

related articles, the one in the far left pharmaceuticals and chemicals related articles, the two at the top of articles with authors from all three disciplines and of cross-disciplinary[22] articles. As it was argued that science convergence should result in an increasing number (or even share) of cross-disciplinarily authored articles, Fig. 5.11 analyzes the development of article authorship over time. While up to the 1970s authors with a pharmaceutical and later chemical background dominated the literature on phytosterols, this picture changed from the 1980s onwards, with an increasing number of authors from the food area. The number of cross-disciplinary articles has peaked around the 1970s.

A similar picture surfaces when looking at the relevant research landscapes over time (Fig. 5.12a–f). Starting with a few dedicated clusters (a) at the beginning of the last century, initially no focus can be identified. The research conducted in that period reveals indeed little interwovenness. Despite the still scattered distribution of articles in the 1960s (b) and earlier 1970s (c), articles from the different disciplines begin to blend and to concentrate on two foci during the 1980s (d) and 1990s (e).

[22] Cross-disciplinary in this case means authors from at least two different scientific disciplines.

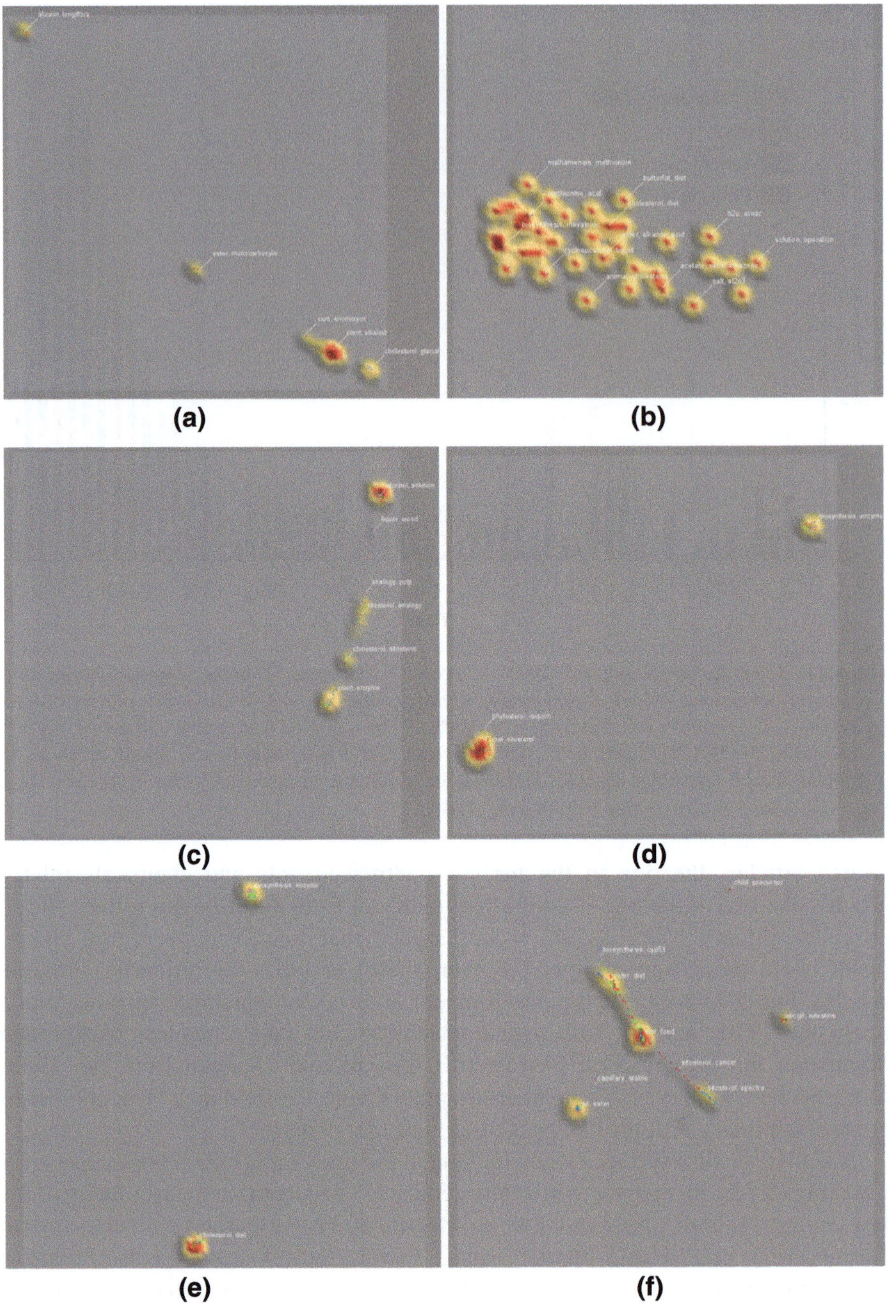

Fig. 5.12 Scientific articles on phytosterols depicted in research landscapes 1904–2008 ($n_{A\_Pub\_A} = 669$; concept frequency 50 %). (The depicted documents add up to only 577 as the years 1909–1958 have been omitted from this figure due to too small numbers of documents in the respective decades. The first landscape contains only articles from 1904 to 1908 and is just presented for illustrative purposes. *Sources* Adapted from [16], p. 648). **a** 1904–1908 ($n_{A\_Pub\_A1} = 20$). **b** 1959–1968 ($n_{A\_Pub\_A2} = 35$). **c** 1969–1978 ($n_{A\_Pub\_A3} = 62$). **d** 1979–1988 ($n_{A\_Pub\_A4} = 54$). **e** 1989–1998 ($n_{A\_Pub\_A5} = 83$). **f** 1999–2008 ($n_{A\_Pub\_A6} = 323$)

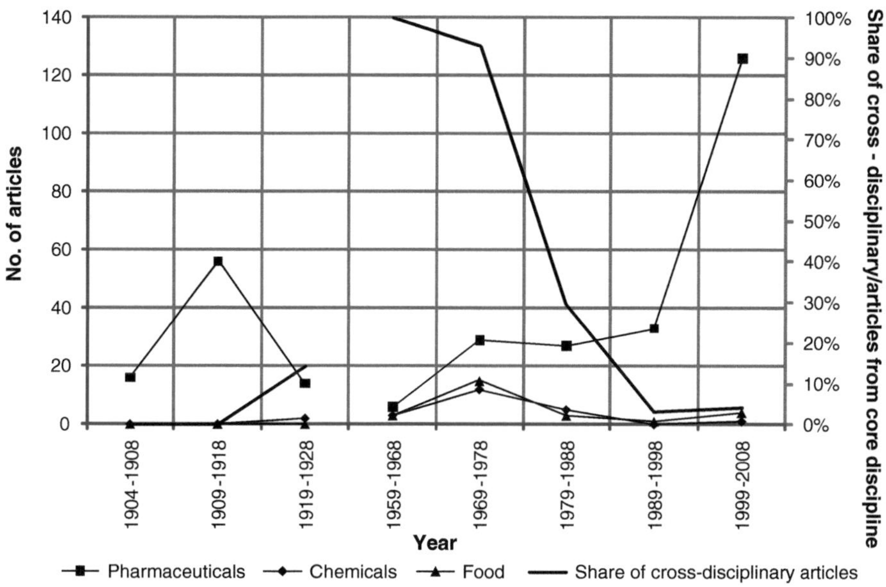

Fig. 5.13 Number of articles in each subject and share of cross-disciplinary articles in pharmaceuticals 1904–2008 ($n_{A\_Pub\_A\_Ph} = 307$; concept frequency 50 %). *Source* Adapted from [16, p. 650f]

While the cluster of documents at the top of the landscapes concentrates on biosynthesis and connected questions more in the biological, biochemical and agricultural scientific fields, the cluster at the bottom[23] entails mainly documents in the area of NFF (also reflected in the broader mixture of disciplines). Strikingly, this focused distribution is being split up again in the last decade under scrutiny (f), leading to more specialized researchers and institutions. In contrast to the middle of the twentieth century, this specialization does not lead to only very loosely related topics. In fact, the five central areas are covering topics such as biochemistry and chemical preparations (the two peaks at the bottom), pharmaceuticals (the peak on the right) and NFF (the two most central areas).

In a last step, the development of numbers of articles and especially co-authorship patterns were explored for the three disciplines. To this end, it was assessed for each of the 669 articles, whether they were authored solely by authors from one discipline, or whether there were authors from any of the other two disciplines as well. Accordingly, Figs. 5.13, 5.14 and 5.15 show the developments over time of the number of articles and the share of articles authored across discipline boundaries for pharmaceuticals, chemicals and food, respectively. When comparing the three figures, the dominance of authors from pharmaceuticals at the

[23] In both decades this cluster is the denser one with more articles (38–16 in 1979–1988 and 52–31 in 1989–1998) focusing on the application of phytosterols.

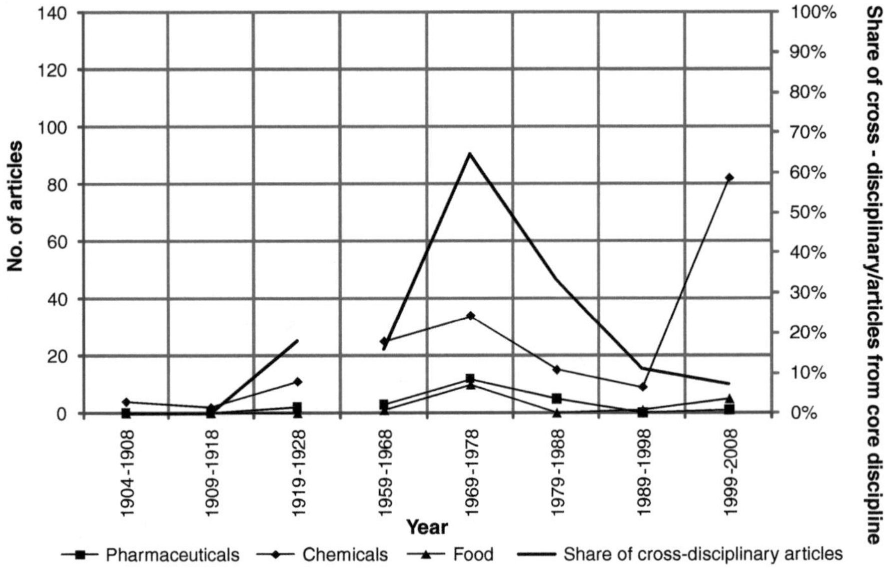

Fig. 5.14 Number of articles in each subject and share of cross-disciplinary articles in chemicals 1904–2008 ($n_{A\_Pub\_A\_Ch}$ = 182; concept frequency 50 %). *Source* Adapted from [16, p. 650f]

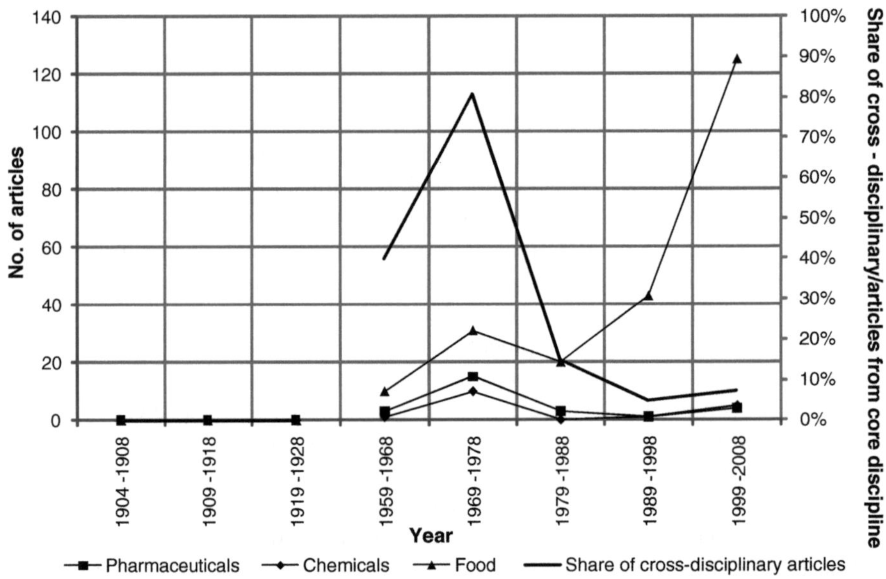

Fig. 5.15 Number of articles in each subject and share of cross-disciplinary articles in food 1904–2008 ($n_{A\_Pub\_A\_Fo}$ = 229; concept frequency 50 %). *Source* Adapted from [16, p. 650f]

beginning of the twentieth century is striking. In the second half of the century authors from pharmaceuticals and food are about evenly productive, whereas publications from the chemicals group lag somewhat behind. In general, the numbers of cross-disciplinary articles are quite small in all of the three groups. Prior to the 1950s and from the late 1980s onwards only little cross-disciplinarity can be observed, with shares mainly below 20 %. Contrastingly, a sharp peak is demonstrated between 1969 and 1978. Articles from pharmaceuticals, chemicals and food were authored across discipline boundaries in 93, 65 and 81 %, respectively. Between the three groups the most decisive difference lies in the shares in the 10 years before these *maxima*. While on the one hand pharmaceuticals had an even higher share of 100 % and food of at least 40 %, the share for chemicals was at only 16 %. On the other hand, absolute numbers were very small for pharmaceuticals and food in that period (6 and 10 articles, respectively).

5.1.2.4 Patents

After the analysis of phytosterol related scientific articles two routes were explored to evaluate progress in technology convergence as mirrored in patents. First, parallel to the approach with the SciFinder sample, the same clustering by firm affiliation was undertaken, yielding a dataset of 391 patents ($n_{A\_Pat\_C}$). This assignment of patents by industry affiliation was used for the results in Table 5.7 and Fig. 5.17. Second, all 785 patents were grouped according to their IPCs. The assignment to food, chemicals, pharmaceuticals and personal care was done on the basis of WIPO's IPC catchword index [cf. 17]. Table 5.5 provides an overview of the applied IPCs. Due to the co-classification of patents and the abolishment of main and secondary IPCs it cannot be assessed to which of the subjects a patent is per se belonging. Therefore, numbers in the following will mostly add up to more than 100 % of the sample. Any patent was counted as a chemicals patent as well as a pharmaceuticals patent, for instance, if it bore IPCs from both domains.

Figure 5.16 presents the development over time of these IPC groups in the three decades between 1978 and 2007. While the first patent on phytosterols had been published in 1916, it was not until the late 1970s that interest in phytosterols triggered more substantial numbers of patents (hence, the preceding 61 years are omitted from the Figure as they carry very little informative value). Especially in respect to patents with pharmaceuticals, food or personal care focus, the number of patents has multiplied since 1998. Personal care and food are at about the same level in 2007 with 38 and 41 patents, respectively. Partly because of the overall quite low numbers, each of the graphs fluctuates substantially. Interestingly, patents on chemicals and on the remaining IPCs exhibit trends deviating considerably from the other three. The number of chemicals patents has halved between 2003 and 2004 and only gradually increased since then. Inventors/assignees have filed for about five patents in the remaining IPCs for nearly 10 years, before this number has risen to 13 in 2007.

Table 5.5 Grouping of IPCs into four industry-related clusters in patents on phytosterols

Subject	IPCs[a]	Meaning of IPC
Personal care	A61K 7	Cosmetics or similar toilet preparations
	A61K 8	Cosmetics or similar toilet preparations[b]
	A61Q	Use of cosmetics or similar toilet preparations
Food	A21D	Treatment, e.g., preservation, of flour or dough for baking, e.g., by addition of materials; baking; bakery products; preservation thereof
	A23	Foods or foodstuffs; their treatment, not covered by other classes
Pharmaceuticals	A61K (except 7&8)	Preparations for medical, dental, or toilet purposes [...]
	A61P	Therapeutic activity of chemical compounds or medicinal preparations
Chemicals	A01N	Preservation of bodies of humans or animals or plants or parts thereof [...]; biocides, e.g., as disinfectants, as pesticides or as herbicides [...]
	C07	Organic chemistry
	C08	Organic macromolecular compounds [...]
	C09	Dyes; paints; polishes; natural resins; adhesives [...]
	C11	Animal or vegetable oils, fats, fatty substances or waxes [...]
	C12	Biochemistry; beer; spirits; wine; vinegar; microbiology; enzymology; mutation or genetic engineering
Rest	All remaining IPCs	...

[a] Provided is the highest hierarchical level possible
[b] IPCs with the same content appear (e.g., *A61K 7* and *8*) as the classification scheme is regularly updated and then all patents are reclassified (if necessary). Thus A61K 7 is the IPC of an older version than A61K 8.

A parallel (but potentially independent) growth of patents across the different areas does not necessarily point at convergence. Consequently, a closer look at co-classification of the 785 phytosterol patents would demonstrate how strongly interwoven the field is. The IPC co-classification depicted in Table 5.6 shows an about even distribution of patents across all four industry-related groups. As this analysis does not allow for deciding which of the IPCs is the main one, the sum of all five groups (including the patent with codes from the remaining IPCs) adds to 1,169—384 more than the actual number of documents. Nevertheless, there is a considerable overlap especially between pharmaceuticals and food/personal care. In personal care, 26 % of all patents are also classified as pharmaceuticals-related. This link is even stronger for food and pharmaceuticals—40 %. In line with the findings in Fig. 5.16, the number of patents with a (additionally) strong chemical focus is smaller but still decisive (17 % of personal care and 19 % of food patents).

The results on industry/IPC co-classification shown in Table 5.7 are meant to test whether the IPC co-classification of patents resulted from all industry groups. When looking at their individual patenting behavior, differences between them are

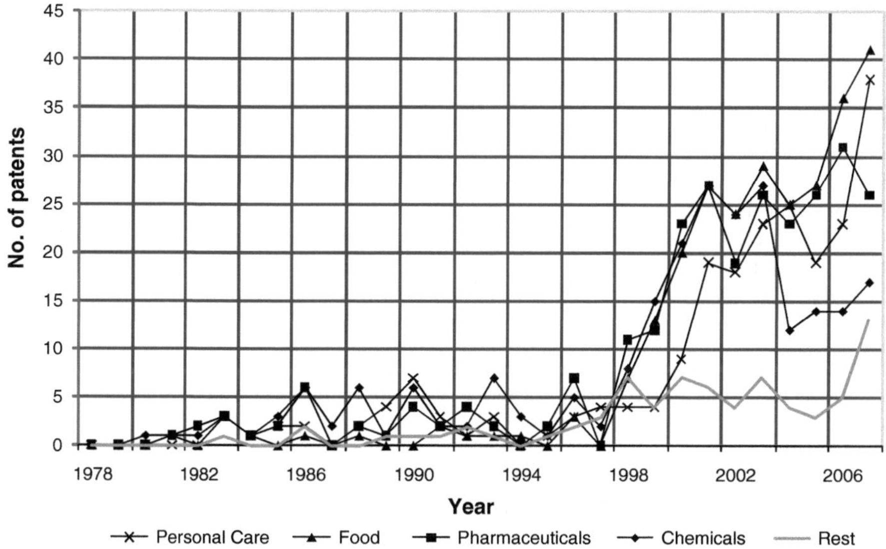

Fig. 5.16 Development over time of IPC clusters 1978–2007 ($n_{A\_Pat}* = 670$). (The figure is based on the patents in n_{A_Pat} with 785 patents, but only 670 in the years 1978–2007. *Source* Adapted from [18, p. 389]

Table 5.6 Co-classification of patents according to IPCs ($n_{A\_Pat} = 785$)

IPC group	Personal care	Food	Pharmaceuticals	Chemicals	Rest
Personal care	**234**				
Food	15	**291**			
Pharmaceuticals	60	115	**283**		
Chemicals	40	83	87	**279**	
Rest	12	17	29	40	**82**

The table is based on n_{A_Pat} with 785 patents, but due to the co-classification the total number of patents in the five groups is 1,169. *Source* Adapted from [16, p. 652]

obvious. Whereas 92 % of personal care patents are at least also with personal care IPCs, this concentration is considerably lower for food, pharmaceuticals and chemicals (48, 52 and 35 %, respectively). A caveat has to be taken into account regarding food patents, as 35 % of them do not contain IPCs.[24]

Finally, all of the aforementioned results constitute static assessments of the degree of interwovenness as a proxy for convergence, but not necessarily whether there is a development over time towards greater convergence. Therefore, $n_{A\_Pat\_C}$ was taken as the basis for a series of research landscapes exhibited in Fig. 5.17a–e. The 338 of the 391 patents published in the decade from 1999 to 2008 evolve from

[24] This particularly high percentage of unassigned patents will be discussed in Sect. 6.1.2.1 in greater detail.

Table 5.7 Patents by companies according to IPCs ($n_{A\_Pat\_C} = 391$)

No. of patents in	Companies in group				Sum
	Personal care	Food	Pharmaceuticals	Chemicals	
Each company group	88	188	52	63	391
Personal care IPCs	**81**	22	12	20	135
Food IPCs	4	**90**	13	26	133
Pharmaceuticals IPCs	19	30	**27**	23	99
Chemicals IPCs	6	35	24	**22**	87
Remaining IPCs	0	7	4	6	17
Not containing IPCs	3	66	5	11	85
SUM	**113**	**250**	**108**	**108**	**556**

The table is based on $n_{A\_Pat\_C}$ with 391 patents, but due to the co-classification the total number of patents across the four company clusters is 556. *Source* Adapted from [18, p. 390]

a landscape of scattered hotspots in 1999/2000 (a) into a landscape with two foci in 2007/2008 (e).[25] These two foci are a NFF-related bimodal hotspot in the lower left corner and a Cosmeceuticals-related peak in the upper right. Furthermore, by taking a closer look at individual documents, this judgment is supported by the tendency of all four industries to increasingly concentrate on common research topics in NFF and Cosmeceuticals.

In contrast to these observations, an analysis of IPC co-classification over time did not exhibit an increasing share of co-classified patents. The results depicted in Fig. 5.18 vary decisively, but all four show an overall decreasing trend. As too little data was available for the first eight decades, this value is provided as well for informative purposes.

5.2 Classification-Based Analysis: Indications of Convergence in NFF and ICT

5.2.1 Searching INPADOCDB with STN Express

Until now, all results were derived from keyword dependent data collection. This does not necessarily pose a problem. In the present case the term 'phytosterols' and its derivatives are with very limited application and employed independently of possible convergence trends. The term has been in use in biology and chemistry already since a time when industry convergence was far from being even discussed. As already discussed above, this may be different in other settings. There,

[25] Due to the small number of patents prior to 1998, outcomes were not expected to be reliable up to then and consequently eliminated from this analysis. In addition, 2 years were added each to one landscape to provide numbers of references above the threshold for a compilation of such landscapes.

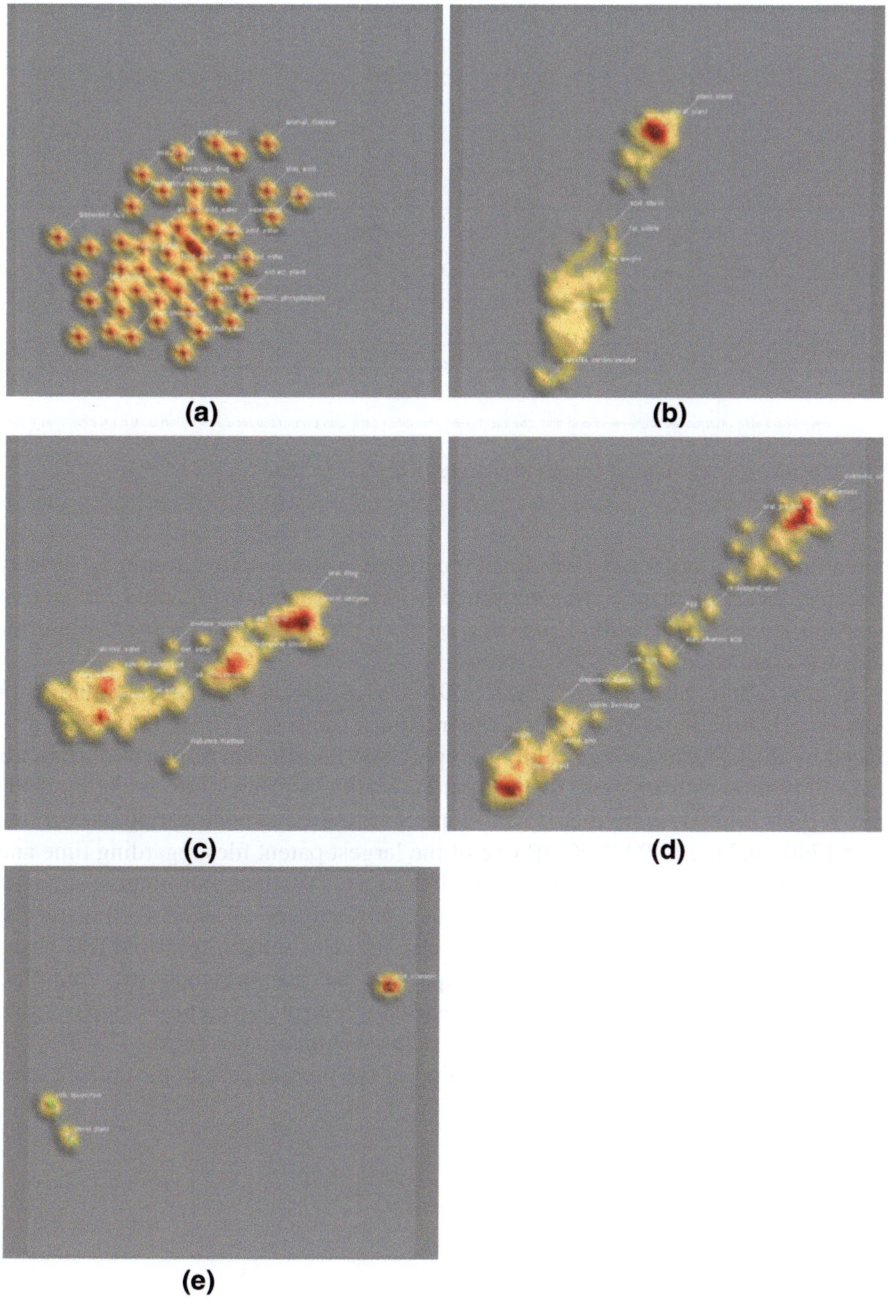

Fig. 5.17 Patents on phytosterols depicted in research landscapes 1999–2008 ($n_{A\_Pat\_C}{}^* = 338$; concept frequency 50 %). The figures are based on the patents in $n_{A\_Pat\_C}$ with 391 patents, but only 338 in the years 1999–2008. *Source* Adapted from [16, p. 653]. **a** 1999–2000 ($n_{A\_Pat\_C1} = 48$). **b** 2001–2002 ($n_{A\_Pat\_C2} = 61$). **c** 2003–2004 ($n_{A\_Pat\_C3} = 80$). **d** 2005–2006 ($n_{A\_Pat\_C4} = 66$). **e** 2007–2008 ($n_{A\_Pat\_C5} = 83$)

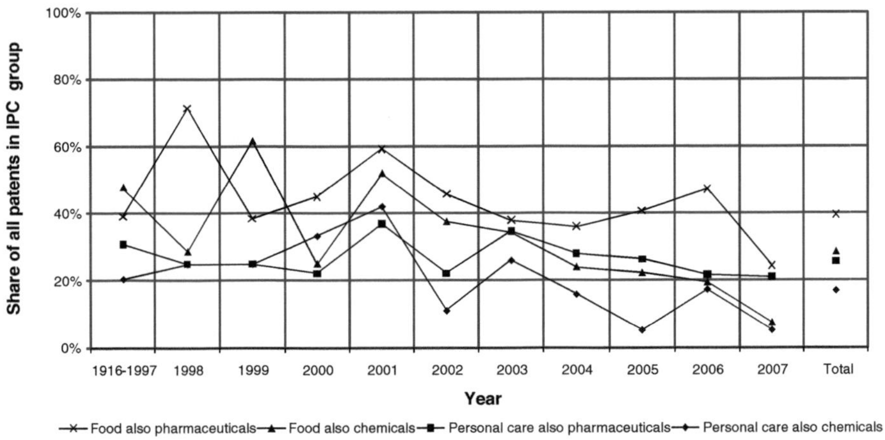

Fig. 5.18 Co-classification of patents according to IPCs over time ($n_{A\_Pat} = 785$)

potential search-terms may be either too broad to be employed or may already describe the convergence product/process itself. Consequently, this last set of analyzes is based on classification data, or more precisely on IPCs as search-terms and was conducted using STN Express.

In contrast to Sect. 5.1, *INPADOCDB* was accessed for the compilation of datasets. INternational PAtent DOCumentation DataBase (INPADOCDB) is produced by the EPO and contains bibliographic and family data of patent documents and utility models from more than 90 patent-issuing organizations, including EPO and WIPO. In total, more than 55 million records are available dating back to the year 1790, making INPADOCDB one of the largest patent files regarding time and country coverage [cf. 19, p. 5f].[26] As INPADOCDB does not contain data on scientific articles (in difference to CAplus), no analyzes of science convergence were conducted in the remainder of this chapter. The suitability of the IPC based approach should be assessed in comparison to the previous two samples. Furthermore, different countries were tested to control for country specifics and finally relating these results to a widely acknowledged convergence—ICT—was meant to enable an approximate positioning of the present case in the convergence continuum.

[26] Time coverage does actually differ considerably from country to country. Even for the few countries with more than 500,000 records in INPADOCDB, this coverage varies between 1790 (USA) and 1985 (France).

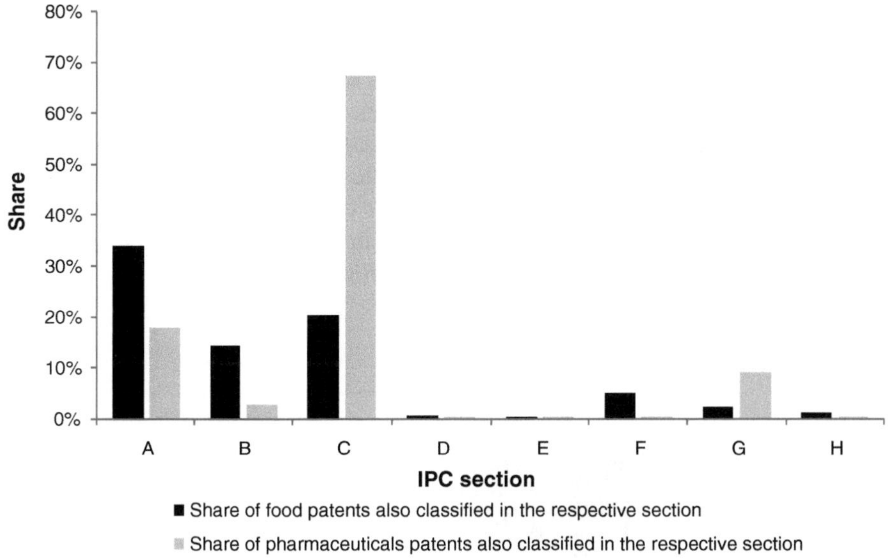

Fig. 5.19 Co-classification of US patents on food and pharmaceuticals with the remaining sections ($N_{NFF\_US} = 341{,}846$). *Source* Adapted from [11, p. 269]

5.2.2 Samples

IPC co-classification as a proxy for technology convergence only makes sense if this co-classification is not an overall trend with even distribution across all IPC sections. Accordingly, a first analysis was carried out before compilation of the samples. As is obvious in Fig. 5.19, the co-classification is indeed varying very much between the IPC section and can thus be judged to be dependent on the technological closeness of IPCs. Food patents show a considerable co-classification with IPCs from section A, C and B, in that order.[27] Contrastingly, approximately two thirds of pharmaceuticals patents are also classified as chemicals patents and are sometimes also classified in section A and G. Hence, this finding indicated the applicability of the proposed approach.

In the course of these analyzes four datasets were generated. The first three contained patents from the food and the pharmaceuticals field and are thus named NFF samples. They were constructed in INPADOCDB on the basis of the IPCs listed in Table 5.5. All patents were searched with the respective IPCs, then limited to one of the three countries USA, UK and Japan as 'patent country' before

[27] Food and pharmaceuticals IPCs are both in section A. The cluster's own IPCs were excluded when calculating the co-classification, of course. Sections have been briefly introduced in Sect. 3.3.1. They are: *(A) human necessities; (B) performing operations, transporting; (C) chemistry, metallurgy; (D) textiles, paper; (E) fixed constructions; (F) Mechanical engineering, lighting, heating, weapons, blasting; (G) physics; (H) electricity.*

being split up into yearly samples based on publication year. Two effects could be achieved by limiting the sample to one country each. First, country specifics could be controlled for as each of the three countries was analyzed separately. Second, the sample was not implicitly biased by patents belonging to one patent family. Finally, for each year the number of co-classified patents in food as well as pharmaceuticals IPCs was derived. A co-classification was deemed to be a sign of convergence in this setting. The construction of samples for NFF analysis is depicted in Fig. 5.20. Based on the analyzes in Sect. 5.1 and due to the small numbers, patent counts were only conducted from 1950 onwards.

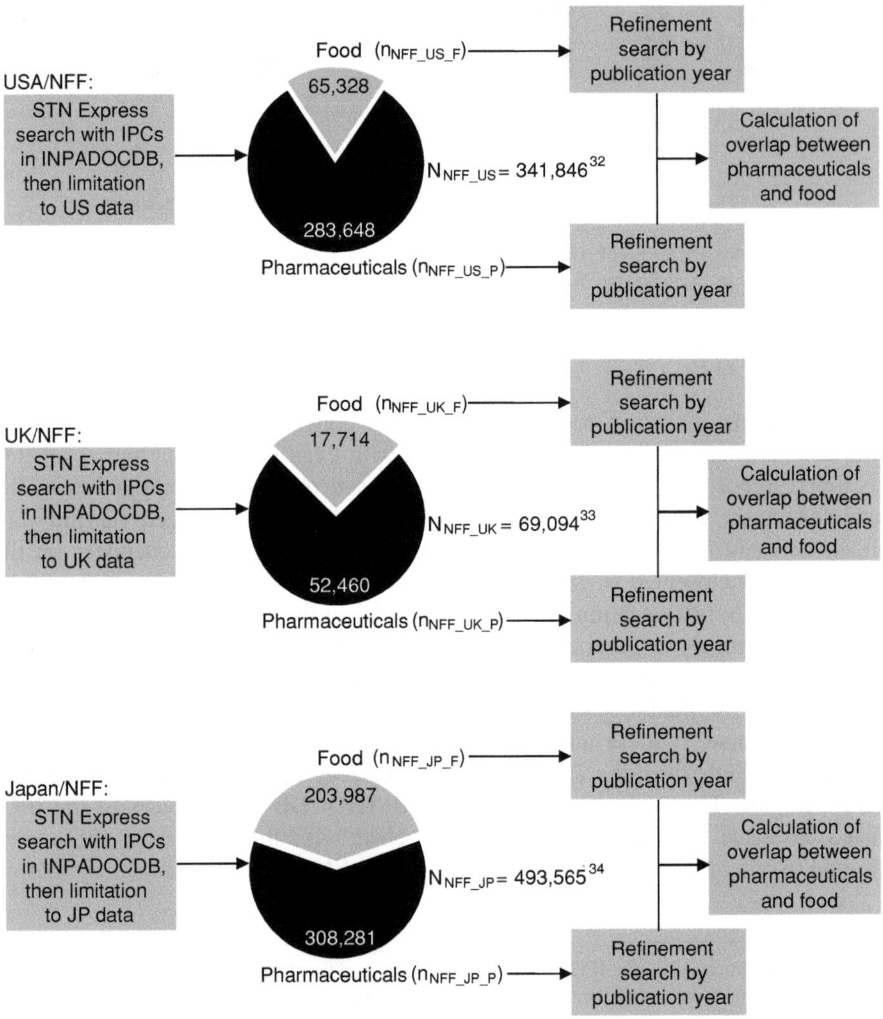

Fig. 5.20 Construction of STN Express samples on NFF in the USA, UK and Japan. (Graph prepared using think-cell® chart)

Table 5.8 Grouping of IPCs into five smartphone-related subjects

Subject	IPCs[a]	Meaning of IPC
Telecommunication	H04M	Telephonic communication [...]
Computer	G06	Computing; calculating; counting [...]
Music	G10	Musical instruments, acoustics
Camera	G03B	Apparatus or arrangements for taking photographs or for projecting or viewing them [...]
Navigation	G01C	Measuring distances, levels or bearings; surveying; navigation; gyroscopic instruments; photogrammetry or videogrammetry [...]

[a] Provided is the highest hierarchical level possible

To compare the assessment of convergence in the area of NFF a last sample was generated on ICT developments. Accordingly, five IPC subjects were selected focusing on technology regarding telecommunication, computer, music, camera and navigation. These IPCs are representing the broader ICT progress regarding the computer and the telecommunications industry, as well as the narrower example of smartphones. In analogy to the derivation of suitable IPCs from WIPO's catchword index in Sect. 5.1.2.4, the respective IPCs for the ICT example are presented in Table 5.8. As this is only an illustrative example, a sample was only created for one country—the USA. All the remaining actions were conducted analog to those in the aggregation of the three NFF samples; with the difference that co-classification was assessed one-on-one for each pair of the five subjects.[28,29,30,31] The construction of this sample is shown in Fig. 5.21.

5.2.3 Analysis of NFF Patents in the USA

The number of US food and pharmaceuticals patents has seen only modest growth rates up to the 1990s (Fig. 5.22). Since then, pharmaceuticals patents have increased sharply from about 4,000 to more than 25,000, especially since 1995. In 2003 they appear to have reached a steady state, at least for the time being. Around 1,000 food patents were filed per year during the 1990s, before being multiplied to nearly 4,000 in 2004. Since then their number has decreased again to

28 Due to the co-classification the complete sample is 7,130 patents smaller than the sum of both groups.

29 Due to the co-classification the complete sample is 1,080 patents smaller than the sum of both groups.

30 Due to the co-classification the complete sample is 18,703 patents smaller than the sum of both groups.

31 Due to the co-classification the complete sample is 31,211 patents smaller than the sum of the five groups.

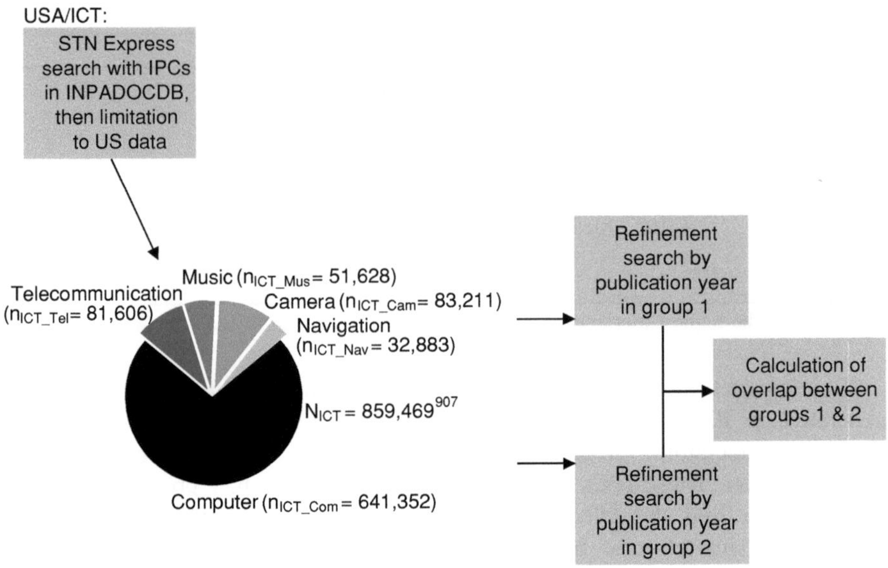

Fig. 5.21 Construction of STN express sample on ICT in the USA. (Graph prepared using think-cell® chart)

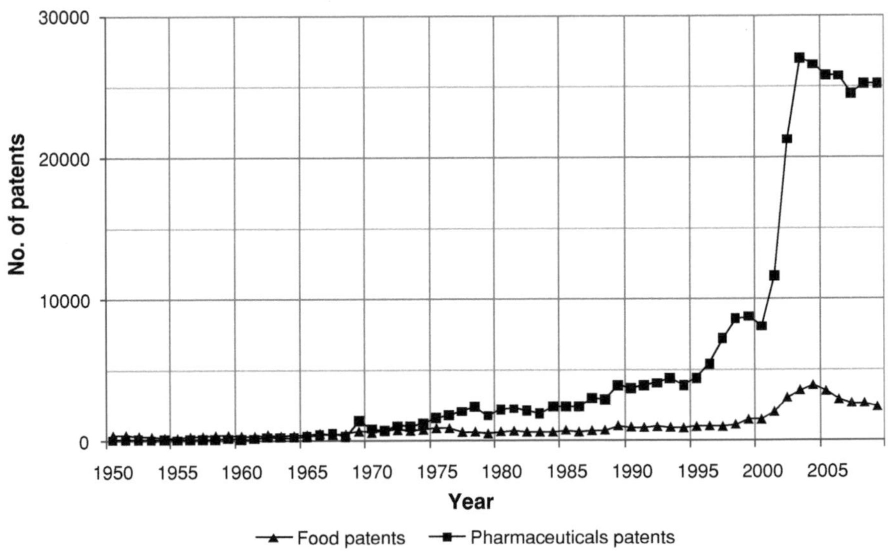

Fig. 5.22 Number of US food and pharmaceuticals patents between 1950 and 2009 (N_{NFF_US} = 341,846)

Table 5.9 Share of co-classified patents on NFF in the USA according to IPCs ($N_{NFF\_US} = 341{,}846$)

IPC group	Food ($n_{NFF\_US\_F} = 65{,}328$)	Pharmaceuticals ($n_{NFF\_US\_P} = 283{,}648$)
Food	–	2.5 %
Pharmaceuticals	10.9 %	–

Sources Adapted from [20, p. 12] and [11, p. 268]

approximately 2,500. In total, the 60 years under evaluation saw the publication of more than four times as many pharmaceuticals as food patents.

As the number of co-classified patents is an intersection of both quantities, the co-classification share differs accordingly (Table 5.9). While 10.9 % of all food patents are also classified as pharmaceuticals patents, the number is only 2.5 % vice versa.

Against the background of convergence's process character, these numbers only mirror a certain degree of interconnectedness. If this analysis is not to be destined to fall short of showing a growing corresponding trend, this needs to be demonstrated over the six decades. Consequently, Fig. 5.23 reports the shares of co-classified patents. For pharmaceuticals patents, the share has remained relatively constant well under 5 %. Numbers had risen slightly between the mid 1970s (approximately 1 %) and the mid 2000s (approximately 3.5 %) but have decreased again to 1.5 % in 2009. A similar development can be observed from the food patents' perspective. From approximately 3 % in 1977, shares had reached 26 % in 2003 before decreasing to 2009s 16.3 %.

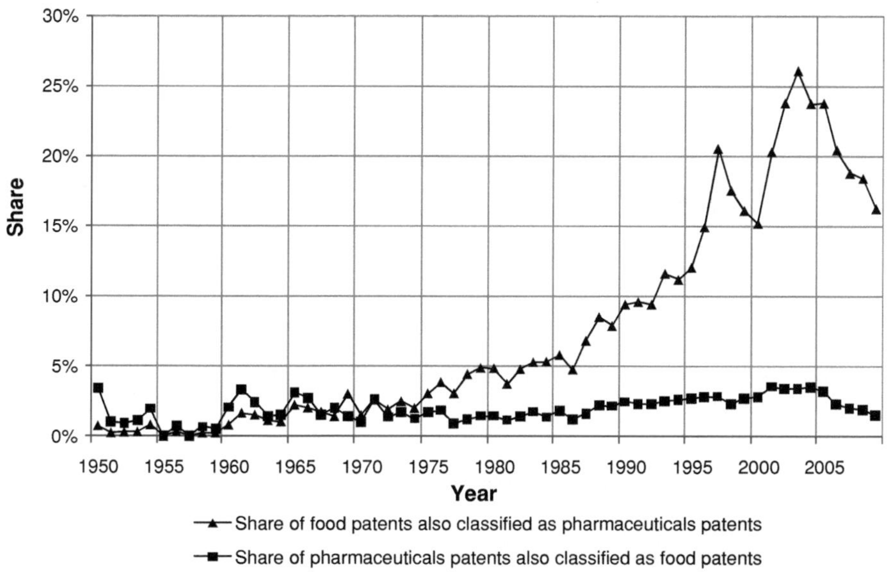

Fig. 5.23 Share of co-classified US food and pharmaceuticals patents between 1950 and 2009 ($N_{NFF\_US} = 341{,}846$). *Sources* Adapted from [20, p. 13] and [11, p. 269]

5.2.4 Analysis of NFF Patents in the UK

The nearly 70,000 patents in the same time period in the UK display a considerably different pattern. Again, there are many more pharmaceuticals patents—about three times as many as food patents. Both graphs peak in 1983 at nearly 2,500 and 600 patents, respectively (Fig. 5.24). Despite local *maxima* around 2000, both numbers have decreased massively since then to 228 and 60 in 2009.

Albeit different in absolute numbers, the co-classification pattern evolves quite similarly. Overall, 6.1 % of all food patents have been also classified as pharmaceuticals patents (2.1 % for pharmaceuticals/food patents; Table 5.10). With the exception of a much steeper decline of the share in food patents, the development over time is virtually parallel to that in US patents (Fig. 5.25). A difference lies in the generally lower shares for food patents and the slightly higher ones for pharmaceuticals patents. The maximum overlap was reached in 2001 with nearly 22 % of food and about 4.5 % of pharmaceuticals patents. These numbers have plummeted to fewer than 7 and 2 %, respectively, in 2009.

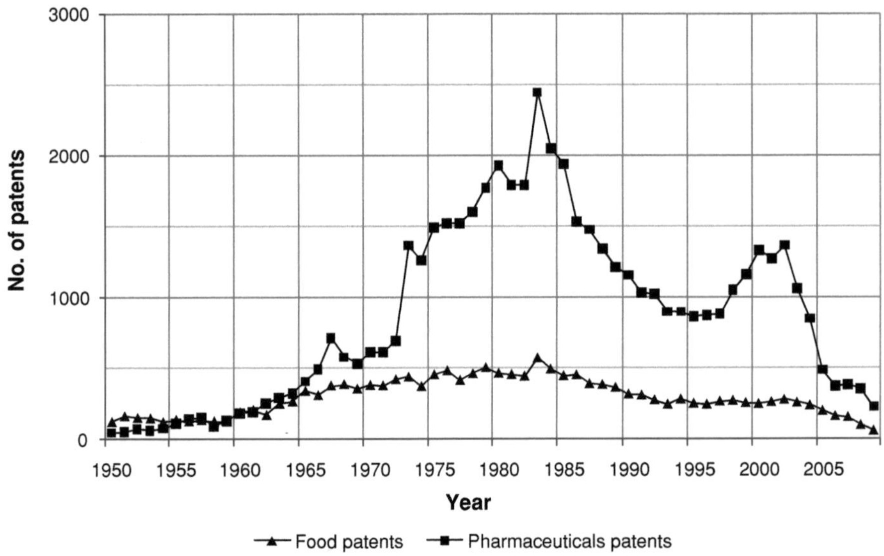

Fig. 5.24 Number of UK food and pharmaceuticals patents between 1950 and 2009 ($N_{NFF\_UK} = 69{,}094$)

Table 5.10 Share of co-classified patents on NFF in the UK according to IPCs ($N_{NFF\_UK} = 69{,}094$)

IPC group	Food ($n_{NFF\_UK\_F} = 17{,}714$)	Pharmaceuticals ($n_{NFF\_UK\_P} = 52{,}460$)
Food	–	2.1 %
Pharmaceuticals	6.1 %	–

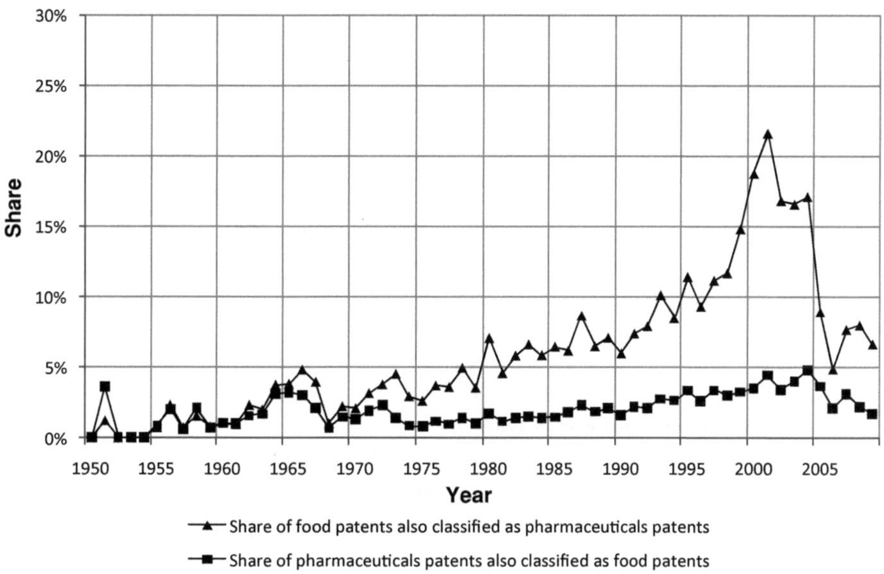

Fig. 5.25 Share of co-classified UK food and pharmaceuticals patents between 1950 and 2009 ($N_{NFF\_UK} = 69{,}094$)

5.2.5 Analysis of NFF Patents in Japan

Japan has been the forerunner in NFF developments since their early beginnings. Furthermore, patent counts in general are traditionally high in Japan—partly because of the Japanese patenting system but also because of the long history in a high propensity to patent. This is also reflected in the sample size. The Japanese sample is the largest of all three ones, with nearly half a million pharmaceuticals and food patents in the last 60 years. Strikingly, numbers are considerably less different than in the USA or the UK. In the light of multiples of three to over four in the UK and the USA, a multiple of 1.5 appears to be rather modest. Both graphs are nearly identical until the mid 1980s, when the number of food patents saw little to negative growth after nearly 15 years of decisive growth (Fig. 5.26). Having reached approximately 7,500 patents in 1991, they have oscillated around 5,500 since then. Contrastingly, the number of pharmaceuticals patents has continued to rise reaching an all time high of over 15,000 in 2008. Interestingly, there were virtually no patents at all in the database between 1950 and 1973 due to database restrictions (and IPC classification of Japanese patents in it).

The lower difference in the number of patents is also reflected in the share values (Table 5.11). 6.1 % of all pharmaceuticals patents and 9.2 % of all food patents have been assigned to both groups in the preceding 60 years. While the share for pharmaceuticals patents is consequently exceptionally high, the share for food patents is in between the US and the UK values.

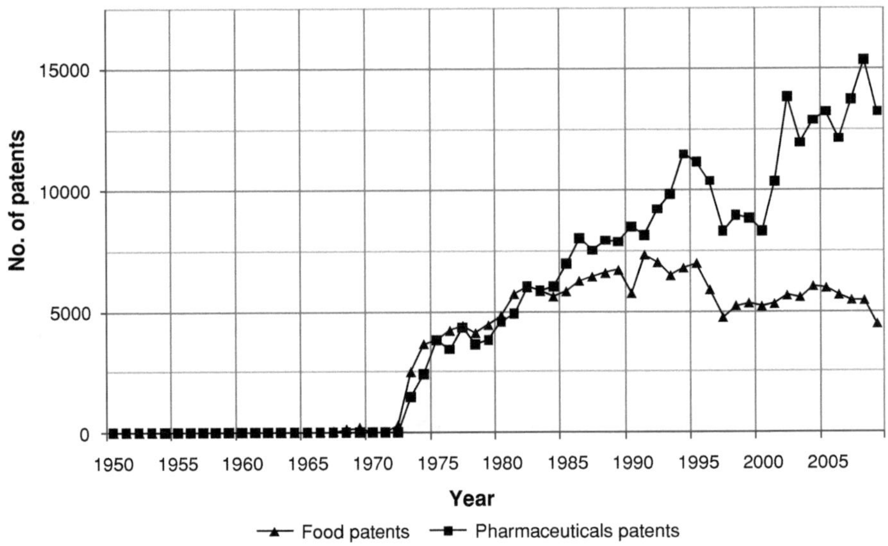

Fig. 5.26 Number of Japanese food and pharmaceuticals patents between 1950 and 2009 ($N_{NFF\_JP} = 493{,}565$)

Table 5.11 Share of co-classified patents on NFF in Japan according to IPCs ($N_{NFF\_JP} = 493{,}565$)

IPC group	Food ($n_{NFF\_JP\_F} = 203{,}987$)	Pharmaceuticals ($n_{NFF\_JP\_P} = 308{,}281$)
Food	–	6.1 %
Pharmaceuticals	9.2 %	–

Figure 5.27 displays the co-classification development over time. Due to the virtual lack of data for the first 20 years, no reliable numbers[32] could be calculated until 1973. After a sudden increase to around 2 %, shares did not change much until the late 1980s. Subsequently, the shares have risen continuously for both groups (steeper for the share in food patents and less steep for pharmaceuticals patents) until 2007 (26.8 %) and 2006 (11.8 %), respectively. Notwithstanding a slight decrease in the years afterwards, this begun much later and has (up to now) been much weaker than in the USA or the UK.

[32] Where they could be calculated without violating mathematical rules, they were 0 % for the first 22 years.

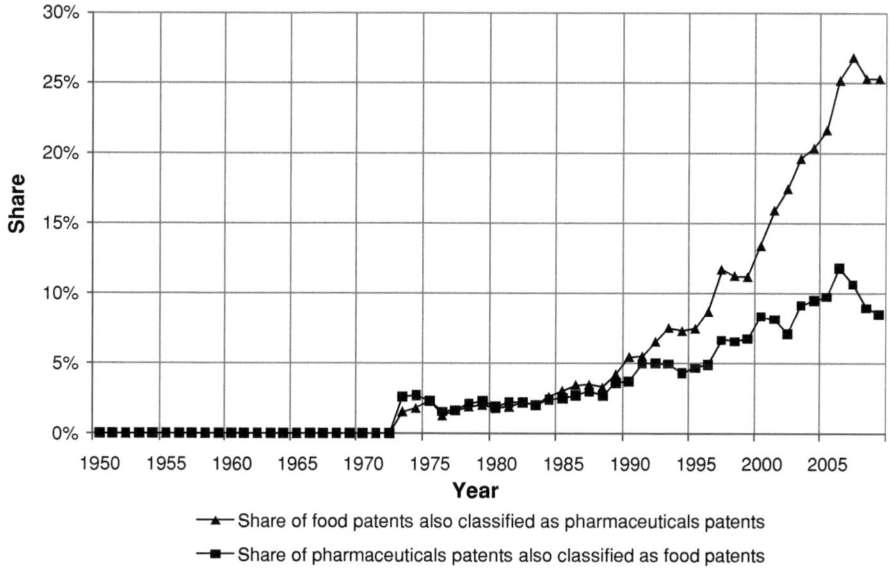

Fig. 5.27 Share of co-classified Japanese food and pharmaceuticals patents between 1950 and 2009 ($N_{NFF\_JP} = 493{,}565$)

5.2.6 Analysis of ICT Patents in the USA

When considering convergence in ICT, the clearest tendency towards blurring boundaries should be expected in respect to telecommunication and computer (both, technologies and industries).[33] Consequently, this is the focus of the analysis presented in the last paragraph of this chapter.

Figure 5.28 reports on the number of patents in both groups. Computer patents had more or less steadily increased since the 1950s (under 300) until their massive growth from 1997 (approximately 10,000) onwards, reaching over 87,000 in 2008. In contrast, telecommunication patents had varied from about 200 to just over 500 between the 1950s and the late 1980s. For another 15 years numbers multiplied to over 9,000 in 2004. Since then assignees seem to have lost interest, reducing the number of newly published patents to approximately 6,500 by 2009.

Co-classification around ICT-related technologies varies decisively (Table 5.12). While 15.9 % of telecommunication patents have been also classified as computer patents, only 0.2 % of them have been co-classified as camera patents. These numbers change to 2.0 and 0.2 % (i.e., remains the same for telecommunication/camera), respectively, when taking the other perspective. Many of the shares are below 1 % and therefore probably negligible. Contrastingly, 17.0 % of navigation patents and 12.1 % of music patents are also classified

[33] For a discussion of convergence patterns in ICT see Sect. 2.3.1.

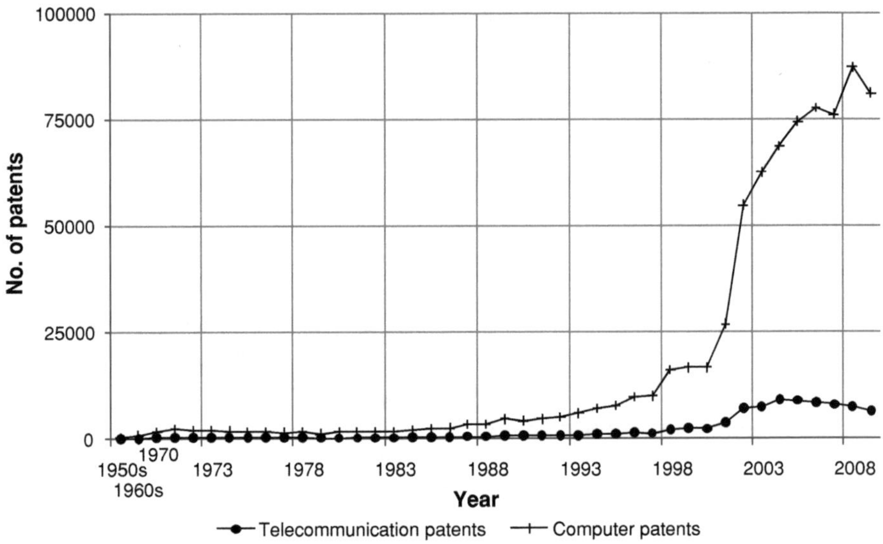

Fig. 5.28 Number of US telecommunication and computer patents between 1950 and 2009 (n_{ICT_TC} = 844,951) (This sample was aggregated later than the overall sample used in Table 5.12. Therefore, the numbers for telecommunication patents (n_{ICT_Tel}* = 93,009) and computer patents (n_{ICT_Com}* = 768,736) were slightly higher than in the original N_{ICT} dataset. Due to an increased overlap in the last year, also the overall share of co-classified patents rises to 18.1 % for telecommunication patents and 2.2 % for computer patents). (The first two peaks are the 10 year averages for the two decades between 1950 and 1969)

Table 5.12 Share of co-classified patents on ICT according to IPCs (N_{ICT} = 859,469)

IPC group	Telecommunication (n_{ICT_Tel} = 81,606) (%)	Computer (n_{ICT_Com} = 641,352) (%)	Music (n_{ICT_Mus} = 51,628) (%)	Camera (n_{ICT_Cam} = 83,211) (%)	Navigation (n_{ICT_Nav} = 32,883) (%)
Telecommunication	–	2.0	4.5	0.2	0.8
Computer	15.9	–	12.1	4.1	17.0
Music	2.9	1.0	–	0.2	0.6
Camera	0.2	0.5	0.3	–	2.4
Navigation	0.3	0.9	0.4	0.9	–

Sources Adapted from [20, p. 11] and [11, p. 267]

as computer patents, underlining the central role of computer technology in ICT developments.

In spite of the substantial difference of importance of co-classification between telecommunication and computer, both graphs run nearly completely simultaneously when considering (local) *minima* and *maxima* (Fig. 5.29). While telecommunication and computer patents both exhibited an increase of co-classification from the early 1950s to the early 2000s (about 2 and 1 % to about 27 and 3.5 %, respectively), the shares have again decreased to 18 and 1.5 % in 2009.

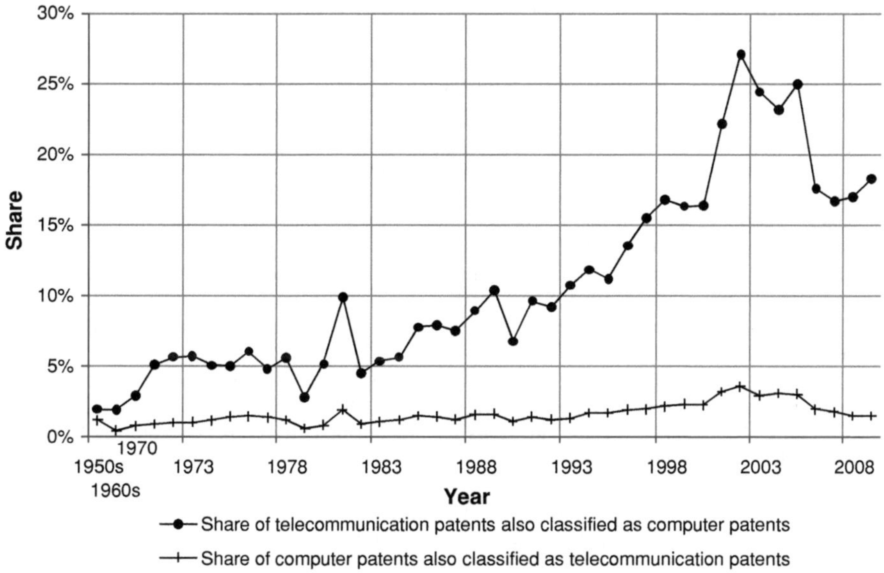

Fig. 5.29 Share of co-classified US telecommunication and computer patents between 1950 and 2009 ($n_{ICT\_TC} = 844{,}951$) (See Fig. 5.28). *Source* Adapted from [20, p. 12]

Overall, there are considerable similarities between each of the samples. First, in most cases[34] there is one technological field (and a closely related industry) which is much more affected by common developments due to smaller numbers of patents. Second, most potential convergence cases reached a maximum overlap in the area of 25 % and often this number decreased again in later years. Third, overlap between fields began to substantially grow in the 1970s to 1980s in all samples. Finally, the general shapes or patterns of progress resembled much of one another. Possible explanations for these similarities and differences will be developed and discussed in the following chapter. Additionally, the discussion will also evaluate to what degree the results presented in this chapter can answer the research questions derived in Chap. 4.

References

1. van Raan, A.F.J.: Advanced bibliometric methods to assess research performance and scientific development: Basic principles and recent practical applications. Res. Eval. **3**(3), 151–166 (1992)
2. Watts, R.J., Porter, A.L., Newman, N.C.: Innovation forecasting using bibliometrics. Compet. Intell. Rev. **9**(4), 11–19 (1998)

[34] In part with the exception of Japanese patents in the area of NFF.

3. CAS: SciFinder®: Part of the process™. http://www.cas.org/products/scifindr/index.html (2010). Accessed 13 Apr 2010
4. CAS: SciFinder: Part of the research process for universities around the world. http://www.cas.org/products/sfacad/index.html (2010). Accessed 13 Apr 2010
5. CAS: IPC7 List I: Guaranteed coverage. http://www.cas.org/expertise/cascontent/caplus/patcoverage/ipcguar.html (2009). Accessed 13 Apr 2010
6. CAS: SciFinder content at a glance: Biomedical information. http://www.cas.org/expertise/cascontent/ataglance/medlinecontent.html (2010). Accessed 13 Apr 2010
7. CAS: SciFinder content at a glance. http://www.cas.org/expertise/cascontent/ataglance/index.html (2010). Accessed 13 Apr 2010
8. CAS: SciFinder scholar™ how to guides. http://www.cas.org/support/academic/sf/htg/index.html (2009). Accessed 13 Apr 2010
9. Curran, C.-S., Bröring, S., Leker, J.: Anticipating converging industries using publicly available data. Technol. Forecast. Soc. Change **77**(3), 385–395 (2010)
10. Curran, C.-S., Bröring, S., Leker, J.: Patent analysis for the anticipation of converging industries. In: R&D management conference 2008, Ottawa, Canada, 17–20 June 2008
11. Curran, C.-S., Leker, J.: Patent indicators for monitoring convergence—examples from NFF and ICT. Technological Forecasting and Social Change **78**(2), 256–273 (2011). doi:10.1016/j.techfore.2010.06.021
12. Fischer, G., Lalyre, N.: Analysis and visualisation with host-based software: The features of STN® AnaVist™. World Pat. Inf. **28**, 312–318 (2006)
13. Yang, Y., Akers, L., Klose, T., Barcelon Yang, C.: Text mining and visualization tools: Impressions of emerging capabilities. World Pat. Inf. **30**, 280–293 (2008)
14. STN International: STN AnaVist: Analysis and visualization software for information professionals. http://www.stn-international.de/stn_anavist.html (2010). Accessed 16 Apr 2010
15. STN International: STN express with discover! http://www.stn-international.com/uploads/tx_ptgsarelatedfiles/INPADOCDB_07.pdf (2010). Accessed 16 Apr 2010
16. Curran, C.-S., Leker, J.: Employing STN AnaVist to Forecast Converging Industries. Int. J Innov. Manag. **13**(4), 637–664 (2009). doi:10.1142/S1363919609002455
17. WIPO: International patent classification. (2009)
18. Curran, C.-S., Leker, J.: Seeing the next iPhone coming your way: How to anticipate converging industries. In: Kocaoglu, D.F., Anderson, T.R., Daim, T.U., Jetter, A., Weber, C.M. (eds.) PICMET '09, Technology management in the age of fundamental change, Portland (OR), USA, 02–06 Aug 2009, pp. 383–395
19. STN International: INPADOCDB: The international patent DOCumentation DataBase, INPAFAMDB: The international patent FAMily DataBase: Handbook. pp. 1–242. Karlsruhe, Germany, (2010)
20. Curran, C.-S., John, G.M., Leker, J.: Finding the Convergence-Needle in the Patent-Haystack. In: Hosni, Y. (ed.) 19th International conference for the international association of management of technology: Technology as the foundation for economic growth, Cairo, Egypt, 08–11 Mar 2010, pp. 1–20

Chapter 6
Discussion of Findings and Implications for Theory and Practice

After presentation of the various results in the preceding chapter, the discussion will provide guidance through the often elaborate but noisy results and direct toward the most important findings. At the same time, shortcomings, limitations, and potential sources of error will be clearly laid out—especially the non-obvious ones. By answering the research questions and while discussing alternative explanations as well, the following chapter therefore aims at distilling a clear picture of the major findings. The remainder of this chapter is structured as follows. First, the main results will be briefly summarized and discussed in respect to immediate explanations and limitations. Second, based on these results answers to each of the research questions will be provided . Third, the different approaches will be compared to each other, including their overall limitations. Subsequently, the theoretical and practical implications of the study at hand will be outlined and suggestions for potential future research *foci* will conclude the discussion.

6.1 Discussion of Results

6.1.1 SciFinder Scholar

6.1.1.1 Scientific Articles

Evaluating the signs of convergence identified through SciFinder analyses, the first interesting result is the development of articles in different subject areas (Fig. 5.2; $n_{SF\_Pub\_i}$). The first 60 years saw only little interest in phytosterols, followed by only modest growth in the 1960s and exponential growth from the mid-1990s. It is now mainly driven by articles from food subjects and since 2000 to a lesser extent from pharmaceuticals. The driving force prior to that were the remaining subject areas and here predominantly articles dealing with biological or physiological topics of plants and their production of phytosterols. Later articles increasingly focused on the applicational value of this class. Strikingly, while publishing activities on pharmaceuticals and food aspects of phytosterols gained decisive

C.-S. Curran, *The Anticipation of Converging Industries*,
DOI: 10.1007/978-1-4471-5170-8_6, © Springer-Verlag London 2013

momentum in the last 10–15 years, personal care does not. It remains a negligible factor in scientific publications. This is in line with the fact that there are only few articles dealing with Cosmeceuticals, many of them being highly critical in regard to the real value of these products.[1] Also, Cosmeceuticals are much more consumer-driven than, e.g., pharmaceutical products and there are only very few publicly funded research institutions on cosmetics (which would be the most likely affiliations of scientific authors).

Considering also the number of scientific articles in the sub-sample of selected firms (Table 5.1; $n_{SF\_Pub\_C}$), it is interesting to notice that the personal care and the food subjects are decisively overrepresented (6 % in sub-sample vs. 1 % in sample and 66 vs. 31 %, respectively) whereas pharmaceuticals and the remaining subjects have a smaller share than in the complete phytosterols sample ($n_{SF\_Pub\_i}$; 4 vs. 12 % and 24 vs. 55 %, respectively). While caution should be generally exerted in light of the small number of articles, this finding indicates a higher degree of 'research consolidation'[2] exists in personal care and food, as companies were selected on the basis of their patent portfolio size. The companies at the top of the list of patentees were mostly large multinational corporations highly active in the businesses around NFF and Cosmeceuticals. Accordingly, they would have a natural (commercial) interest in pushing publications on positive health effects of foods or cosmetics. Excluded from the sample were research institutions, government agencies, and universities. The distribution of articles would thus indicate a lower likelihood of researchers from these institutions to report on food and especially personal care applications. As articles are more often positive in tone than negative, this would also be in line with the fact that so few articles on Cosmeceuticals have been published—as health effects are vigorously debated.[3] Due to tighter regulation around foods[4] there is, probably at the same time, also a greater demand for scientific discussion, which is also reflected in food industry's comparably (71 % of the articles by the sub-sample's firms are authored by the food industry) high propensity to publish scientific articles. Not surprisingly, articles on more basic research questions in pharmaceuticals and other subject areas are less likely to be published by the firms in the sub-sample. Lastly, firms related to chemicals and pharmaceuticals published the majority of their very few articles around food subjects or in none of the core subject areas. Bearing in mind

[1] See Sect. 4.3.5 for a short discussion of the literature on Cosmeceuticals.

[2] This research consolidation refers to intellectual property in the form of patents and scientific articles.

[3] As already discussed in Sect. 4.3, health claims of NFF are also in the center of attention. But more generally, their effects are not per se as much debated as in cosmetics, scientific articles focus rather on potential negative side effects and dose questions (and do not question an effect at all).

[4] This tighter regulation, when compared to cosmetics, is largely caused by the greater importance of nutrition and its effects on people's health, thus calling for much more science-based testing. Undoubtedly, most substances will have much stronger effects on the human body when incorporated orally than when topically applied.

the importance of their core scientific disciplines as ancillary sciences to NFF and Cosmeceuticals, this is far from surprising.

6.1.1.2 Patent Documents

The SciFinder analysis of patenting complements this picture. The share of patents has grown substantially, especially since the late 1980s (Fig. 5.3; N_{SF}). With its first local maximum in the 1940s, it follows a pattern similar to Ernst's 'technology life cycle' seen in patents (Fig. 6.1). He discerns between three phases of the life cycle of a technology. In the first '*emerging*' phase, patent applications will see a sudden increase. In the second '*consolidation*' phase, the number of applications is decreasing for a short time. In the third '*market penetration*' phase, the number of applications as well as the number of patenting entities grows sharply, before eventually interest in the analyzed technology may decrease again [cf. 1, pp. 367ff].[5]

While the growth in the patent ratio until, approximately, 1950 resembled the *emerging phase* with a modest sudden growth, the second phase until the early 1980s reminds of the *consolidation phase* with a limited decrease. The third phase with exponential growth bears analogy to the *market penetration phase*. It definitely corroborated a strong increase in commercial interest and one should expect the reaching of a limit soon, when most of the potential inventions around phytosterols would already have been patented or protected otherwise.

Looking at the distribution of patents according to originating industries (Fig. 5.4; $n_{SF\_Pat\_C}$) as well as targeted subject areas (Fig. 5.5; n_{SF_Pat}), very similar overall trends can be examined. While the sub-sample entails less than half the number of patents, it still shows very similar patterns, supporting the argument of unbiased company selection. The main difference between both figures is the more prominent role of patents from personal care and food in the sub-sample ($n_{SF\_Pat\_C}$). They are the most important contributing industries to the strong growth in patents since 1997. Their considerably larger share in the sub-sample points at the important role, pharmaceutical topics play in the patenting activity of the two more consumer-driven industries. Hence, this could serve as an indication for at least technology convergence. Analysis of co-classification of industries and subject clusters (Table 5.2; $n_{SF\_Pat\_C}$) demonstrates the dichotomous distribution of activities across the four industries. While the more active industries of personal care and food (184 and 152 patents, respectively) file for patents in mainly their core technological areas, the two less active industries pharmaceuticals and chemicals (58 and 57 patents, respectively) spread their activity far more across all subjects. Based on the fact that patents are granted on their commercial applicability and since especially the chemical companies find their customers mainly in

[5] Where next to the theoretical introduction also empirical data on developments around computerized numerical control (CNC) machine tool technology is presented. See also [2, p. 983], discussing that "patent growth generally follows a similar trend that can resemble s-shaped growth".

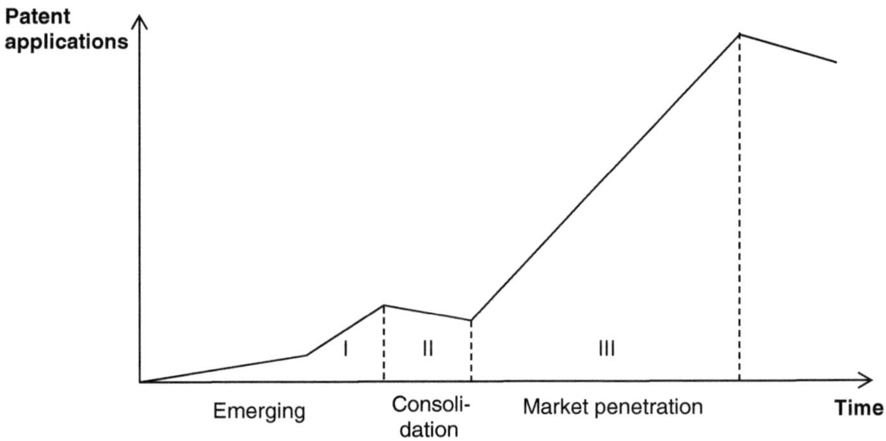

Fig. 6.1 Theoretical technological life cycle as measured in patent applications according to Ernst. *Source* Adapted from [1, p. 367] with kind permission from Springer Science+Business Media B.V.

the other three industries, this is easily understandable. Albeit these results could to some extent be interpreted as weak signals of a convergence process, the small role of, e.g., pharmaceuticals patents for the personal care and food industries is surprising. This deviation from the expected importance could be rooted in an assignment of a food industry's patent mainly to food subjects and only in very clear cases also pharmaceutical subject areas.[6]

A central line of reasoning in describing the convergence process was that it could be expected to experience science convergence first and only later technology convergence.[7] In consequence, signs of this science convergence (most importantly scientific publications) should be preceding signs of technology convergence (mainly patents). This would mean that also the scientific articles on phytosterols should on average be older than the corresponding patents. The findings on the *WAY* (weighted average year of publication) depicted in Table 5.3 could in principle support this. While across all industries and subject clusters the respective average patent was younger than the average article (between 0.8 and 24.7 years), the only exception was the chemical industry with a time lag of -3.0 years. A possible explanation may be the special role of the chemical industry in phytosterols. It is much less active regarding potential market applications, its competencies focus rather on topics such as methods for analysis, separation, or processing (of phytosterols). But as outlined before, this largely procedural

---

6 As this assignment is not conducted by patent authorities but by CAS, it cannot be fully excluded that there is a bias towards less cross-classifications.

7 See Sect. 2.4 for this study's reasoning on a sequential process of convergence. According to this reasoning, industry convergence does not necessarily have to be based on technological convergence. If it is, then it will be preceded by science convergence (Fig. 2.6).

knowledge is more difficult and less likely to be patented. Strikingly, the overall time lag (N_{SF}) is eight times higher than in the sub-sample ($n_{SF\_Pat\_C}/n_{SF\_Pub\_C}$) (15.9–2.0 years, respectively), indicating a high interest in commercialization and probably also a holding back of research results until promising findings are filed for patenting. One additional reason for the considerably higher age of the publications by organizations not within the company sub-sample could be based on academic papers focusing on general characterizations and descriptions of phytosterols, e.g., in their natural environment of plants (as presented in Table 5.1, 55 % of all publications are not from any the three dedicated subject clusters). In the personal care as well as in the food industry, the (second) oldest patents are from the other of the two subject areas. This might speak for a development from basic patents (subject cluster 'rest') over the respective other area and pharmaceuticals to the primary subject of the industry sector itself. This would be in line with the reasoning of a trend of convergence, as this means companies have seen possible applications of 'their' products in the other sector and have deemed it interesting enough to have it patented. Apparently, both industry groups have identified these cross-sectional applications, pointing to a certain degree of fading industry boundaries.

Although the WAY seems to yield useful results for this study's approach, the findings have to be interpreted with caution. Most of the sample sizes are comparably small and the weighted average year of publication might be the same for very different patent distributions. For instance, a set P_1 with only two patents (one in 1950 and one in 2000) would have a WAY of 1975.0. A second set P_2 with 51 patents (one in each year between 1950 and 2000) and a third set P_3 with 20 patents (ten in 1974 and 10 in 1976) would both yield an identical WAY of 1975.0. All three samples would yield the same WAY, although the distributions and thus the underlying patenting activities would be undeniably extremely different. Two approaches could potentially heal this problem. First, a look at the distributions instead of just the WAY would be more informative—but at the same time also considerably more difficult to handle and feed into a combination of different measures. Second, using the median age of publications would shed more light on skewed distributions (although that would help little in the abovementioned example). Such an approach is taken with the TCT indicator used in patent analyses. It is defined as the "median age of patents" (normally those cited in a patent document). The line of argument is that a higher TCT indicates a slower replacement of one technology by its successor [cf. 3, pp. 127ff]; [4, pp. 47ff]. The median was not used in this study, as this would not reflect the long-term history of a development. A set P_4 including three patents (one in 1950, one in 1999 and one in 2009) would have a WAY of 1986.0 and a median age of 1999. Its median age would remain unchanged, if the first patent had been published in 1980 instead of 1950, but its WAY would increase to 1996.0—at least indicating a different distribution. In summary, the WAY was not calculated as a proof for convergence, but it does serve as a supplementary measure shedding more light on the development of activities over time.

6.1.2 STN AnaVist

6.1.2.1 Scientific Articles

While similar in the sampling approach, the analysis functionalities of the second software tool in use were decisively different. Considering the STN AnaVist-based results, they can again be grouped into results from analyses of scientific articles and from patents. The first research landscapes including the complete sample serve well to provide a first overview and to identify general hotspots. But for an assessment of convergence they are of little use. This is different with the series of landscapes in Fig. 5.12 ($n_{A\_Pub\_A}$). Moving from a scattered plot of about two dozen in the 1960s, publications were more and more concentrated around two areas of a predominantly medical related peak and a mainly chemical/biological related peak. Although these peaks were to a lesser extent the domain of only one group than earlier ones, this growing interconnectedness changed into a specialist approach in subsequent decades. This is in accordance with other studies, reporting on increasing specialization and a growing necessity for interdisciplinary research approaches, as single research groups or organizations will be less likely to possess all of the required knowledge itself.[8] Specialization is even reflected in the names of affiliations. Most are a lot less general or broad today than they were some decades ago. This tendency could be expected to be generally valid especially in science—if a new field is formed due to increasing collaboration it is often followed by more specialized sub-fields.[9]

On the limitation side of this kind of research landscapes, the way they are created has to be carefully considered as a potential source of errors. The underlying algorithm can be adjusted very decisively, based on:

- Inclusion or exclusion and grouping of clustering concepts
- Concept frequency thresholds (10–70 %)
- Sources of clustering concepts (title, claims, abstracts, IPCs)
- Inclusion or exclusion and grouping of individual documents.

Naturally, there is a benefit to such a possibility but a danger as well. On the one hand, these adjustments can be used to increase the discriminatory power of this analysis tool. On the other hand, it may be tempting to try out different combinations until a desired outcome is achieved. And even unwilling influence may be exerted through the abovementioned parameters or through the mere

[8] For a similar argumentation see e.g., [5–7].

[9] Future studies should analyze whether such a specialization-convergence-specialization pattern can also be found in other converging areas. Generally, convergence should not be viewed as a static situation, but rather as a highly dynamic process with changing momentum and impact. Most likely will convergence in NFF and in ICT also be followed by a later separation of different industrial *foci*. This is in line with the reasoning in Chap. 2 and bears more analogy to a reshuffling (of industry segments) than a melting into one unified industry.

reorganization of landscapes. If a landscape is created again it will most likely bear resemblance, but may be different in parts. The present study did not use any of the possibilities to alter the basis for clustering and clearly pointed out all necessary user adjustments. Wherever possible, no further manual adjustments or refinements were done to limit the investigator's influence.[10] At the same time, this study aims to develop an approach that is less susceptible to (un-) intentional manipulation. Within certain limits, this makes an uncovering of cloaked patterns necessary.

Turning to co-authorship patterns, Figs. 5.13, 5.14 and 5.15 represent a general trend in the natural sciences. Whereas at the beginning of the twentieth century many scientists were generalists across a large part of the natural sciences, the specialization has grown, paralleling the exponentially increasing amount of knowledge. At the beginning of research into phytosterols, most scientists concentrated on either their chemical characteristics or their effects in a medical sense. But even the articles by researchers from the medical field often dealt with chemical properties of single phytosterols. With realization of the possible applications in the area of medical or nutritional entities came an awareness of the necessity of a knowledge combination. Neither of the specialists was in most cases able to discuss and explore possible applications at the intersection of all three disciplines without the aid of other sources. Thus, the content of an average article as well as the (co-) authorship structure changed. With a tendency toward more applied research and the demand for more focused solutions, this changed again to the more specialized picture illustrated in the last two decades. Hence this study argues that scientific convergence has occurred in the NFF sector and has moved on to represent a research focus in its own right, as represented in scientific publications, specialized journals, and research institutions. A caveat needing to be mentioned exists in the low number of articles, especially in the decades with many cross-disciplinary articles. These low numbers may have influenced the observed pattern. Albeit the existence of an artifact cannot be 100 % denied, it remains highly unlikely as most other decades did not see many more articles. So the question would remain unanswered why this one decade should have been different across all three areas.

6.1.2.2 Patent Documents

Noticeably, all four IPC clusters do not only run nearly parallel until the early 2000s (Fig. 5.16; $n_{A\_Pat}*$) but also amount to very similar numbers (Table 5.6; n_{A_Pat}). A potentially unrelated but parallel development does not necessarily point at convergence. Contrastingly, as co-classification is based on technological links,

---

[10] Nevertheless, it would be interesting to analyze the effect of such adjustments. As is done in the optimization of, e.g., fermentation media for bacteria, ideally *design of experiments* should be used to find the maximum difference caused by different combinations of the parameters.

it may refer to convergence, especially when analyzed longitudinally over time. The co-classification pattern reported in Table 5.6 exhibits strong technological links between personal care and pharmaceuticals and particularly between food and pharmaceuticals. Remarkably, this indication of convergence is thwarted by the longitudinal analysis depicted in Fig. 5.18. All four of the co-classification shares vary decisively over the course of time but exhibit negative trends. While the static analysis thus leads to the impression of increasing convergence, the process perspective transmits an opposite picture. A caveat in respect to the development over time exists in the fact that yearly sample sizes are very small, ranging from four patents (personal care in 1998 and 1999) to 41 (food in 2007). In consequence, Fig. 5.17 does not negate Table 5.6 per se, but it does serve as a clear danger signal and stresses the importance of analyzing convergence predominantly from a process perspective.

The industry/IPC co-classification results in Table 5.7 suffer from a similar problem. While numbers in each section were too small to conduct an analysis over time, the static breakdown yields an unsurprising concentration of industry groups on their core technological field[11] for personal care and food. This holds true at least as long as it can be assumed that the large number of patents by food companies not containing IPCs could be proportionally assigned to the IPC groups. Then food companies' concentration on food IPCs would reach approximately 75 %. Numbers are considerably lower in pharmaceuticals and chemicals, which again points to the emphasis of the most central applications in NFF and Cosmeceuticals.

Finally, the graphical analysis of research hotspot development between 1999 and 2008 as done in Fig. 5.17 does constitute an indication of convergence. Both, the evolution of two main areas related to NFF and Cosmeceuticals and the growing mix of patents from different industries point in this direction. The landscapes further add to the impression of a strong correlation between (a) pharmaceuticals and personal care and (b) pharmaceuticals and food. This is in line with the hitherto discussed results from co-classification. Allowedly, this is neither a quantifiable measure nor very close to a proof of convergence. In accordance, the next paragraph will discuss the results derived from a simple and large-scale calculation of co-classification shares.

6.1.3 Classification-Based Analyses: STN Express

6.1.3.1 Discussion of Sample Selection

Most of the aforementioned, keyword-based analyses pointed toward a convergence in NFF and Cosmeceuticals, at least in respect to the limited example of

[11] Concentration rates are: 92 % in personal care, 48 % in food, 52 % in pharmaceuticals, and 35 % in chemicals.

phytosterols. At the same time several serious concerns surfaced, some particular to individual analyses but many also applying more to the general approach. Thus, the last set of results was derived from analyses of IPC co-classification in NFF and ICT. One central advantage of this approach is the fact that investigators do not need expert knowledge in the area under scrutiny. Knowing which broad technological area is of interest to them, or to assess the range of IPCs a certain firm or industry is most active and interested in, is sufficient for a first evaluation of convergence influence. The degree of co-classification (as a proxy for convergence) can be easily calculated for one point in time and even as a longitudinal examination. The corresponding analyses were conducted on the basis of three NFF samples from the USA (N_{NFF_US}), the UK (N_{NFF_UK}), and Japan (N_{NFF_JP}), and one ICT sample (N_{ICT_US}). Altogether, 1.76 million patents were analyzed in respect to pairwise co-classification patterns to demonstrate the applicability of this measure in different countries and fields.

Grounding the IPC selection on WIPO's catchword index [cf., e.g., 8, p. 296].[12] (as done in this study) limits investigators' influence on sample creation, which could potentially lead to some kind of bias. Three individual national patent authorities were chosen to exclude family patents from the analysis but still be able to compare country characteristics. Chosen were three of the most important countries for NFF—Japan, the USA, and the UK. The three patent offices also use three different specific patent classifications (which are then translated into the IPC classification). This was done to ensure maximum applicability of the convergence measurement across country (and industry) boundaries.

6.1.3.2 Share of Co-classified Food Patents in the USA, the UK, and Japan

Remarkably, all three countries exhibit a similar pattern when it comes to the central results from the time-dependent analysis of co-classification shares (Fig. 6.2). The most important deviation from this common line lies in the fact that Japan seems to have experienced the weakest decrease in co-classification, while it is most obvious in the UK. Moreover, the Japanese graph runs nearly exactly parallel to the US graph, only with a delay of 2 to 10 years. For instance, the share in the USA reached 5 % around 1980, in Japan in 1990. Later this time distance decreased to 2 years (e.g., 15 % reached in 1999 in the USA and in 2001 in Japan) and increased again to about 4 years after 2000. Moreover, the maximum share was very similar in both countries with 26.1 % for US, 26.8 % for Japanese, but only 21.7 % for UK patents. Patenting activity in the UK appeared to be a little further ahead of US developments, resulting in the (lower) UK maximum to be reached in 2001, 2 years prior to the US maximum and even 6 years earlier than

[12] This approach has been suggested by other scholars as well.

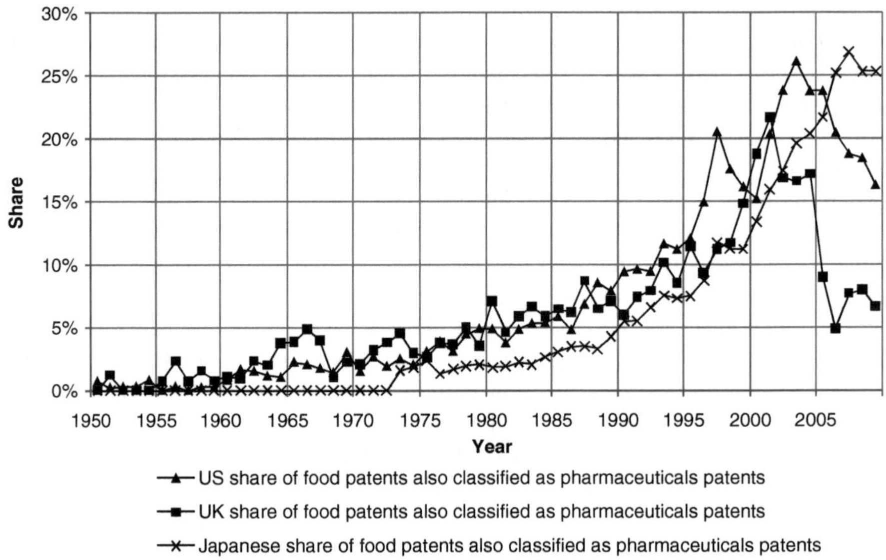

Fig. 6.2 Share of co-classified US, UK and Japanese food and pharmaceuticals patents between 1950 and 2009 (The used samples are the original samples: $N_{NFF\_US} = 341{,}846$, $N_{NFF\_UK} = 69{,}094$ and $N_{NFF\_JP} = 493{,}565$)

the Japanese maximum. The following characteristics may explain the similarities and differences in the comparison of the three countries.

First, developments in respect to an internalization of research and market activities, especially by large multinational corporations,[13] lead to very similar and broadly applied patenting activities in the most important markets. As adumbrated earlier, the majority of pharmaceuticals-related inventions will be filed for patent protections at the trilateral offices (and additional important offices). Accordingly, developments should be very similar in a defined technological area. Dissimilarities should only result from differences in the respective patent systems, such as classification rationales or patentability differences.[14] Accordingly, the graphs for all three countries exhibit a considerable degree of similarity.

Second, a difference in the maximum share between Japan and the USA on the one hand and the UK on the other hand could have resulted from higher patentability hurdles in the UK or from slightly lower acceptance of functional foods in the UK. This might have resulted in some functional foods not being patented there, partly also due to tighter regulations in the European Union.[15] Also, there

[13] These companies play a specifically important role in the increasingly consolidated food and pharmaceutical industries. See, e.g., [9, p. 13].

[14] See Sect. 3.3.1 for a description of differences between Europe, the USA and Japan.

[15] This could be tested in future studies by comparing UK results to patenting patterns from other European countries.

might be more other food-related patenting activities in the UK than in the other two countries, leading to a lower ratio of NFF-related patents over all food patents. Additionally, a general reservation in respect to UK patent activities lies in the uncommon development over time. In contrast to other countries (and the general worldwide trend), the number of patents does not increase but rather decrease since it is maximum in 1983. This is not only true for the IPC groups under scrutiny in the present study, but it is indeed a general pattern for patent applications and especially granted patents [cf. 10, 11]. Both experienced a maximum level in the 1970s, which will have led to the described shapes of the graphs in the study at hand. The small time lag will probably be attributable to the differences between publication year (in our samples) and filing dates (used in WIPO's statistics), but also due to differences between IPC clusters. There are two causes for this development. European Patents (EPs) are only 'designated' to a selection of the member states and do thus not count as a national application. This is in contrast to international PCT applications, which have to be filed through any of the national authorities and are thus also counted as one application of this respective patent office [cf., e.g., 12, pp. 6ff; 13, pp. 20ff]. Hence, the growing importance of the EP is reflected in a reduction in national applications, including the ones in the UK. Furthermore, the employed database and software tool do not provide the patents *designated* to a national office if the search field "patent country"[16] is used, but rather only the ones *filed* at this office. As the co-classification analysis is based on relative and not absolute measures, this should not have posed a problem—as numbers were still large enough for a reasonable analysis.

Third, the time lag of US and especially Japanese patents over UK patents will most likely be the result of differing paces at which patents are handled in each of the trilateral offices. While the EPO has the on average shortest handling times for patents as well as smaller backlog, the US system and especially the Japanese system are decisively slower with larger backlog. It should be kept in mind that, for instance, average first-action pendency periods for examination are 9, 21, and 26 months, respectively. While Japan has recently introduced measures to counter resulting potential negative side effects by speeding up the handling of patent applications, this only applies to a limited number of patents.[17] Reconsidering Ernst's technology life cycle (Fig. 6.1) and the results of the patent share in phytosterols, all three present graphs do not follow a pattern similar—unless the drop in share in the most recent years would be indicative of the beginning of the consolidation phase in NFF. With shares already reaching 25 %, it seems unlikely that the future should bring a far more drastic increase in shares.

---

[16] To generate references for both types of patents the *super search field* 'PCS' (patent country group, which is a combination of PC = patent country and DS = designated states) instead of 'PC' (patent country) has to be used additionally. As explicitly all applications and granted patents in each of the respective countries was to be sampled, this was not done in this study.

[17] See Sect. 3.3.1.

6.1.3.3 Share of Co-classified Pharmaceuticals Patents

The differences in developments over time also explain the differences in the overall proportion of food patents co-classified as pharmaceuticals patents. These numbers deviate considerably, but without changing the general distribution. In the UK the share is only 6.1 %, whereas it is 9.2 % in the somewhat lagging Japanese patent collection and 10.9 % in the USA (Tables 5.9, 5.10 and 5.11). According to the line of reasoning, the share should be growing in Japan over the next few years and thus approach the US number. Interestingly, this situation is different when it comes to the share from pharmaceuticals' point of view. The overall share for the last 60 years is considerably higher in Japan (6.1 %) than in the USA (2.5 %) and the UK (2.1 %). The same holds true when looking at the development over time, very similar developments the USA and the UK with maximum shares of just under 5 %, but a much more decisive increase in Japan resulting in up to 11.8 % of all pharmaceuticals patents being co-classified as food patents (Figs. 5.23, 5.25 and 5.27). This unusually high share in pharmaceuticals is caused by the exceptionally large number of food patents in Japan, driven by the very high acceptance of (functional) food innovations in the Japanese market. As discussed in Sect. 4.3, this rarely challenged acceptance has led to a highly innovative market and a well-established industry structure in Japan.

Figures 5.22, 5.24 and 5.26 are the only figures which are each decisively different from the other two. The USA saw an explosion in published documents between 2001 and 2004 (introduction of the publication of applications) and more or less steady numbers since then. The UK graph follows a completely different pattern, reaching its peak in 1983 and falling drastically since then. While in both countries the food patent trends were very similar (just on a lower level) to the pharmaceuticals patents, Japanese food patent numbers have been slightly shrinking since the early 1990s when at the same time pharmaceuticals patents overall retained their strong growth rate. This finding is surprising, as co-classified pharmaceuticals/food patents continues to strongly grow in numbers as well as in shares during the same period (Fig. 5.27). Moreover, the relative growth was much stronger in food patents, but also clearly observable in respect to pharmaceuticals patents.[18] Apparently, the importance of NFF has risen substantially within food as well as pharmaceutical technologies. This can probably again be attributed to the unique Japanese NFF market where customers appear to have traditionally been much more convinced regarding the health-related effects than in other markets. Accordingly, due to this stronger competition the pharmaceutical industry had more reason to engage in NFF activities.

[18] The share grew from 5 % in both IPC clusters at the beginning of the 1990s to over 25 % at the end of the 2000s in food patents and around 10 % in pharmaceuticals patents.

6.1.3.4 IPC Co-classification as a Sign of Convergence

As argued earlier, a growth in IPC co-classification over time is seen as a proxy for technology convergence, thus pointing at potential industry convergence. On that account, the discussed results from NFF (verified in three different countries) would confirm a technology convergence between the areas of food and pharmaceuticals. The results exhibited in Fig. 5.19 ensured that co-classification is no random or evenly distributed phenomenon across all different sections and groups. Another line of reasoning could be based on some underlying pattern leading to a growing co-classification between any two closely related technological fields. As both fields, pharmaceuticals and food, are also closely related to invention in chemicals, this study also analyzed the co-classification patterns between chemicals and food/pharmaceuticals for each of the three countries. Again, Japanese numbers are zero (or close to it) until the early 1970s [cf. 8, pp. 296f; 14].[19] Each of the cases yields different results but presents largely similar results across countries.

Looking at the co-classification of chemicals and food, co-classification is negligible from the chemicals' viewpoint, as these IPCs are listed on between 14 (Japan) and 22 (UK) times as many patents. Moreover, the shares remain comparably steady between 1 and 2 %. This picture changes from food patents' perspective. As shown in Fig. 6.3 shares grew to about 40 % in 1975 in the USA and the UK, falling to slightly lower levels in the following. In Japan numbers, reached are only about 20 % and they have remained to slowly grow up to the present day.

It could be argued from these results that there was a process of technology convergence around food and chemicals up to the 1970s. Undisputedly, both areas are strongly linked as the food industry uses many chemical processes and substances to produce especially processed food products. Several facts speak against such an argument. First, there were no further signs of industry convergence in the 1970s. No signs of massive industry regroupings surfaced, nor were many new food-chemical firms or even a whole industry created.[20] Second, also patent classification has not changed to account for such a newly converged field, which could have been another indication of convergence. Third and most importantly, chemical firms are supplying more and more ingredients to the food industry, explaining a growing overlap of IPCs beyond a technology convergence. The chemical industry is indeed a very important supplier of the food industry, without engaging in food production or marketing.

---

[19] IPC classification is contained in the INPADOCDB file only as of the early 1970s for most countries. Hence, more Japanese patents exist on food and pharmaceuticals in the years prior to 1970, but will not be returned when searching INPADOCDB based on IPCs. However, there is no reason to believe in a decisively different pattern if all patents could have been included in the analyses.

[20] Although it has to be borne in mind that technology convergence does not have to be followed by complete industry convergence, it is still a likely connection. See also Sect. 2.4.

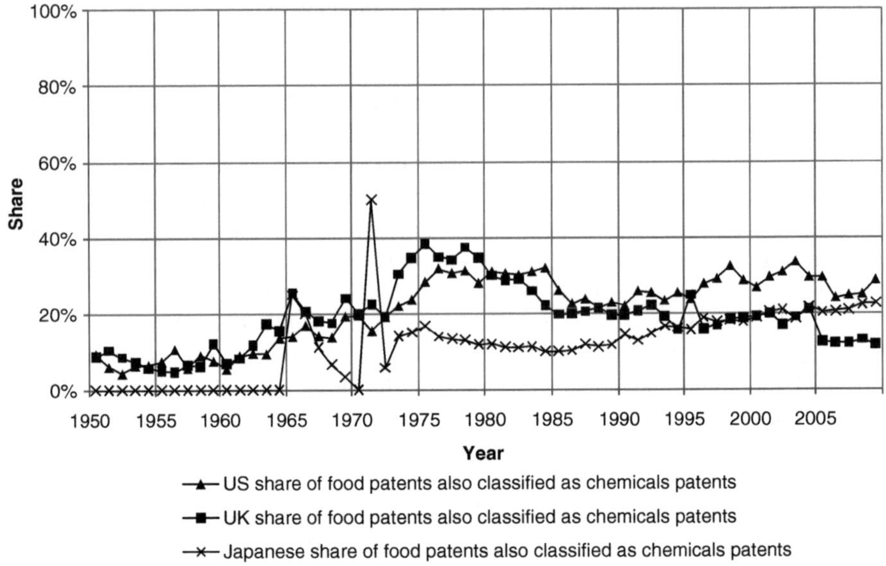

Fig. 6.3 Share of co-classified US, UK and Japanese food and chemicals patents between 1950 and 2009 (The used samples are: $N_{FC\_US} = 1{,}358{,}609$, $N_{FC\_UK} = 395{,}976$ and $N_{FC\_JP} = 2{,}969{,}582$)

Despite an also much larger body of patents in chemicals (between four (USA) and ten (Japan) times as many chemicals-related patents than pharmaceuticals-related), the changing developments are depicted from the pharmaceuticals (Fig. 6.4) and the chemicals (Fig. 6.5) point of view. Besides the very low numbers in Japan, the patterns are nearly identical for all three countries when looking at co-classified pharmaceuticals patents. Starting at about 20 % in the 1950s, shares rose to approximately 90 % by 1975 and afterward decreased to around 60 % today.

The increasing importance of chemistry for pharmaceutical developments can be easily explained by the evolution of pharmaceutical drugs from using largely biological components to being based nearly solely on chemical entities. Over the last decades biotechnological drugs have been introduced to a greater extent, which are capable of mimicking biochemical processes and substances in the human body. Thus, they make it possible to combat and sometimes even heal illnesses, which could not be fought with purely chemical drugs—leading to a little less dominant role of chemistry in the pharmaceutical sciences.

Contrastingly, the development of pharmaceutical drugs and diagnostics constitutes an increasingly appealing outlet for (particularly new) chemical products (Fig. 6.5). Rooted in the ever growing size and importance of the healthcare market and combined with the high returns in the pharmaceutical industry, many chemical companies have been putting a lot of attention on their role as supplier to the pharmaceutical industry. This is even more so the case, where other potential

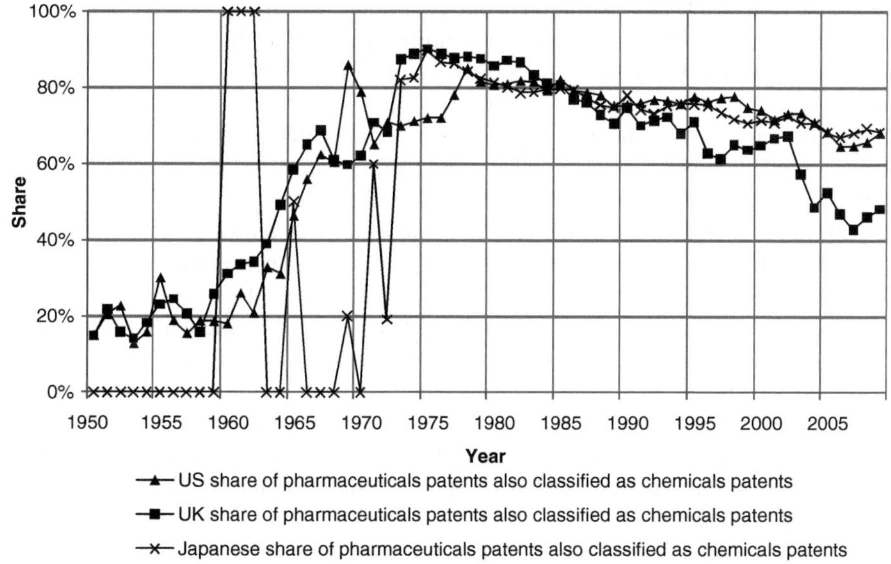

Fig. 6.4 Share of co-classified US, UK and Japanese pharmaceuticals and chemicals patents between 1950 and 2009 (The used samples are: $N_{PC\_US} = 1{,}404{,}569$, $N_{PC\_UK} = 395{,}185$ and $N_{PC\_JP} = 2{,}871{,}894$)

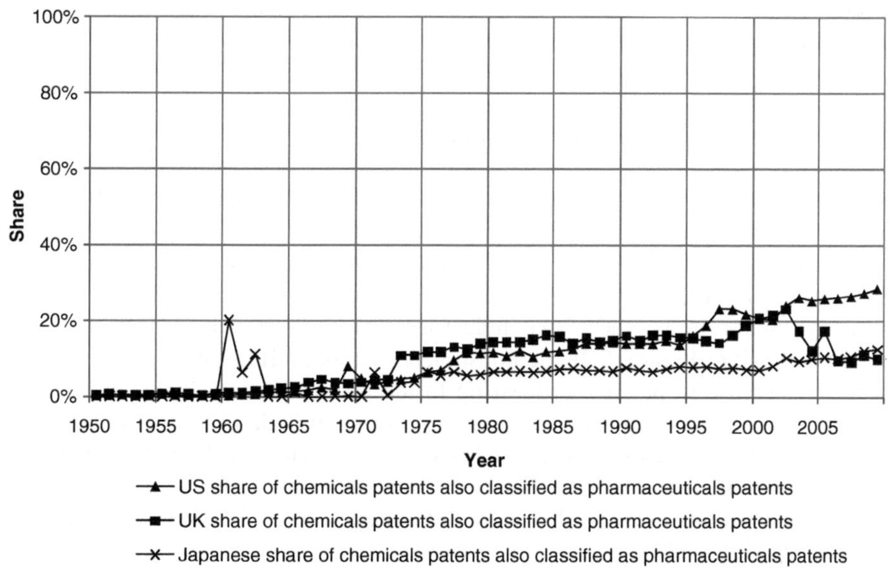

Fig. 6.5 Share of co-classified US, UK and Japanese chemicals and pharmaceuticals patents between 1950 and 2009 (The used samples are: $N_{PC\_US} = 1{,}404{,}569$, $N_{PC\_UK} = 395{,}185$ and $N_{PC\_JP} = 2{,}871{,}894$)

customer industries have suffered from economic downturns. These phenomena could explain the relatively slow but steadily growing (with the exception of the UK since 2003) numbers of co-classified chemicals patents, approaching 30 % in the USA and 20 % in Japan.[21]

In summary, both industries share a long common history with many chemical firms having being pharmaceuticals' producers as well for decades. Most companies have later focused on their core chemical businesses and divested their pharmaceutical activities. However, it could again be reasoned that there were (based on pharmaceuticals patents) or still are (based on chemicals patents) tendencies of convergence. Three observations negate such a line of reasoning. First, there appears to be no real activities in either of the two industries pointing at convergence. Second, the supposed convergence trends in Fig. 6.5 are unusually slow and steady for the more radical changes found in the light of converging technologies and industries. Third, while shares of co-classified patents differ distinctively between food and pharmaceuticals in NFF, both graphs evolve with a similar pattern. Contrarily, these developments are completely different in chemicals and pharmaceuticals.

In consequence, it is argued that the presented results on NFF do present support for convergence between pharmaceuticals and food. A further analysis was incorporated in the present study, however, to rule out another possibility of an alternative explanation. The characteristics observed in NFF could theoretically represent a single anomaly or a pattern unique to the involved technologies. Therefore, the most widely accepted example of convergence was examined in the final analysis presented in Chap. 5.

6.1.3.5 Share of Co-classified ICT Patents

Section 2.3.1 analyzed in greater details signs of convergence in ICT. Being the industry probably most affected by a convergence to date, many actions on all levels and by all stakeholders were implemented to react to this convergence. In the very focus of these processes were the telecommunication and the computer industry. Accordingly, these two industries were analyzed regarding their patenting behavior. Figure 5.28 demonstrated that both IPC groups have seen a decisive increase in patents, at least up to the mid-2000s. At the same time, the number of patents is larger by about an order of magnitude in respect to computer inventions. The time-dependent analysis exhibits wide analogy between both graphs, but while more than 25 % of telecommunication patents were co-classified as computer patents in 2002, this share did not even reach 5 % from the computer perspective (Fig. 5.29).

[21] The considerable interest in the pharmaceutical industry is also reflected in M&A activities between the chemical and the pharmaceutical industry.

While the developments over time covered only the central hotspot of ICT convergence, Table 5.12 presented also other IPC groups related to product developments in ICT, particularly smartphones. Thus three further groups were added, covering music, camera, and navigation patents. All three groups are substantially smaller than the computer group and in a similar range as telecommunication. Consequently, computer patents exhibit very low co-classification shares with the remaining four groups (between 0.5 and 2.0 %). Besides telecommunication and computer and two other combinations, all pairwise overlap is in the lower one-digit percentage reason. The other two exceptions are music and computer and navigation and computer (12.1 and 17.0 %, respectively). Unsurprisingly, recent developments in both fields have often (or always as in respect to navigation) relied on the application of computer functionalities. Based on the static overlap of IPCs alone, it is difficult to tell the difference between potential convergence and mere utilization of computer electronics. Furthermore, future research would have to clarify whether these two links also exhibit clear tendencies toward increasing convergence.

All in all, these results again stress the asymmetry in convergence effects. From the abovementioned results one should expect much more decisive changes in the telecommunications and the food industry than in the computer and pharmaceutical industry. Outstandingly, these ICT results are very similar to the results obtained in NFF, which is again taken as clearly mirroring convergence between telecommunication and computer technologies. Even a decline in co-classification at the end of the timeline is included. Complemented by convergence of products, markets, and in part even industries as discussed in Sect. 2.3.1, this underlines the explanatory power of the discussed NFF results.

6.2 Answering the Research Questions

Grounded on the previous discussion of results and aiming at increasing clarity, the following paragraphs will condense these results and their discussion into brief answers to the research questions posed in Sect. 4.2 of this study.

Q.1: Can convergence be anticipated on the basis of weak signals in publicly available data?

Yes, it could be demonstrated that convergence can be anticipated by scrutinizing publicly available data.

Besides discussing developments around NFF, Cosmeceuticals and ICT qualitatively, this study has only used publicly available data to assess degrees of convergence in different industry settings. Based on the findings introduced in Chap. 5 and discussed in the preceding paragraphs, convergence could be confirmed *ex post* for NFF as well as ICT. However, the question is more difficult to answer when concentrating on the *anticipation* perspective. It is relatively easy to proclaim convergence when looking at the whole picture, but would it been possible to decide on a convergence pattern already 10, 20 or even 30 years ago?

This would necessitate either a specific pattern of developments or a threshold at which convergence would be destined to happen. Such a *critical threshold* has been identified as the *tipping point* in many complex dynamical systems, from ecosystems to financial markets. Scheffer et al. argue that in such systems often a *critical slowing down* precedes the reaching of the critical threshold [cf. 15, pp. 53ff].

Strikingly, a slowing down was observed in most of the time-dependent analyses in this study. The numbers of phytosterol patents in the SciFinder analyses (Figs. 5.4 and 5.5) as well as in the STN AnaVist analysis exhibit a slowing down between 2002 and 2004. IPC co-classification shares in NFF (Figs. 5.23, 5.25 and 5.27) decreased (slightly) between 1997 and 1999.[22] Furthermore, ICT developments also experienced a local plateau (Fig. 5.29) from 1998 until 2000. All of these findings would suggest that a slowing down after a longer phase of growth could be followed by an exponential phase of growth. For the IPC co-classification shares this plateau was reached after crossing the 10 % (NFF in the UK and Japan), 15 % (NFF in the USA), or 20 % (ICT) level. In summary, while this reasoning would need more evidence to be substantiated it appears to be feasible for at least the cases analyzed in this study. Especially when monitoring possible convergence cases over time by IPC co-classification, the reaching of a plateau after crossing the 10 % level should be taken as a serious indication for convergence.

Q.2: Can the assessment of different levels of convergence be based on analyzing samples derived from different data sources?

Yes, scientific publications could be shown to represent science convergence, while patents mirror more application oriented technology convergence.

Indications of science convergence were appearing in the phytosterol-restricted samples in the late 1960s and 1970s (Figs. 5.11, 5.12, 5.13, 5.14 and 5.15), about 30 years earlier than corresponding developments in patents. This would point toward the possibility of differentiating between science and technology convergence on the basis of these two sources. However, science convergence still lacks an easily and robustly applicable quantitative measure. Whereas this could be achieved for technology convergence, [23] the approaches tested in this study appear to be only of limited use in an *ex ante* setting. Accordingly, future research should try to develop more easily applicable quantitative measures which can be keyword-independently derived from larger samples.

Q.3: Is a keyword-based anticipation of convergence possible and can this also be achieved employing broader classes of articles or patents?

Yes, keyword-independent analyses appear to be even much better applicable in anticipating convergence—especially in patents—than keyword-based approaches.

---

[22] Interestingly, this difference was found for both phytosterols samples. This would indicate an intrinsic difference between developments in phytosterols (apparently appearing a little later) and NFF as a whole.

[23] See Sect. 6.3 for a thorough assessment of the applicability of the different measures and data sources.

While the same holds true for science convergence as discussed in answering Q.2, the IPC co-classification employed in this study could be shown to be the most easily applicable measure. At the same time, it yielded more stable results than those derived from the massively smaller sample of phytosterols. There, only convergence indications could be derived and only by using different more complicated and less transparent handling procedures.

Q.4: Are the food and the pharmaceutical industries converging in the area of NFF?

Yes, each of the samples yielded considerable support for a convergence of both industries in the area of NFF.

While especially the results from IPC co-classification provide substantiation for convergence, the results derived from the limited area of phytosterols support this finding. Interestingly, the convergence effects can be expected to be asymmetrically distributed, as the food industry seems to be much more concerned with the introduction of NFF than the pharmaceutical industry. This is in line with the expectations voiced in Sect. 4.3.4.1 on the basis of differences in perspectives for growth of revenues and profits. At the same time the findings indicate that there is no unified pharmaceutical/food industry to be expected, which would be replacing the two hitherto distinct industries. Instead of such a *substitutive new inter-industry segment*, a *complementary* is being created [cf. 16, p. 83]. While pharmaceutical companies appear to be focusing more on the introduction of Nutraceuticals (which are closer to drugs), the food industry is more involved in the market of Functional Foods (being closer to conventional food products).[24]

Q.5: Are the personal care and the pharmaceutical industries converging in the area of Cosmeceuticals, perhaps even into one broad NFF/Cosmeceuticals area?

Yes, there is some reason to expect a convergence into Cosmeceuticals, but no there is no indication for a common NFF/Cosmeceuticals industry.

Considering the results from both phytosterol examples, there are indications of convergence around Cosmeceuticals. On the basis of the low interconnectedness of personal care and food scientific articles and patents, there is no evidence hinting at one common NFF/Cosmeceuticals industry. Instead, there are several signs of two distinct areas of NFF on the one hand and Cosmeceuticals on the other hand. While these results were restricted to the functional ingredients class of phytosterols, this question can only be answered completely when looking at the general overlap of pharmaceuticals and personal care (for convergence into Cosmeceuticals) and the overlap of both with food (for convergence into one broad NFF/Cosmeceuticals area). Accordingly, Figs. 6.6 and 6.7 present the results from IPC counts and co-classification analysis, respectively.[25]

---

[24] See Sect. 4.3.1 for a definition of Nutraceuticals and Functional Foods.

[25] As the study at hand is focusing on developments in NFF and has only reported on developments around Cosmeceuticals where they were an inherent part of NFF developments, these results were not part of Chap. 5. As they can shed some light on the structure of convergence in NFF, they are presented here.

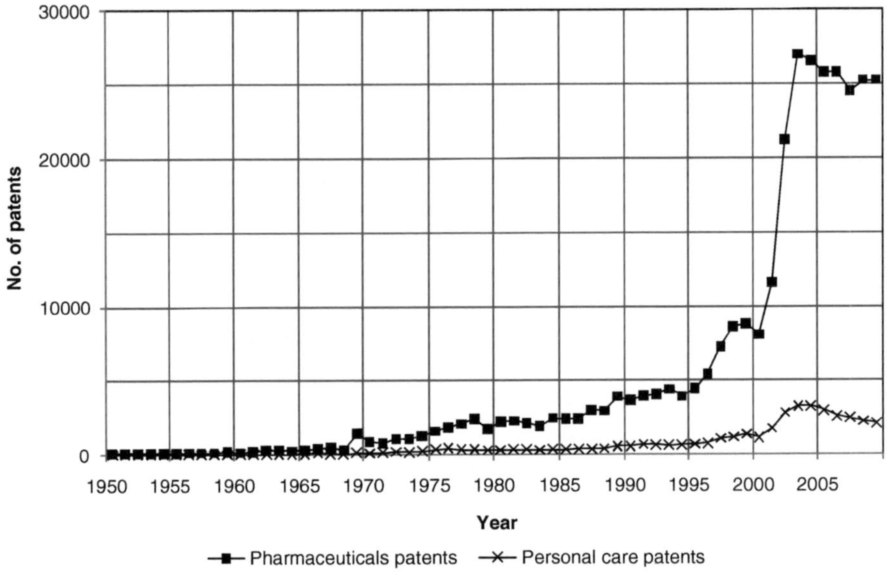

Fig. 6.6 Number of US pharmaceuticals and personal care patents between 1950 and 2009 ($N_{CM\_US} = 357{,}325$)

The numbers and general pattern of personal care were similar to those in food patents. Consequently, the overall number of personal care patents ($n_{PC\_US} = 39{,}615$) is much closer to the number of food patents than those of pharmaceuticals patents. Looking at the co-classification over time (Fig. 6.7), a substantial overlap of personal care and pharmaceuticals surfaces, but by and large with considerably lower shares than in respect to NFF (the respective graph is added in grey for informative purposes). Interestingly, developments were similar until the early 1990s, when the Cosmeceuticals share (from the personal care perspective) remained at around 10 % for about 15 years. This share has grown massively to about 20 % from 2007 to 2008, before slightly decreasing to 18 % in 2009. While results up to 2007 would indicate much weaker convergence trends than in NFF, the sudden surge in the last years may point at a renewed interest in functional cosmetics. In total, the share of pharmaceuticals patents also co-classified as personal care was 1.3 % in the 60 years under scrutiny and 10.4 % from the personal care viewpoint. These comparably high numbers are caused by the substantial increase in the number of patents in more recent years (exhibiting larger co-classification shares). Accordingly, *unweighted* average shares would be 0.7 and 4.9 %, respectively.

The results of the triple co-classification of patents on food, pharmaceuticals, and personal care do substantiate the findings from the phytosterols examples as well. While trends of convergence can be identified in both individual areas of NFF and Cosmeceuticals, there is little support for the expectation of one common

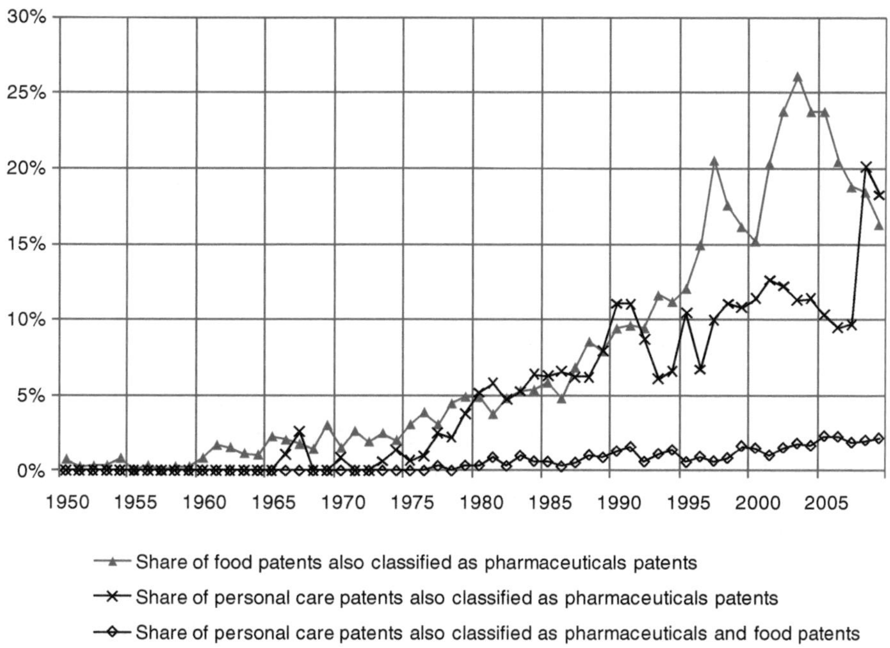

Fig. 6.7 Share of co-classified US personal care, pharmaceuticals and food patents between 1950 and 2009 ($N_{NFF\_CM\_US} = 410{,}466$)

area. This is despite the fact that the share (from a personal care viewpoint[26]) has actually grown to 2.2 % over the last 20 years.

In addition, entering the cosmetics market is most likely also less appealing for the pharmaceutical industry than the NFF market. It is even more about brand names and resulting products would be even closer to prescription free consumer products already being sold by pharmaceutical firms. This could provide an explanation for the overall lower interest and the slower pace of developments around Cosmeceuticals.

Q.6: Does the analysis of activities in different countries lead to different results?

Partially, while the results differ slightly regarding the exact numbers, their overall patterns are very similar.

As discussed above in greater detail, the general patterns in respect to pairwise co-classification in food, pharmaceuticals, and chemistry are very similar across all three countries. However, a closer look and more detailed comparison do reveal differences as well. These differences are probably rooted in differences of the

[26] Due to the similar numbers, the order of magnitude would be similar from a food patents' perspective.

respective patent laws, the activities of patent authorities, and also the overall mentality and buying behavior of customers. Consequently, the degree of convergence may be different; or, for instance, developments in Japan seem to trail those in the UK and the USA.[27] Nevertheless, the general patterns are very similar and thus there is no reason to believe that convergence could not be anticipated when choosing a 'wrong' patent country.[28]

Q.7: Are patterns of convergence generalizable across industry boundaries?

Yes, patterns seem to be very similar across industry boundaries and can, for instance, also be found in ICT.

On the one hand this corroborates technology convergence patterns in ICT. At the same time it becomes obvious that these developments have been increasingly intensive over the last few decades, explaining the longstanding discussion about the reality of ICT convergence. Adding the example of the intersection of personal care and pharmaceuticals to this discussion, it remains unclear how strong the trend toward convergence is in that area. However, with the exception of an outstandingly long 'lag phase' of nearly 20 years, the general pattern resembles developments in NFF and ICT. Naturally, three examples will be insufficient to prove the existence of a general pattern. Future studies should compare results to the findings in this study, ideally especially in connection with different degrees of convergence or convergence on different levels. Thus, with more data on different cases the pattern could not only be described but also explained.

6.3 Limitations and Comparison of the Different Approaches

6.3.1 Keyword-Based Analyses

6.3.1.1 General Limitations

After discussing the individual results of the SciFinder and STN AnaVist analyses of scientific articles and patents on phytosterols, a few caveats regarding sample construction and processing have to be considered. First, the search-terms used were chosen on the basis of a broad class of substances—phytosterols. Although

[27] Although caution should be exerted in this assessment, as the samples were created on the basis of publication years not priority years. So a 'trailing' Japanese development may be caused by more recent patenting activities or by the fact that patent applications and granted patents need longer to be published at the JPO than at the UK or US patent office. See also Sect. 3.2.1, for a description of the different backlogs.

[28] This, of course, only refers to the larger and more important patent countries. When employing patent samples from very small patent offices, they might indeed differ significantly.

different alternative forms of naming and spelling were incorporated into the search, it is very likely that not all documents dealing with phytosterols were included. For instance, documents dealing only with a few of the over 250 different substances may have not been found, if they did not also mention the class name. While inclusion of every possible spelling and each subclass or individual substance would have been technically possible, this would have led to a multiplication of search procedures and therefore time, effort, and costs incurred.[29] Also, there is no reason to believe that this procedure led to any bias in the sample toward or against patterns of convergence.[30] In fact, especially for patents an inclusion of all the individual substance names might have led to more 'noise' in the sample. It was not the aim of this search to generate all patents where phytosterols were not a central part of the content but, for example, only a chemical byproduct. If all such patents were also included, an identification of underlying patterns would have only been all the more difficult.

Second, this stresses the importance of a careful selection of keywords in general. Indeed, in most cases there will not be an ideal keyword or set of keywords. It will rather be a question of tradeoffs and the lesser of two 'evils'. This is a major weakness of keyword-based searches; and one of the strongest assets as well. As long as even an imperfect search strategy will yield sufficiently robust *weak signals*, it will be an easy to use way of evaluating the degree of convergence trends around one firm or industry. In the case of phytosterols, about 40 % of all patents were classified as both, pharmaceuticals and food related. This figure, which is well above the numbers calculated in the STN Express samples (between 6.1 and 10.9 %), underlines the role of phytosterols as an appropriate example of NFF.

Third, in contrast to INPADOCDB, which encompasses virtually all patents,[31] neither CAplus nor MEDLINE entail all important publications. While their coverage of patents (at least in the chemical field) is very comprehensive, their coverage of scientific journals is extensive but limited.[32] Patent assignees can be expected to file for patents at the most important patent offices (either in the first place or within the priority year), especially if holding a valuable patent. Contrastingly, authors of scientific articles normally only have one chance to report

[29] Especially in light of the emphasis on practical applicability this would have been counterproductive.

[30] For patents, it is on the one hand highly doubtable that any assignee would voluntarily limit his claims to a certain subclass of phytosterols. On the other hand, patent examiners would surely regard a patent on one subclass as potentially harmful to the patent newness of a patent on a different subclass. Although such a limitation to a subclass is slightly more likely in scientific articles, even there most authors would probably refer to the more general class as well or point out potential applications based on related substances.

[31] As outlined in Sect. 5.2.1, INPADOCDB does of course not encompass *all* published patents in the world. But building on over 90 patent-issuing organizations and with by now more than 55 million documents in its records, it does cover *nearly all* patents and especially all that can be expected to be of considerable value.

[32] See Sect. 5.1.1.1 for more information on the coverage of CAplus and MEDLINE.

their findings, as most journals will rule out the publication of non-original articles. If—for any reason—this first journal is not in the set of journals sampled for the respective databases, an article might remain unnoticed. Most high quality articles can be expected to be published in the included journals (which are mostly of higher quality and standards themselves), but sometimes the author may have chosen a different journal of lesser reputation.[33] Again, this will probably lead to incompleteness of the sample but there is no convincing argument why this should bias the sample. It is likely that scientific articles will be less completely covered than patents, which is another case in point for favoring patent data over scientific publications. Furthermore, if convergence is still in an early phase where these few highly specific articles would matter for the overall judgment, signals would probably be generally too weak to detect them.

A further caveat to keep in mind is the assignment of companies and subject areas to broader subject clusters. While this step was done very carefully and based on a clear reasoning, a certain probability of misclassification cannot be denied. On the one hand, even the IPC-based classification will include a certain number of erroneous assignments, as no classification is ever 100 % accurate. On the other hand, due to manual assignment by the investigator, the probability will be considerably higher than when relying on standard classifications (like IPCs). However, the central point is that there is no evidence or even indication for a systematic misclassification and thus a bias in sample selection and handling. Whenever using IPCs it has to be born in mind that the IPCs are not necessarily evenly distributed representatives of industrial sectors, but *per se* only of technological and applicational fields within the patent literature.

Another potential source of distortion of results is the employment of different thresholds in this study. Most of the analyses in SciFinder and STN AnaVist were based on sub-samples reduced in size. Firms (top 50 patentees), authors (at least ten articles on phytosterols) and even subject areas (representing at least 1 % of all publications) to be included were cut off based on some form of threshold. This was necessary because of the plentitude of items on each of the lists. Arguably, thresholds were decided on arbitrarily to a certain degree and may have led to a bias toward larger firms and research institutions with more resources. On the one hand, some might argue that larger organizations were more likely to try to use their positions to stretch out into new domains. On the other hand, others might respond that often larger entities are less flexible and more likely to remain in their beaten tracks. In summary, it is not clear how the picking of thresholds should have led to a distortion of the findings regarding the anticipation of converging industries, thus it should not have biased the findings.

Finally, although neither of the two phytosterols samples was based on US patents alone, one particularity of the US system should be considered. Many US

[33] While there may be many different reasons for such a choice, including a shortened time horizon until publication, a too novel approach for established journals, the wish to assist in installing a new and more focused journal or a topic (which had been too specialized for the established journals). For further discussion of reasons and rationales in publishing see Sect. 3.2.

patent applications do not carry assignee names, potentially leading to an artificially lessened impact of certain firms, which might have been excluded from the firms sample because of their perceived low level of patenting activity. This might have led to an overall underrepresentation of US patents (which could in addition also have been caused by the fact that until 2001 no US applications were excluded). However, these characteristics will most likely not have led to less reliable results, as the STN Express analyses have demonstrated the high similarity between different countries. In addition, most patents will not have been applied for and granted solely in the USA. Important developments in patents should hence appear in other countries' patents as well.

6.3.1.2 SciFinder

Recollecting the results presented in Sect. 5.1.1 and their discussion in Sect. 6.1.1, SciFinder could be demonstrated to be able to yield indications for a convergence. These results stretched from scientific publications pointing at science convergence to patents representing technology convergence. Besides the one-by-one discussed, overall limitations and caveats do not support adequacy of this approach. The two most central drawbacks are the necessity to use keyword-based search-terms and SciFinder's architecture which is not constructed to be used for advanced statistical analyses. Considerable effort was necessary to unravel the underlying structure of the sample, or at least a part of it. Based on the already difficult selection of keywords, investigators would have to invest a lot of time and thought to derive not even compelling results. Therefore, if used in the anticipation of converging industries at all, SciFinder should rather be used to delve deeper into a certain (limited) area of a possible convergence field. Building on CAplus and MEDLINE, it can furthermore only be used around the chemical or pharmaceutical industries.

6.3.1.3 STN AnaVist

Contrastingly, STN AnaVist could also be used in additional industries or scientific disciplines, depending on the databases used to generate the visualized[34] sample. Still the same limitations affect this program as far as keyword-based search-terms are concerned. But as datasets can be employed which were generated in standard databases, all kinds of different search strategies can be used to create the target sample. The results presented in this study are particularly appealing due to the possibility of interacting with the sample and combining

[34] Initial visualization in STN AnaVist includes the importing of the sample into the program to allow for all subsequent visualizations and analyses.

quantitative[35] as well as qualitative[36] means of analysis. At the same time they could only be derived in many time consuming operations, hence being of limited value to researchers trying to assess the degree of convergence in any given area.[37] STN AnaVist generally provides the opportunity for various examinations. While these are very useful in assessing convergence patterns, the applicability is further limited by the maximum number of importable documents. Not more than 20,000 documents can be used to create an importable sample. Accordingly, only very limited areas can be analyzed for convergence patterns and it does thus seem more sensible to use this program for a more thorough analysis of certain aspects of a convergence process.

6.3.2 STN Express

In accordance with this study's reasoning of convergence being a process rather than a static situation, the co-classification shares' variations in time have clearly corroborated the applicability of this measure as an instrument for anticipating convergence. This was achieved by comparing different country samples and different co-classifiable IPC groups, with similar results across all discussed convergence hotspots. Notwithstanding these encouraging substantiations, a few limitations call for attention.

Especially on the basis of the similarities of convergence patterns in ICT and the three NFF samples a question arises from the decisive decrease in co-classification shares (with the exception of NFF patents in Japan) in more recent years. The easiest explanation would be that this indeed is the general pattern of technology convergence as seen in patents. Then future studies around different fields should also exhibit this drop in shares. Another explanation would be that this is the individual pattern of convergence in NFF and ICT, which should also be identifiable in time-dependent depictions of other measures and indicators. A last reasoning could be built on patent characteristics. If such characteristics would lead to a decrease in most recent patents, the share should be different for the most recent years if analyzed again in a few years. A, however unlikely, cause for such an artifact could be rooted in the computation of the most recent documents and their indexing with IPCs etc. Contrastingly, it would be difficult to explain the already earlier and more decisive decrease in the UK sample and the sudden increase between 2007 and 2008 in the Cosmeceuticals example, if this last reasoning was to be true.

[35] For instance calculations of co-authorships or co-classifications.

[36] For instance the creation of research landscapes.

[37] This is reinforced by the incurred costs, which will be too high for most academic researchers and also potentially deterrent for practitioners, especially when thinking of a set of visualizations recurring regularly.

A few caveats on IPC co-classification have already been discussed.[38] In addition, not only the question of potentially erroneous classification is an important one, but also the completeness of classification and especially the degree of reclassification. Many countries' patents are included in databases with only incomplete coverage of IPCs [cf. 8, pp. 296f; 14]. Furthermore, especially with changes in the IPC structure (introduction of new IPCs or regrouping of older ones) old references do constantly need reclassification. Based on resources of the different patent authorities as well as of the database providers, this reclassification may take considerable amounts of time. Hence, any analysis has to consider the influence of incomplete classification. As long as no systematic differences are neglected (like only using newly introduced IPCs in a search), this should not lead to skewed results.

Originally invented as a shelf-location tool, Adams discusses the relevance of IPCs in patent analyses altogether. In summary, he argues that IPCs are particularly useful when used in parallel with other search-terms and when classification coincides exactly with the field or object of interest. In difference to keyword-based search-terms and indexing terms, classifications are designed to represent a certain concept with one class [cf. 8, pp. 291f]. Thus, the present study has used IPCs as they could be used without the risk to also include unwanted content—as would have been the case when using wider arrays of indexing terms or keywords. A further challenge is the correct use of the *hierarchical*, *non-expressive notation* of the IPC. As such, the length of the class mark is not reflecting the relative position of a term within the overall hierarchy of the IPC. This can lead to distortions in results when searching with incorrect truncation [cf. 8, pp. 291f]. Within this study, searches were conducted nearly exclusively on the class and sub-class level and were hence not confronted with truncation problems. In addition, where main groups were used the fact that STN Express loads IPCs as single alphanumeric sequences enables an easy correct use of the chosen IPCs.

Many of the analyses conducted on the basis of IPC co-classification covered only US patents. Consequently, it should be discussed whether US data is actually a good representative of global developments. As argued earlier, the inclusion of all possible countries could lead to an overemphasizing of especially larger companies being more likely to file for patent protection in a range of countries. Accordingly, if not corrected for potential patent families, this approach would lead to more patents than actually relevant in respect to a technological analysis. The concentration on particularly US data was done for two central reasons. First, the USA is a dominant country when turning to technological as well business developments. This holds true especially (but not only) for NFF[39] and also ICT. Second, Fai and von Tunzelmann state: "In view of the consistency of US screening and the importance of the US market, it has frequently been asserted that

[38] See, e.g., Sect. 6.3.1.1.

[39] While other countries also play decisive roles in selected technological areas (especially Japan in NFF), the USA can undoubtedly be asserted to be overall a dominant economic player.

US patent data provide the most useful basis for international comparisons of technological change".[40] When conducting IPC searches in US patents the fact that the structurally different USPC[41] is translated into IPCs via a concordance list has to be considered. A translation may lead to systematic differences in numbers or distribution of patents, especially when comparing countries.[42] Such problems are less likely when using larger quantities of patents and broader technological fields, as differences in individual cases will to a lesser degree influence the overall evaluation. Furthermore, to control for such limitations this study has also included data from Japan and the UK. Based on the general similarity of the derived findings, US sample results should not suffer from a higher quantity of erroneous classifications. This is especially the case, as the sheer numbers of patents were not objects of investigation.

Despite all potential limitations and caveats, the use of patent analysis in respect to technological fields is a commonly used approach. Most industries tend to patent largely in their corresponding technological fields [cf. 17, pp. 141ff]. Accordingly, an analysis of developments across technological and industrial boundaries can serve well as a proxy for convergence. In addition to the empirical findings presented in Fig. 5.19, Fai and von Tunzelmann's respective findings support the reasoning that co-classification is not a general pattern in today's patenting activities [cf. 17, pp. 141ff].

STN Express should be interpreted as only a wildcard for any program able to work with standard patent and scientific literature databases.[43] Although only simple counts and co-classification calculations were conducted with it in this study, it could also be used to carry out various other analyses. This includes most prominently different search operations around citations. Once investigators accustomed themselves to the retrieval language, standard scripts can be used to, e.g., perform routine searches at certain points in time. Consequently, this approach appears to be ideal for the anticipation of convergence as it can handle nearly unlimited numbers of documents and calculate convergence-related measures in comparably short times. Furthermore, all procedures are very transparent and can be saved in transcripts for later analysis. Finally, IPCs as a basis for patent analyses have been welcomed by others, leading to the conclusion that "Increasing access to the IPC will give skilled searchers new avenues to explore in coping with the ever-increasing volume of patent literature".[44]

---

40 [17, p. 148]. See also, e.g., [18, p. 106], stating that studies often refer to US data due to availability and importance of US R&D activities.

41 See Sect. 3.3.1 for a description of differences between USPC and IPC.

42 One way to heal potential translation illnesses would be to refrain from using IPCs and instead use USPC. While the USPC has other weaknesses, such as the re-introduction of discontinued notations or the lack of a concordance list for notation changes over time, the main problem would be a further decrease in comparability between different patent countries. [cf. 8, p. 292].

43 The same, in principle, holds true for SciFinder and STN AnaVist, as long as other programs with a comparable architecture and resulting functionalities are concerned.

44 [8, p. 300].

6.3.3 Overall Limitations

While the preceding paragraphs have dealt with limitations and caveats of the various approaches used in the study at hand, some overall limitations have to be taken into account as well.

All the findings have been derived and discussed from an ***ex post* perspective**. As argued earlier, such an approach is necessary when trying to develop methods and tools and in particular when dealing with long time horizons. There the past is necessary as a reference point as well as a training ground to test and refine theorized models. At the same time, it carries the danger of merely trying to find patterns in the mass of data that would support a theorized model. Moreover, it is hard to resist the temptation to only select supportive data and results and to argue why all negative findings are exceptions from the rule or negligible due to other limitations. Where empirical data is missing, the theoretical concept has to be solely convincing and then to prove what it is worth in the light of future empiricism. However, in the light of convergence in ICT being anticipated, discussed and 'executed' for decades but still being far from finalized for likely another few decades, using the past as a reference is the only sensible approach. The only alternative would have been to make rather arbitrary predictions now—and then leave it to following generations of researchers to validate or falsify them in the more distant future. But because of this dilemma, great emphasis was put on developing a sound concept and underlying theory for an anticipation of converging industries prior to testing it with empirical and historical data.

A further limitation lies in the question whether the observed developments can be really coined 'convergence'. While the presented results clearly indicate a blurring of boundaries between separate scientific and technological fields, a complete **industry convergence** would also have to include changes in industry structure.[45] In spite of some examples discussed in the analysis of the ICT and the NFF/Cosmeceuticals cases, such structural changes were not the focus of the study at hand. Hence, it does fall short of providing tools for *forecasting* industry convergence. Yet it does not fail to introduce means to *anticipate* converging industries.[46] Building on the nature of the sequential process of industry convergence and the significance of science and technology convergence for it in research-intensive industries, the determination of one of them would be sufficient to justify preparations for a convergence of industries. The anticipation concept is based on the detection of *weak signals* and then the consideration as well as potentially the acceleration of convergence. A monitoring of each of the steps of the sequential process would be of mainly informative value and is thus of subordinate importance to this study.

---

[45] See Sect. 2.4 for a discussion of the sequential nature of industry convergence and its various forms and aspects.

[46] See Chap. 3 for differences between forecasting, foresight, and anticipation activities.

Finally, Chandler speaks of the importance of the scope of a technology for the effectiveness of organizational structures. He discerns between industries like the chemical industry with a wide selection of different products and processes and much more focused industries like the steel industry [cf. 19]. The first one's success would be grounded on so called **extensible technology**, with the chemical industry extending into pharmaceuticals for example [cf. 20]. The latter concentrated on cost reductions and product improvement because of its non-extensible technology. Adhering to this terminology, one could, on the one hand, surely speak of the use of phytosterols and other technologies within NFF as some kinds of extensible technologies. On the other hand, this would neglect the decisive alterations and adjustments firms have to undertake when extending their activities from foods into pharmaceuticals or vice versa. Thus, changes in industry structure would be too massive an event to be referred to as just a result of 'extensible technology'. Future developments will have to show which description is more to the point. Furthermore, a discussion about the correct terminology and the underlying concepts does not diminish the practical importance of monitoring these kinds of developments in industries.

6.4 Implications and Future Research Opportunities

6.4.1 Theoretical Implications and Future Research Opportunities

The present study may contribute to three distinctive research streams. First, the thus far rather scarce literature on convergence was enriched with some fundamental definitions and classifications, as well as the more detailed description of the two industry convergence examples of ICT and NFF. Second, the research stream on forecasting and foresight was supplemented with a concept for the anticipation of converging industries and its different steps. Third, the RBV was employed in this study's setting to lay a theoretical foundation and derive the aforementioned research questions.

Despite the massive effects of convergence on science, technology, industries, and even society at large, the existent research on **convergence** lacks a sound theoretical foundation. Although it has been discussed in a plethora of cases[47] from the machine tool industry to the cognitive sciences, no consensus has been achieved on definitions, delineations, and classifications. This hampers a fruitful discussion among researchers occupied with different aspects of convergence as methods and findings can hardly be compared and refinements on the basis of prior research becomes virtually impossible. More importantly, it decisively complicates the development of successful strategies for all stakeholders to cope with

[47] See Sects. 2.3 and 4.3 for examples of converging industries.

converging industries. Consequently, the study at hand has consolidated as much of the body of literature as possible. Wherever suitable, it tried to build on already employed or even widely accepted definitions and approaches. However, most researchers have thus far concentrated on only selective parts of industry convergence, necessitating the introduction of a new and comprehensive definition of *convergence* and its delineation from the often synonymously used term *fusion*.[48] Grounded on these definitions a differentiation between *triggers* and *drivers* was derived that grouped the causes of convergence into the five classes of technological, managerial, industrial, political, and societal triggers/drivers.[49] As only very few of the hitherto published studies on convergence have tried to classify these causes at all, this study could serve as a seed crystal for a discussion of the importance of the individual triggers/drivers. Furthermore, stakeholders will fail to slow down or accelerate convergence trends as long as they are unaware of its causes. As this study set out to build a concept for anticipating convergence, the question 'where to look' was of essential importance. Without a distinction of different steps, levels or phases of convergence it would have proven impossible to deduce dependable and clearly delimited measures. Developing further the few fragmented and non-disjunct taxonomies, a sequential process was proposed on the basis of the four steps science, technology, market, and industry convergence.[50] On their basis the anticipation concept was derived as an addition to the literature on forecasting and foresight. Albeit this study has added further aspects to the theoretical foundation of converging industries, there are still decisive research gaps. Analyses of further convergence examples will have to assess the applicability of the definitions, delineations, and classifications. However, the most decisive gaps refer to questions of management in the light and in anticipation of convergence: How can managers and policy makers deal with convergence and use it to the advantage of their organizations—be it firms or whole countries? What strategies have evidenced to be successful in the past and under which parameters? How can these strategies be adjusted to different degrees or forms of convergence and under different regimes?

In comparison to the scarce and scattered literature on convergence, a few clearly marked research streams exist in regard to **predictions of the future** of technologies, markets or industries. Thus, this study positioned its anticipation concept closer to foresight with its more active and preparatory stance. Differing from many of the established foresight/forecasting tools, this concept contains a substantial monitoring perspective without the use of exact mathematical functions and calculated probabilities or margins of error. Grounded in the *futures analysis'* underlying theory of detecting weak signals, it aimed at using simpler models' assets (such as robustness, validity, and especially transparency) to enable other

---

48 See Sect. 2.1.4 for this study's definition of *convergence* and its delineation from *fusion*.

49 See Sect. 2.2 for the different aspects around the classification of *triggers* and *drivers*.

50 See Sect. 2.4 for an explanation of this sequential process of industry convergence.

scholars to change and adjust the concept to better serve their purposes.[51] Starting from the sequential convergence process model, different sources for scrutiny of the level of convergence were discussed in greater detail. The main focus lay on the possibilities of the use of bibliometrics (in respect to science convergence) and patent analyses (regarding technology convergence) under the umbrella of an anticipation approach and considering established forecasting principles as much as possible.[52] The study at hand could influence the underlying theory mainly in two ways. First, it opens another potential field of application for different future-oriented research streams, where different approaches should be tested and compared and then adjusted to the specific environments and needs of converging scientific disciplines or industries.[53] Second, the anticipation of future events under high uncertainty could broaden the focus of the future-oriented literature and serve as a discussion trigger around topics like sample size, significance testing, and combination of different approaches.

Only very rarely have organizational theories been used to explain patterns in the light of convergence.[54] Where they were discussed, authors agreed on the applicability of the **RBV** as a theoretical foundation for convergence. This study followed the corresponding line of reasoning and used the resource-based perspective of firms to derive general and theory-oriented research questions. A focus on resources and their asymmetrical distribution was deemed particularly useful as firms will in most cases of industry convergence not know which resources will be useful or necessary. Resources are defined as inimitable, non-substitutable, valuable, and rare in the RBV. Hence, the importance of an anticipation possibility correlates well with the RBV. Further research should evaluate the importance of resources in convergence more thoroughly and also elaborate on the lessons firms facing converging industries can learn from the RBV and its application across different industries and settings.

6.4.2 Practical Implications and Future Research Opportunities

Recollecting the research questions and the main findings, this study yields several practical implications for three different groups. First, policy makers who strive to adjust incentives and regulation to changing industry dynamics. Second, firms and their managers who try to prepare for an uncertain future. Third, academics who

---

[51] See Sect. 3.1.2 for a reasoning of the anticipation approach.

[52] See especially Sects. 3.2.3 and 3.3.3 for a brief description of these analyses can be used in the current study.

[53] Of course, this in principal also applies to converging technologies and markets. However, these two steps are not as immediately represented by human actors and their interests and are thus of subordinate importance in this context.

[54] See Sect. 4.1 for theories applied in the convergence literature.

conduct research in connected streams or who are active in scientific fields involved in science convergence.

Especially the examples from ICT and NFF, but also the classification of political drivers, underlined the importance of regulation for the development of industries. Convergence may be the driver for the necessity to introduce new or overhauled regulation,[55] or may itself be driven by new regulation. In light of often lengthy procedures to adjust the 'regulation landscape', the anticipation of converging industries could provide a time advantage to enable the undertaking of steps in the desired direction. This refers not only to technological and industrial developments, but explicitly also to developments in scientific disciplines. There, **public authorities** often try to allocate their funding according to shifting necessities and interests. The identification of new cross-disciplinary collaboration or interests may thus serve as a warning signal for new developments. Accordingly, new departments or study programs could be installed at universities and funding incentives could invigorate research in a special direction. As for the other groups, one decisive advantage of this concept in the present form is rooted in the relatively simple analyses to be conducted even without the need for expert knowledge. At the same time, most analyses are depending on patents as a source. Patents in turn are widely accepted as proxies for technological developments and can hence work well as a foundation for public discussion on politicians' or public authorities' decisions. Future research should scrutinize the impact of regulation on convergence and what kind of actions could accelerate or decelerate it. Furthermore, bearing in mind the massive backlogs particularly at the JPO and the USPTO, the impact for planning of human resource allocations to subject areas at patent offices could be another benefitting research direction.

Companies[56] are largely path-dependent[57] in their technologies and decisions and thus struggle with adjusting to new environments as caused by industry convergence.[58] Accordingly, the earlier they can identify such decisive changes to industry structure, the more time they will have to make the necessary preparations. These can refer to various aspects at science, technology, market or industry level. While scientific collaborations or technological developments (either in-house or in collaborations as well) may be necessary to source know-how hitherto alien to the firm, the installation, and maintenance of effective and efficient collaborations is an intricate task.[59] To this end, companies could use the discussed

[55] Regulation in this context explicitly refers to all kinds of, e.g., laws, policies, incentives or funding, as are decided on and introduced by public authorities.

[56] Naturally, it is not the companies themselves who are making the decision but rather the managers. For increased legibility, this study uses firm or company as a short form of 'managers of a firm'.

[57] For a detailed discussion of how path-dependent internal rigidities influence firms' decision making especially in settings of convergence, see [16].

[58] See, e.g., Sect. 3.1.1 for the impact of convergence on companies.

[59] For important aspects regarding collaboration, especially in research-intensive industries see, e.g., [21].

anticipation concept not only to become aware of potential areas of convergence but at the same time to get a deeper understanding for the respective areas, to identify central trends and important researchers. Accordingly, they would already generate a list of potential collaboration partners and fields.[60] At the intersection of technologies and markets often the availability of specific intellectual property but also of physical resources such as production facilities or raw materials is often limited.[61] Hence, it is important to secure access to such key success factors before it may be banned by competitors. This is even more challenging where it is not a consequence of beaten tracks in the core industry of a firm, but in an additional field of convergence.

On the market level, companies are constantly struggling to stay abreast of the latest developments in respect to market structures and customer preferences. Customers' increasing trend to buy products which bundle certain electronic products into one device (such as in the case of smartphones) poses opportunities and threats to companies in each of the affected industries. Consequently, anticipating such a development could provide companies with a first mover advantage related to not only technological components, but also to consumers' perceptions.[62] Especially in respect to analyzing market trends and buying behavior, the current concept needs to be broadened considerably. Future research should thus include suitable measures for assessing convergence trends in different (later stage) data sources.[63]

Finally, convergence anticipation may have effects for companies at the industry level, in so far as they will have to react to a new 'power distribution' within a converged industry. For instance, thinking of the food and the pharmaceutical industries, both industries are dominated by large multinational companies with a deep knowledge of their customers, markets, and regulation. While there is a considerable degree of competition between the companies in each of the

---

[60] This would be a perfect example of how the different approaches to anticipate convergence could be employed within one specific case. While an IPC co-classification and subsequent analyses on the basis of tens or hundreds of thousands of patents (and partly scientific articles) could serve as an early warning signal, this could be followed by more detailed and content oriented analyses of smaller samples to identify key research streams, institutions and individual researchers.

[61] While this may be important in any setting, these availabilities are particularly crucial in areas of ICT, the pharmaceutical and the chemical industry. See Sects. 2.3.1 for a description of the setting in ICT and 2.3.4 and 4.3 for the pharmaceutical industry and NFF. See also [22], for a brief discussion of the setting in the chemical and materials industry and the role of access to renewable raw materials.

[62] Undeniably have three smartphones shaped consumers' preferences and expectations at different points in time. The Nokia Communicator was one of the first smartphones, which proved the possibilities of this new category to mainly a limited number of business customers. Later RIM's BlackBerry introduced typical business functionalities to a large group of business managers and started to get more and more private customers interested. Finally, the Apple iPhone was adopted by the mass market based on design, reputation and functionalities.

[63] See Sect. 3.4 for a short overview of potential measures and data sources for each step of the sequential convergence process.

industries, certain patterns of activities[64] have evolved that change only slowly. In the convergence area of NFF these incumbents in their respective industries are now competing with the other industry's incumbents and may have to fundamentally adjust their business processes accordingly. This will often lead to collaborations, strategic alliances, and M&A activities within the hitherto disjunct industries but specifically across industry boundaries. As potential partners for such actions are limited, a reluctance to engage early in these activities may prove to be detrimental.

The study at hand has implications for **two groups of academics**. First, implications and potential future research activities for academics conducting research around convergence and forecasting/foresight will be discussed in the next chapter. Second, academics which might be affected by trends of especially science convergence could utilize the discussed approach in planning future research *foci* as well as university-university or university-industry collaborations. Again, the different approaches in this study could serve as a means to identify potential streams and partners [cf. 21].[65] Academics in areas affected by developments around NFF, Cosmeceuticals or ICT should be expecting a still growing demand for more solid research to line technological, regulatory, and industrial activities.

6.4.3 Implications and Future Research Opportunities for Researchers

Recollecting the different approaches and the main findings, this study yields several implications for **researchers[66] in anticipating convergence**. Scholars have identified a clear demand for ways to predict convergence, while at the same time identifying the negligence of the forecasting literature to provide or even consider means to do so [cf., e.g., 23, p. 157]. This research gap has been attempted to be filled with the current study, leading to implications for the research streams of convergence and forecasting/foresight.[67] Moreover, scientists conducting research on the anticipation of converging industries could use this study as a reference point for a broadening, deepening and, of course, critical review of the proposed approach to anticipate converging industries. While all in

[64] This refers to, e.g., addressed therapeutic areas in the pharmaceutical industry or certain product ranges in the food industry. While there is a constant flow of activities in respect to diversification or concentration, these developments are largely more incremental than radical changes.

[65] See, e.g., Niedergassel who reports on the difficulties to align academics' interests/research foci/procedures with those of other academics and especially industrial partners.

[66] While they will in most cases be academic researchers, they may be also affiliated with companies, research institutions or public authorities. Although not explicitly mentioned, they are implicitly addressed as well in subsequent paragraphs.

[67] See Sect. 6.4.1 for theoretical contributions of this study.

all, a recommendation to anyone interested in anticipating convergence would always include a multi-step approach with more quantitative (like the keyword-independent searches) and more qualitative (like the keyword-dependent searches) analyses, its detailed design should be carefully scrutinized.

The **breadth of research** refers mainly to possible fields of application or research settings. While convergence in ICT and NFF served well as widely accepted examples of converging industries, many more industries of potential interest exist. The approach would in many cases have to be adjusted to the individual settings, but could serve as a kind of blueprint for various other industries or areas. Future studies could assess the degree of convergence in other areas of the chemical industries concerning, for instance, the blurring of boundaries between the chemical industry and the agricultural industry (in respect to renewable resources in materials productions), the biotechnological industry (in respect to production of intermediates for consumer goods or pharmaceutical production) or the various industries of energy production and storage. Further industries would entail less research-intensive industries, which would pose the challenge to most likely focus less on patents as a source of information. Progress in the service sector would present additional research opportunities, e.g., around the banking and the insurance sector or the different parts of the health care sector.[68] Especially in service industries a stronger use of information on customers'/clients' needs and preferences would have to be considered.

In comparison, the **depth of research** refers to different data sources and indicators. The present study relied mainly on patent data and to a lesser extent on scientific publications. Naturally, other databases or targeted patent offices would be possible, but there appears to be no compelling necessity for such a step. Contrastingly, other indicators should be seriously considered. Notwithstanding the caveats voiced in this study, researchers should employ citation-based analyses to assess the degree of interconnection from that point of view. This could be done in scientific publications as well as in the more cautiously citing patents.[69] An additional possibility would be the application of text mining approaches to generate co-word usage descriptors or similar indicators. Although complex measures may be tempting, all measures should be carefully selected, keeping in mind the aim to develop a concept which would ideally be applicable to different industries and in different countries without the need to substantially alter the applied methodology. Considering the various interested stakeholders, it would as well be reasonable to refrain from too complicated and specialized measures in order to retain an approach not only applicable by experts.

Furthermore, especially the remaining two steps in the sequential process (market and industry convergence) should be evaluated in respect to their suitability in anticipating convergence. In spite of the argument of less practical and

68 See also Sect. 2.3 for further examples of fading industry boundaries in different industries.

69 See Sects. 3.2 and 3.3 for a discussion of the benefits and drawbacks of citation-based analyses.

more academic value of an assessment of later stage convergence, it would be interesting to validate (or falsify) this reasoning. Additionally, analysis of these steps would yield a more complete picture of convergence processes and could be of particular benefit to industries without a strong information base in scientific publications or patents. To this end, products like *LexisNexis®* or *Capital IQ* could be employed to aggregate data on product launches, collaborations and M&A activities, construct industry samples based on industry classifications, and partially even for discussing a connection between convergence events, and financial performance.

Several minor points regarding potential limitations of this study due to the indicators and data sources chosen have been already discussed. A **critical review** might also consider differentiating between national and international patents, between applications and granted patents and probably also evaluate the validity[70] of patents. If using a sample with more than one country, the influence of patent families should be controlled for. The development of patents over time can be assessed based on different sources for a patent's date. In this study the publication date was used, as this is also reflecting potential differences in patent authorities' handling of applications. Alternatively, the filing date or the priority date could be used. While analyses based on priority dates appear to be using less recent data, they are closer to the original date of invention and construction of the patent application. In contrast, filing dates will be more recent if referring to subsequent filings of other patent family members and might indicate a prolonged interest of assignees in protection of their inventions. Whether a change in respect to patent dates would have a considerable impact on an anticipation approach based on weak signals is questionable but cannot be decided without the employment of empirical testing.

This leads to a final opportunity for future research. Within the present study's concept more isolated measures to analyze the convergence of industries in NFF and ICT from different perspectives were employed, without combining them into some kind of holistic measure. For researchers more interested in a broader concept that would be leaning more to a quantifiable **monitoring concept** and less toward an early warning system based on weak signals, the development of a comprehensive distance measure should be of interest. Such a distance measure could be based on the discussion of the change in distances between converging areas in Sect. 2.1 and will be outlined briefly in the following.

In order to be able to assess to which degree a set of scientific fields or industry segments have converged, an appropriate measure would need to be defined. This approach would lead to three main challenges. First, how could this distance measure be operationalized? As most likely the distance should be measured for all stages of the convergence process, different measures would be needed to, e.g., measure the distance between scientific areas in comparison to industrial segments.

---

[70] A granted patent may not be valid anymore due to, e.g., the non-payment of fees or appeals by other stakeholders.

Second, where should the distance be measured? Even if a depiction of an area could be generated similar to those in Fig. 2.1 or 2.2, should the distance be measured between the (weighted) focal points or between the closest two points of two areas? In the anticipation of industry fusion it is of minor interest whether areas as a whole shift their focal points in a certain direction. More importantly, one has to take into account the outposts of two fields. Where those begin to move toward each other a possible area of convergence appears. If the complete areas move toward each other, this would only mean that a substitutive new segment would be more likely than a complementary. Albeit this is of high strategic weight in the long run, it could be neglected for a first assessment. Hence this study argues that the distance should be measured as the shortest distance between areas.

Third, would it be possible to aggregate an absolute distance or would it be more appropriate to employ relative distances? This would be of special importance as a set of different measures for science, technology, market, and industry convergence should be combined. If trying to compile an absolute convergence distance, this would pose manifold challenges. How should, e.g., a co-authorship count be weighted and connected with data on patent assignee clusters? Only when comparing relative distances, they could also be more easily aggregated. Furthermore, this study would propose to (initially) use binary measures whenever possible. Although one could rightly argue that there might be a difference in the degree of interrelatedness between citing one reference from a different area and citing many, for a thorough assessment one would then also have to judge on the importance of the other area's contribution. In contrast to this, a reduction on just "interrelated or not" would undoubtedly mean a simplification as well. Notwithstanding this downside, it would at least stress that on the basis of this information the depth of a connection between both areas could not be made. Additionally, for many of the other measures (like co-authorship) a binary measure should be particularly seriously considered.[71]

Such a monitoring concept would also be in line with Ansoff's priority assessment of strategic issues. He distinguishes between low, significant, and pressing urgency as well as low, significant and major impact. For strategic issues of major impact but low urgency he calls for a continuous monitoring [cf. 24, p. 140]. In line with this study's reasoning of industry convergence having a major impact on the different stakeholders its potential impact can be judged to be major. On the other hand, especially the results discussed in the preceding chapter have highlighted the low urgency caused by long time horizons applicable to convergence processes thus far.

[71] For instance, a publication is either co-authored by authors from two areas or not. Consequently, a larger number of co-authors will (in most cases) not reflect a closer collaboration but rather involvement of more researchers resulting from, e.g., more complex methods..

References

1. Ernst, H.: The use of patent data for technological forecasting: The diffusion of CNC-technology in the machine tool industry. Small Bus. Econ. **9**(4), 361–381 (1997)
2. Daim, T.U., Rueda, G., Martin, H., Gerdsri, P.: Forecasting emerging technologies: Use of bibliometrics and patent analysis. Technol. Forecast. Soc. Chang. **73**(8), 981–1012 (2006)
3. Kayal, A.A., Waters, R.C.: An empirical evaluation of the technology cycle time indicator as a measure of the pace of technological progress in superconductor technology. IEEE Trans. Eng. Manage. **46**(2), 127–131 (1999). doi:10.1109/17.759138
4. Daim, T.U., Jordan, S.: A foresight based on scientific indicators: A framework drawn from the case of laptop battery alternatives. Foresight **10**(3), 43–54 (2008)
5. Chesbrough, H.W.: Open innovation: The new imperative for creating and profiting from technology. Harvard Business School Press, Boston (2003)
6. Gassmann, O.: Opening up the innovation process: Towards an agenda. R&D Manage. **36**(3), 223–228 (2006)
7. Howells, J., James, A., Malik, K.: The sourcing of technological knowledge: Distributed innovation processes and dynamic change. R&D Manage. **33**(4), 395–409 (2003)
8. Adams, S.: Using the international patent classification in an online environment. World Patent Inf. **22**(4), 291–300 (2000)
9. Gassmann, O., Reepmeyer, G., von Zedtwitz, M.: Leading pharamceutical innovation: Trends and drivers for growth in the pharamaceutical industry, 2nd edn. Springer-Verlag, Heidelberg (2008)
10. WIPO: Statistics on patents: patent applications by patent office (1883–2008). http://www.wipo.int/export/sites/www/ipstats/en/statistics/patents/xls/wipo_pat_appl_from_1883_table.xls (2009). Accessed 27th Apr 2010
11. WIPO: statistics on patents: patent grants by patent office (1883–2008). http://www.wipo.int/export/sites/www/ipstats/en/statistics/patents/xls/wipo_pat_grant_from_1883_table.xls (2009). Accessed 27th Apr 2010
12. Greenhalgh, C., Longland, M., Bosworth, D.: Protecting intellectual property: British, European and American patents and trade marks of selected UK companies 1986–95, (2001)
13. Khan, M., Dernis, H.: Global overview of innovative activities from the patent indicators perspective. In: STI Working Paper 2006/3: Statistical Analysis of Science, Technology and Industry, vol. DSTI/DOC(2006)3. OECD, (2006)
14. EPO: Contents and Coverage of the DOCDB Bibliographic file. http://documents.epo.org/projects/babylon/rawdata.nsf/0/B4C9ED03518F26C5C12577060031AE28/$File/PFS_1016.xls (2010). Accessed 26th Apr 2010
15. Scheffer, M., Bascompte, J., Brock, W.A., Brovkin, V., Carpenter, S.R., Dakos, V., Held, H., van Nes, E.H., Rietkerk, M., Sugihara, G.: Early-warning signals for critical transitions. Nature **461**(3), 53–59 (2009)
16. Bröring, S.: The front end of innovation in converging industries: The case of nutraceuticals and functional foods. DUV, Wiesbaden (2005)
17. Fai, F., von Tunzelmann, N.: Industry-specific competencies and converging technological systems: evidence from patents. Struct. Change Econ. Dyn. **12**(2), 141–170 (2001)
18. Soete, L.: The impact of technological innovation on international trade patterns: The evidence reconsidered. Res. Policy **16**(2–4), 101–130 (1987)
19. Chandler, A.D.: Scale and scope: the dynamics of industrial capitalism, 1st Harvard University Press pbk. ed. Belknap Press, Cambridge (MA), USA (1994)
20. Tidd, J.: Development of novel products through intraorganizational and interorganizational networks—The Case of Home Automation. J. Prod. Innov. Manage **12**(4), 307–322 (1995)
21. Niedergassel, B.: Knowledge sharing in collaborative R&D projects. University of Münster (2009)

22. Preschitschek, N., Curran, C.-S., Leker, J.: Are plastics turning organic? Antecedents of convergence in the chemical industry. paper presented at the The XXI ISPIM conference—The dynamics of innovation, Bilbao, Spain, 06–09 June 2010
23. Schnaars, S., Thomas, G., Irmak, C.: Predicting the emergence of innovations from technological convergence: Lessons from the twentieth century. J. Macromarketing **28**(2), 157–168 (2008)
24. Ansoff, H.I.: Strategic issue management. Strateg. Manag. J. **1**(2), 131–148 (1980)

Chapter 7
Summary and Conclusion

Starting on Darwin's notion of a combination of sound scientific methods and human combinatory and judgmental power, this study set out to develop a comprehensive approach for the anticipation of converging industries. Despite the plentiful literature on different methodologies to predict the future—or rather a set of 'possible futures'—the scattered literature on convergence is lacking a concept for a science-based prediction of convergence. This is even more surprising when considering the massive impact that converging industries have on the hitherto disjunct industry segments. Especially the ICT industry has been widely impacted by convergence during the last few decades. From AT&T's purchase of a part of Olivetti in 1983 to HP's acquisition of Palm in 2010, many strategic decisions have been made to use the blurring of boundaries for an extension of firm's sales and profits. While many scholars, managers, and politicians had expected convergence to be much faster in pace, it was experienced as a long-lasting process on different levels and at different *loci*. With developments in nutraceuticals and functional foods (NFF), more importantly since the 1990s, this fading of boundaries is also observable at the intersection of the food and the pharmaceutical industry.

Against this background, the study at hand proceeded towards fulfilling its research objective to identify underlying patterns in converging industries focusing on two differentiating factors. First, the sequential process of convergence was divided into four different steps: science, technology, market, and industry convergence. Second, patterns were unraveled in respect to environmental differences regarding industries and countries. In the course of these analyses a solid theoretical foundation was laid by aggregating the thus far scattered literature and deriving more comprehensive definitions and delineations of the most central terms, to add to the existing body of literature.

Convergence [fusion] is a blurring of boundaries between at least two hitherto disjoint areas of science, technology, markets, or industries. Through this convergence [fusion], a new (sub-) segment is being created in a new spot [the same spot] as a merger of (parts) of the old segments. It is marked by an increase of interchangeability and connectedness between the respective areas, as can be seen in collaboration, licensing, patenting, or publishing behavior.

C.-S. Curran, *The Anticipation of Converging Industries*,
DOI: 10.1007/978-1-4471-5170-8_7, © Springer-Verlag London 2013

A **trigger** denominates an event or situation that **sets off** a certain sequence of reactions, while a **driver** denominates a series of events or a situation that **sustains** a certain sequence of reactions.

Based on these definitions, triggers and drivers to be observed in different settings of convergence were grouped into five main classes of:

- Technological drivers
- Managerial drivers
- Industrial drivers
- Political drivers
- Societal drivers

In the following, the concept of anticipating converging industries on the basis of publicly available data was introduced. 'Anticipating' is distinguished from the more common terms of 'forecasting' and 'foresight' in respect to its early identification of (weak) signals by a central characteristic: its immanent active and preparatory twist. Both, the definition as well as the term's common use refer to the future orientation not so much in the form of an exact mathematical formula, but rather simpler and more intuitive methods, thus paying respect to the difficult task of detecting weak signals.

Building on different methods to analyze approximately 1.8 million patents (and a few thousand scientific publications) several samples were generated based on two different search strategies. First, keyword-based searches were used to assemble two samples on the NFF-related substance class of phytosterols. These were analyzed using the two standard software products SciFinder Scholar and STN AnaVist, employing (semi-)quantitative and qualitative indicators for science and technology convergence. For a further assessment of technology convergence (as an antecedent of industry convergence) in the area of NFF, three large samples were aggregated from patent references on the basis of IPCs in the three countries of USA, UK, and Japan. In a last set of analyses, these results were compared to a sample of US patents on ICT.

Major findings were discussed in respect to this study's research questions. They were derived in part from a brief discussion of the RBV as an underlying theory for converging industries and in part from the discussion of the empirical example of NFF as an area of convergence. The questions and an abridged version of the answers are provided in the following:

Q.1: *Can convergence be anticipated on the basis of weak signals in publicly available data?*

Based on the above-mentioned samples, convergence can be confirmed *ex post* for NFF as well as ICT on a validity assumption of the proposed anticipation concept. However, the question is more difficult to answer when concentrating on the *anticipation* perspective. It is relatively easy to proclaim convergence when looking at the whole picture, but would it have been possible to decide on a convergence pattern already 10, 20, or even 30 years ago? This would necessitate either a specific pattern of developments or a threshold at which convergence would be destined to happen. To this end, a decrease in, for instance, co-classification shares being observed in most of the time-dependent analyses could suggest that a slowing down after a longer phase of growth could be followed by an exponential phase of growth. For the IPC co-classification shares this plateau was reached after crossing the 10 % (NFF in the UK and Japan), 15 % (NFF in the USA) or 20 % (ICT) level.

Q.2: *Can the assessment of different levels of convergence be based on analyzing samples derived from different data sources?*

Indications of science convergence were appearing in the phytosterol-restricted samples about 30 years earlier than corresponding developments in patents. This would point toward the possibility of differentiating between science and technology convergence on the basis of these two sources. However, science convergence still lacks an easily and robustly applicable quantitative measure. Whereas this could be achieved for technology convergence, the approaches tested in this study appear to be only of limited use in an *ex ante* setting. Accordingly, future research should try to develop more easily applicable quantitative measures which can be keyword-independently derived from larger samples.

Q.3: *Is a keyword-based anticipation of convergence possible and can this also be achieved employing broader classes of articles or patents?*

While the same holds true for science convergence as discussed in answering Q.2, the IPC co-classification employed in this study was the most easily applicable measure. At the same time it yields more stable results than those derived from the massively smaller sample of phytosterols. There, only convergence indications could be derived and only by using different, more complicated and less transparent handling procedures.

Q.4: *Are the food and the pharmaceutical industries converging in the area of NFF?*

While especially the results from IPC co-classification provide substantiation for convergence, the results derived from the limited area of phytosterols support this finding. Interestingly, the convergence effects can be expected to be asymmetrically distributed, as the food industry seems to be much more concerned with the introduction of NFF than the pharmaceutical industry.

Q.5: *Are the personal care and the pharmaceutical industries converging in the area of Cosmeceuticals, perhaps even into one broad NFF/Cosmeceuticals area?*

Considering the results from both phytosterol examples, there are indications of convergence around Cosmeceuticals. However, on the basis of the low interconnectedness of personal care and food scientific articles and patents, there is no evidence hinting at one common NFF/Cosmeceuticals industry. Instead, there are several signs of two distinct areas of NFF on the one hand and Cosmeceuticals on the other hand. These findings were supported by an additional sample based on IPCs. While results up to 2007 would indicate much weaker convergence trends than in NFF, the sudden surge in the last years may point at a renewed interest in functional cosmetics. Regarding a triple co-classification of patents on food, pharmaceuticals, and personal care, there was little support for the expectation of one common area.

Q.6: *Does the analysis of activities in different countries lead to different results?*

This question receives a mixed answer as results differed slightly regarding the exact numbers, while maintaining very similar overall patterns. Consequently, when debating which to choose as a patent sampling country, there is no reason to believe that convergence could not be anticipated when choosing a 'wrong' patent country.

Q.7: *Are patterns of convergence generalizable across industry boundaries?*

Patterns are very similar across industry boundaries as well, as could be shown, for instance, in ICT. Furthermore, the observations from the intersection of personal care and pharmaceuticals support this assessment. Naturally, three examples are insufficient to prove the existence of a general pattern. Future studies should compare results to the findings in this study, ideally especially in connection with different degrees of convergence or convergence on different levels. Thus, with more data on different cases the pattern could not only be described but potentially also explained.

Following a detailed discussion of limitations in respect of individual analyses, the three software products and two broad approaches were compared. This comparison highlights the limitation of the study in respect of keyword-/IPC-selection, choice of the patent country, and decisions on clustering and thresholds. While none of the limitations should per se rule out a validity of the major findings, this study does suggest following different routes based on the desired outcomes.

SciFinder should (if at all) only be used to delve deeper into a certain (limited) area of possible convergence. The two most central drawbacks are the necessity to use keyword-based search-terms and SciFinder's architecture which is not constructed to be used for advanced statistical analyses. Considerable effort is necessary to unravel the underlying structure of a sample, or at least a part of it.

Contrastingly, **STN AnaVist** can employ samples which were generated in standard databases, thus all kinds of different search strategies can be used to create it. While the possibility of interacting with the sample and combining quantitative as well as qualitative means of analysis is appealing in anticipating convergence, the same limitations affect this program as far as keyword-based search-terms are concerned. Furthermore, the level of effort to be invested is comparably high, due to the many time-consuming operations necessary. Based on its opportunities for various examinations, it is thus rather recommended for a more thorough analysis of certain aspects of convergence patterns.

By using **STN Express** in accessing a patent database, the co-classification shares' variations in time have clearly corroborated the applicability of this measure as an instrument for anticipating convergence. While this measure is very easy to use, a few caveats on IPC co-classification are discussed. They entail not only the question of potentially erroneous classification, but also the completeness of classification and especially the degree of reclassification. However, as most industries tend to patent largely in their corresponding technological fields, an analysis of developments across technological and industrial boundaries can serve well as a proxy for convergence. Although only simple counts and co-classification calculations were conducted with it in this study, it could also be used to carry out various other analyses. This includes most prominently different search operations around citations. Once investigators accustomed themselves to the retrieval language, standard scripts can be used to, e.g., perform routine searches at certain points in time. Consequently, this approach appears to be ideal for the anticipation of convergence as it can handle nearly unlimited numbers of documents and calculate convergence-related measures in comparably short times.

Finally, in addition to the implications for the respective research streams, the study also presents implications for three groups of stakeholders. Academics, managers, and policy makers dealing with areas of convergence could benefit from this study's approach and findings by understanding patterns of convergence and being able to judge on the degree of convergence. Furthermore, they could use the anticipation approach to prepare in advance for potential impacts of convergence. When trying to identify ideal collaboration partners, to secure access to scarce resources or to gather momentum in directing research and development, time is a very valuable resource.

Consequently, while this study does not claim to present a holistic, exact, and ready-to-use mathematical formula for any setting and to derive one insightful convergence measure, it may serve as a reference point for further refinements of the discussed approach. Additionally, it would be reasonable for the individual groups of stakeholders to adjust the methodology to their environments and to identify potential areas of science, technology, market, or even industry convergence. Bearing in mind the decisive impetus of convergence on various aspects of peoples' professional and private life, this endeavor seems to be worth taking.

Appendix

A.1 Hierarchical Construction of IPCs

For instance, Forbes Medi-Tech's patent application AU002005211536A1 on "Conjugates of phytosterol or phytostanol with ascorbic acid and use thereof in treating or preventing cardiovascular disease" is classified as:

IPC main class[1]: C07J 51/00

IPC secondary class[1]: A61K 31/59
A61P 3/06
A61P 9/10
C07J 17/00

Its main class for example can be hierarchically divided up into:

C	Chemistry; Metallurgy				
C07		Organic Chemistry[…]			
C07J			Steroids		
C07J 51				Normal steroids with unmodified cyclopenta[a] hydrophenanthrene skeleton […]	
C07J 51/00					(/00 means there is only this level of hierarchy)

[1] Main and secondary classes were only used up to the 7th version of the IPC (IPC-7). Since 2006 (IPC-8) there is no differentiation on this level anymore.

C.-S. Curran, *The Anticipation of Converging Industries*,
DOI: 10.1007/978-1-4471-5170-8,

MIX
Papier aus verantwortungsvollen Quellen
Paper from responsible sources
FSC® C105338

If you have any concerns about our products,
you can contact us on
ProductSafety@springernature.com

In case Publisher is established outside the EU,
the EU authorized representative is:
Springer Nature Customer Service Center GmbH
Europaplatz 3, 69115 Heidelberg, Germany

Printed by Libri Plureos GmbH
in Hamburg, Germany